Current Topics in Microbiology and Immunology

172

Editors

Transmissible Spongiform Encephalopathies:

Scrapie, BSE and Related Human Disorders

Edited by B.W. Chesebro

With 40 Figures

Springer-Verlag
Berlin Heidelberg NewYork
London Paris Tokyo
Hong Kong Barcelona
Budapest

BRUCE W. CHESEBRO, M.D.
Chief, Laboratory of Persistent Viral Diseases
Rocky Mountain Laboratories
National Institute of Allergy and Infectious Diseases
National Institutes of Health
Hamilton, Montana 59840, USA

ISBN 3-540-53883-6 Springer-Verlag Berlin Heidelberg New York
ISBN 0-387-53883-6 Springer-Verlag New York Berlin Heidelberg

Library of Congress Catalog Card Number 15-12910
Printed in Germany

Typesetting: Thomson Press (India) Ltd, New Delhi;
Offsetprinting: Saladruck, Berlin; Bookbinding: Lüderitz & Bauer, Berlin.
23/3020-543210—Printed on acid-free paper.

Preface

In spite of a long history of intense investigation the transmissible spongiform encephalopathies remain a poorly understood family of neurodegenerative diseases. This group of diseases has been described in a wide variety of animal species and includes kuru, Creutzfeldt–Jakob disease, and Gerstmann–Sträussler syndrome in humans, and scrapie, bovine spongiform encephalopathy, and related syndromes in ruminants and rodents. In all cases spongiform degeneration and astrocytosis are seen in specimens of brain and a filterable transmissible agent is present in the brain and some other tissues of affected individuals. However, the precise nature of this agent remains unknown. Agent infectivity, which can so far only be assayed by serial transmission to new individuals, has been shown to be remarkably resistant to inactivation by heat, chemicals, and irradiation. These properties create significant biohazard possibilities during exposure to infected tissues. Transmission between humans was originally recognized in the unique epidemiology of kuru in New Guinea tribesmen, and concern about transmission from animals to humans has re-emerged as a result of the current epidemic of bovine spongiform encephalopathy in dairy cattle in Great Britain. Although interspecies transmission has often been achieved experimentally, its efficiency is highly variable. Therefore, the possibility of spread of bovine spongiform encephalopathy from cattle to humans or various animal populations cannot be accurately predicted at this time.

This volume presents a comprehensive update of knowledge concerning the transmissible spongiform encephalopathies. Areas covered include epidemiology, agent properties, relationship of the agent infectivity to the prion protein, properties of the prion protein and its gene, evidence for genetically diverse agent strains, host susceptibility genes, altered expression of host genes, agent propagation in vitro, and possible detection of transmissible agent in Alzheimer's disease. Many of these areas remain controversial, and readers will find significant overlap in the subject matter and strong differences in the interpretation of the facts by different authors. The editor hopes that readers will not be overly frustrated by the lack of unanimity among these experts in the field, but rather will take this as an indication of the challenges to be surmounted.

BRUCE CHESEBRO

Contents

List of Contributors

(Their addresses can be found at the beginning of their respective chapters.)

The Human Spongiform Encephalopathies: Kuru, Creutzfeldt–Jakob Disease, and the Gerstmann–Sträussler–Scheinker Syndrome

P. BROWN and D. C. GAJDUSEK

1 Introduction

Kuru and the transmissible virus dementias Creutzfeldt–Jakob disease (CJD) and Gerstmann–Sträussler–Scheinker syndrome (GSS) belong to the group of virus-induced slow infections that we have described as subacute spongiform virus encephalopathies because of the strikingly similar histopathological lesions they induce (Table 1). Scrapie, mink encephalopathy, chronic wasting disease with spongiform encephalopathy of captive mule deer and of captive elk, and bovine spongiform encephalopathy all appear from their histopathology, pathogenesis, and the similarities of their infectious agents to belong to the same group. The basic neurocytological lesions in all these diseases are a progressive vacuolation in the dendritic and axonal processes and cell bodies of neurons and, to a lesser extent, astrocytes and oligodendrocytes; an extensive astroglial hypertrophy and proliferation; and spongiform change or status spongiosis of gray matter with extensive neuronal loss.

Kuru was the unfortunate result of virus contamination introduced into the mortuary practice of handling the infected brain tissue of deceased relatives as a

* Laboratory of CNS Studies, National Institute of Neurological Disorders and Stroke, National Institutes of Health, Building 36, Room 5B21, Bethesda, MD 20892, USA

Current Topics in Microbiology and Immunology, Vol. 172

part of the ritual rite of mourning with endocannibalistic consumption of the tissues of dead kinsmen. Thus, all cases of kuru appear to have been the result of a direct chain of infection, usually occurring in infancy, and probably resulting from massive contamination of the skin and mucus membranes of infants with highly infectious brain tissue handled by the women, rather than the men.

In CJD and GSS, however, fewer than 1% of the infections can be attributed to a direct chain of infection. In these cases, a high technological neocannibalism resulting from contaminated instruments or stereotactic electrodes in neurosurgery, tissue transplantation (cornea, dura mater) and the preparation of hormones from infected human pituitaries has transmitted the infection, which seems to have no arthropod vector or other natural means of spread.

These atypical infections differ from other slow virus diseases of the human brain in that they do not evoke an inflammatory response: no perivascular cuffing or invasion of the brain parenchyma with leukocytes, and no pleocytosis or elevated protein in the cerebrospinal fluid. Furthermore, they do not stimulate an immune response to the causative virus and there are no recognizable virions in sections of the brain visualized by electron microscopy, whereas in other virus encephalopathies virions have been readily observed. Instead, they show an ultrastructural alteration in the plasma membrane that lines the vacuoles, piled up neurofilaments in some swollen nerve cells, and strange arrays of regularly arrayed tubules that look like particles in cross section in postsynaptic processes.

Table 1. Comparison of epidemiological, clinical, biological, and pathological features typically found in the three varieties of human spongiform encephalopathy

Characteristic	Kuru	Creutzfeldt–Jakob disease	Gerstmann–Sträussler–Scheinker syndrome
Epidemiological pattern	Epidemic	Sporadic, familial, iatrogenic	Familial, sporadic(?)
Clinical features			
Age at onset	5–40 years	50–75 years	35–55 years
Duration	3–9 months	2–5 months	2–8 years
Early signs	Cerebellar shivering, tremor	Dementia, cerebellar, visual	Spinocerebellar ataxia
Later signs	Extrapyramidal, behavioral changes	Extrapyramidal, pyramidal, myoclonus	Dementia, pyramidal, amyotrophy
Abnormal biologic tests	None	PRIP gene mutations periodic EEG	PRIP gene mutations
Neuropathology			
Spongiform change	+ + +	+ + + +	+ +
Astrogliosis	+ + + +	+ + +	+ +
Plaques	+ + +	+/0	+ + + +
System tract degeneration	+ +	0	+ +

The accumulated data of four decades have made it evident that infection proceeds with the slow conversion of a normal host precursor protein to a β-pleated configuration of amyloid, and that we are dealing with a slowly replicating infectious agent that contains no DNA or RNA or non-host protein, which in the majority of cases originates de novo in each individual. We thus have the exciting new paradigm of an agent fulfilling postulates for infectious virus and yet undergoing a virus-like replication that more closely parallels the nucleation or seeding of protein crystal growth and fibril polymerization, such as occurs in the patterned laydown of tropocollagen networks, of which the nucleation requires an oligomer or microfibril of tropocollagen itself. Nucleation must cause an autopatterned transformation of configuration of the normal host precursor into an infectious form of the whole molecule, which is then proteolytically cleaved into the subunit infectious protein that polymerizes into the fibrillar structures and plaques seen in all subacute spongiform encephalopathy brains of kuru, the transmissible virus dementias, and scrapie. This paradigm from crystallography, invoking a posttranslational configurational change of polypeptide winding and fibril polymerization and crystallization, better serves us than any derived from the basic tenets of biology, involving transcription and translation from nucleic acid.

Kuru, the first proved slow virus infection of man is thus caused by a very unconventional virus which is an infectious form of a host-specified precursor protein. Certainly in kuru, and in CJD from receiving injections of CJD-contaminated cadaveric-derived human growth hormone, an infectious chain accounts for the cases. In its worldwide form of CJD (or GSS) it is usually generated de novo in each individual rather than the result of a direct chain of infection. This spontaneous de novo transformation of the normal precursor to an infectious form by conformational change in secondary and tertiary structure normally occurs with a probability of about one per million per annum, which is the worldwide incidence of CJD in diverse communities. Each of several newly identified point mutations in the familial forms of CJD (or GSS) increases the probability of spontaneous occurrence of this new conformation by one hundred thousand- to one million-fold, thereby transforming a rare event into a Mendelian dominant, inherited trait.

Such a heretic model leads to the prediction that a mutated virus would replicate in a new host without "breeding true": all progeny of the virus (infectious polypeptide) will have the amino acid sequence of the new host protein from which it is derived and will not carry the point mutation of the parent virus of familial CJD or GSS that led to its original generation.

2 Kuru

Kuru is characterized by cerebellar ataxia and a shivering-like tremor that progresses to complete motor incapacity with dysarthria and total loss of speech and death in less than 1 year from onset. It was confined to highland New

Guineans living in a number of adjacent valleys in the mountainous interior of Papua New Guinea and occurred in 169 villages with a total population of just over 35 000. Kuru means shivering or trembling in the Fore language. In the Fore cultural and linguistic group, among whom more than 80% of the cases occurred, kuru had a yearly incidence rate and prevalence rate of about 1% of the population. During the early years of investigation, it was found to affect all ages beyond infants and toddlers; it was common in male and female children and in adult females, but rare in adult males. This marked excess of deaths of adult females (compared to males) has led to a male to female population ratio of more than 3:1 in some villages and of 2:1 for the whole South Fore group (ALPERS et al. 1975; FARQUHAR and GAJDUSEK 1980; GAJDUSEK 1963, 1972, 1973, 1976, 1977, 1978, 1957–1989; GAJDUSEK and ZIGAS 1957, 1959; GLASS 1963; KLITZMAN et al. 1984; LINDENBAUM 1979; ZIGAS and GAJDUSEK 1957).

Kuru has been disappearing gradually during the past 25 years (ALPERS 1968; ALPERS and GAJDUSEK 1965; ALPERS et al. 1975; GAJDUSEK 1977, 1978; KLITZMAN et al. 1984). The incidence of the disease in children and adolescents decreased during the 1960s and by 1975 there were no further patients under 20 years of age. By 1985, the disease was no longer seen in persons under 35 years of age. This change in occurrence of kuru appears to result from the cessation of the practice of ritual cannibalism as a rite of mourning and respect for dead kinsmen, with its resulting conjunctival, nasal, skin, mucosal, and gastrointestinal contamination by highly infectious brain tissue mostly among women and small children (FARQUHAR and GAJDUSEK 1980; GAJDUSEK 1976, 1957–1989; GLASS 1963; KLITZMAN et al. 1984; LINDENBAUM 1979). In recent field work on kuru, it has been possible to obtain clear documentation of incubation periods of 30 years and more in human kuru and, because kuru patients are now scarce, to identify small clusters of related patients and to ascertain the exact cannibal feast that caused contamination (KLITZMAN et al. 1984). Over 90% of the infants and children of women present at a contaminating event of cannibalism have already developed kuru (KLITZMAN et al. 1984). Continued surveillance has revealed no alteration in the unusual pattern of kuru disappearance, which indicates the artificial, manmade nature of the epidemic. Kuru virus clearly has no reservoir in nature and no intermediate biological cycle for its preservation except in humans. Furthermore, there is no transplacental or neonatal transmission, even as an integrated genome or as a milk factor from a mother who is incubating kuru or who is already a kuru victim. There were only six to twelve kuru deaths a year since 1988, and the youngest kuru patients are now over 35 years of age (GAJDUSEK 1977, 1957–1989).

2.1 Clinical Course

The clinical course of kuru is remarkably uniform, with cerebellar symptoms progressing to total incapacitation and death, usually within 3–9 months. It starts insidiously without antecedent acute illness and is conveniently divided into three stages: ambulant, sedentary, and terminal. Some patients report prodromal symptoms of headache and limb pains. There is no fever or convulsions (ALPERS et al. 1975; GAJDUSEK 1963, 1973, 1978; GAJDUSEK and ZIGAS 1957, 1959; ZIGAS and GAJDUSEK 1957).

The first, or ambulant, stage is usually self-diagnosed before others in the community are aware that the patient is ill. There is unsteadiness of stance and gait and often of the voice, hands, and eyes as well. Postural instability with truncal tremor and titubation and ataxia of gait are the first signs. Dysarthria starts early, and speech progressively deteriorates as the disease advances. Eye movements are ataxic, but no ture nystagmus occurs. Convergent strabismus often appears early in the disease and persists. Tremors are at first no different from those of hypersensitivity to cold; the patient shivers inordinately. Incoordination affects the lower extremities before progressing to involve the upper extremities. Patients arising to a standing posture often stamp their feet as though angry. In attempting to maintain balance when standing, the toes grip and claw the ground more than usual. Very early in the disease, the inability to stand on one foot for many seconds is a helpful diagnostic clue. For most of the first stage of the disease, the patient usually continues to work in the garden and attempts to pursue his normal activities and take full part in village social life. In the latter part of this first stage, the patient usually takes a stick for support in order to continue to walk about the village unaided by others.

The second, or sedentary, stage is reached when the patient can no longer walk without complete support. Tremors and ataxia become more severe, and a changing rigidity of the limbs often develops, associated with widespread clonus or sometimes shock-like muscle jerks and occasionally coarser athetoid and choreiform movements, especially when the patient is thrown into an exaggerated startle response by postural instability, or even by sudden noise or bright light. Deep tendon reflexes are usually normal; the Babinski response is negative. Ankle clonus is usually present and patellar clonus often. Although muscle activity is poorly maintained, there is no fasciculation or real weakness or muscle atrophy. Emotional lability, leading to outbursts of pathological laughter, is frequent, sometimes even appearing in the first stage of the disease; smiling and laughter are terminated slowly. This feature has given rise to the unfortunate journalistic synonym of "laughing death" for the disease. Most patients show a resignation to, and a light-hearted attitude toward their disease, bordering on a true euphoria. Some patients, especially adolescent and young adult males, become depressed, and a rare patient develops a pathological belligerence to all disturbances by family members or others. Mental slowing is apparent, but severe dementia is conspicuously absent. No sensory changes have been noted. The fundi appear normal.

The third, or terminal, stage is reached when the patient is unable to sit up without support and ataxia, tremor, and dysarthria become progressively more severe and incapacitating. Tendon reflexes may become exaggerated. A grasp reflex may appear, and some patients show characteristic extrapyramidal defects of posture and movement. Terminally, urinary and fecal incontinence develop, and dysphagia leads to thirst and starvation. Flaccidity, inanition, and signs of bulbar involvement develop, and the patient becomes mute and unresponsive. Deep decubitus ulcerations and hypostatic pneumonia appear in this stage, and the patient finally succumbs, usually emaciated, but occasionally quickly enough to be still well nourished.

2.2 Laboratory and Pathological Findings

There are no changes in the peripheral blood smear, and no pleocytosis or elevated protein in the CSF; the erythrocyte sedimentation rate is normal. There is no disturbance, renal or hepatic, and endocrine function tests remain normal (GAJDUSEK 1963, 1973; GAJDUSEK and ZIGAS 1957, 1959; ZIGAS and GAJDUSEK 1957).

There is no pathological process outside the CNS, and no gross brain lesions are noted, except for an atrophic cerebellar vermis in many of the patients. In kuru, the changes in the nervous system are widespread and are characterized by marked proliferation and hypertrophy of the astrocytes throughout the brain, mild status spongiosus of the grey matter, diffuse neuronal degeneration that is most severe in the cerebellum and its different and efferent connections, and minimal demyelination which, though not readily seen in sections stained for myelin, can be demonstrated by the presence of neutral fat. Typical intracytoplasmic vacuolation is usually observed in the large neurons of the striatum (BECK et al. 1966a, b, 1973, 1975, 1982; KLATZO et al. 1959; LAMPERT et al. 1972; LANDIS et al. 1981).

In about three fourths of the cases, there are periodic acid-Schiff (PAS)-positive, birefrigent, amyloid-containing plaques. All patients develop some scrapie-associated fibrils (SAF) seen by electron microscopy in density gradients of brain suspensions (MERZ et al. 1981, 1983a, b, c, 1984). or in touch preparations on electron microscopic grids (NARANG et al. 1987, 1988). Both these SAF and the kuru plaques are composed of the same amyloid. Thus, all cases of kuru are cerebral amyloidoses. The amyloid subunit is totally different from that of the amyloid plaques or vascular amyloid deposits of Alzheimer's disease and normal aging brain as shown by amino acid sequence and immunocytochemistry.

Because of the massive appearance of amyloid plaques structurally, ultrastructurally, and histochemically similar to the amyloid plaques of the normal aging brain and of Alzheimer's disease, we facetiously called kuru "galloping senescence of the juvenile" in the first years of its intensive study. Because 20% of patients failed to show amyloid plaques, we discounted them as not an essential feature of the disease. Only after MERZ et al. (1983c, 1984) demonstrated that even those CJD patients *without* kuru plaques had amyloid fibrils (SAFs) on ultrastructural study of brain suspensions, and we recognized that these were polymers of the same amyloid subunit as were the plaques, did we realize that all of these infections were *transmissible cerebral amyloidoses* (DIRINGER et al. 1986; GAJDUSEK 1985, 1988a, b; GUIROY and GAJDUSEK 1989).

2.3 Epidemiology

Since the beginning of kuru investigations in 1956, over 2500 cases have been recorded. Most have ended in death within less than a year, a few patients have a somewhat more prolonged disease. Kuru mortality has declined continuously over the past 30 years, and the disease no longer appears in children, adolescents, or young adults. Over 200 patients died annually during the early years of investigation, but now only 5–10 patients, all over 35 years of age, still die of the disease each year. Case-finding is complete, as the entire population of all villages in which kuru has occurred has been surveyed at least twice annually (ALPERS 1968; ALPERS and GAJDUSEK 1965; ALPERS et al. 1975; FARQUHAR and GAJDUSEK 1980; GAJDUSEK 1957–1989).

Epidemiological observations have given no evidence for its contagion. No outsiders of the kuru region have yet developed kuru after residence in the region. Disease has, however, developed in many subjects after many years of absence from the kuru region, for work or study. Such persons were always found to be from families in which kuru occurred in the past. No cases of the disease were seen in natives from elsewhere in New Guinea eating and living together in school or in labor compounds with kuru-region persons in whom kuru developed while they were away from home. Furthermore, hundreds of native peoples from elsewhere in New Guinea and from surrounding Melanesian islands and Caucasians have now been residing for long periods in the kuru region since early government penetration almost a half-century ago. In spite of the many thousands of man-years of close contact of these immigrants with the flora, fauna, food, and people of the region, no case of kuru has occurred among them. In the peripheral region of low incidence in the north and northwest of the kuru area, the few cases that have appeared are either immigrants from villages of higher kuru incidence or their family members who have participated in the cannibalistic mortuary rites, although such immigrants from but a small portion of the population of these peripheral villages of low kuru incidence.

Soon after the discovery of kuru and at a very early stage in the disease patients were brought totally outside the kuru region and placed on a regular hospital diet; yet their disease progressed to death in typical fashion. This was also true of patients who acquired their disease after a prolonged sojourn outside the region but who never returned before death to their kuru-affected homeland.

The culture in the kuru resembles those of surrounding kuru-free highland peoples. The pattern of the women, small children, and pigs sharing the women's houses was by no means unique to the kuru region. Pigs, the principal domestic animals throughout the New Guinea highlands, had no unique relation with humans in the kuru region, nor did they show any illnesses not encountered in the pigs in adjacent kuru-free regions. Adult men, who were the group least affected by kuru, are rarely with the women and children but had quite similar diets. There was no item in the Fore diet not also consumed in similar quantity by some neighboring kuru-free groups. Nor was the endocannibalism as a rite of respect for the dead confined to the kuru-affected villages.

The mechanism of spread of kuru is undoubtedly contamination of the population during ritual cannibalistic consumption of their dead relatives as a rite of respect and mourning. Women did the autopsies barehanded and did not wash their hands thereafter. They then wiped their hands on their bodies and in their hair; they picked sores and scratched insect bites; they wiped their infants' eyes and cleaned their noses; and they ate with their hands. Scratching of ubiquitous itching scabies and impetiginous skin eruptions with infected fingernails could have produced hundreds of intradermal and deeper inoculations (ALPERS 1968; ALPERS and GAJDUSEK 1965; ALPERS et al. 1975; FARQUHAR and GAJDUSEK 1980; GAJDUSEK 1973, 1976, 1957–1989; GLASS 1963; KLITZMAN et al. 1984; LINDENBAUM 1979). It is intriguing to suggest that a sporadic case of CJD in a New Guinean may have given rise to the kuru epidemic; such a sporadic case of CJD has been seen in a New Guinean (HORNABROOK and WAGNER 1975).

The liquefying brain tissue, scooped by hand into bamboo cylinders, was infectious at a titer of 10^7 or higher. It is infectious by the peripheral routes of inoculation, and intracerebral inoculation is not necessary (GAJDUSEK and GIBBS 1973). Infection of squirrel monkeys has been accomplished by feeding them kuru-infected brain tissue (GIBBS et al. 1980), but chimpanzees and other monkeys have failed to become infected by the oral route. The cooking of brain and other visceral tissues by steaming in bamboo cylinders at an elevation of 6000–7500 ft would expose them to temperatures of no more than 95°C and would be insufficient to fully inactivate the virus of kuru. Women and the children crowding about or in the arms of women were more exposed to the kuru-infected human tissue than were the men, who let this butchery of dead kinsmen to the women. Men rarely handled or ate the brains of dead kuru victims. We must presume that all children who had a close kinsman who died of kuru and whose mothers took part in the cannibalistic rite were contaminated with kuru. If a boy did not have a relative who died of the disease before he left the women's society to live in the men's house and ate with the men at feasts, he might thereafter escape contamination; but this was not so for the girls (ALPERS 1968; ALPERS and GAJDUSEK 1965; ALPERS et al. 1975; FARQUHAR and GAJDUSEK 1980; GAJDUSEK 1973, 1957–1989; GLASS 1963; KLITZMAN et al. 1984; LINDENBAUM 1979). Kuru is not transmitted to offspring of kuru-infected mothers transovarianly or in utero or during parturition, or from a milk factor or by some other route of infection. We know this from our failure to see kuru in children born to kuru-infected mothers since the cessation of cannibalism; infants born during the mother's incubation period or during active disease remain well.

2.4 Etiology

In 1959 Hadlow pointed out the similarity in the neuropathology of scrapie in sheep with that of kuru in man (HADLOW 1959), as reported by KLATZO et al.

(1959). An intensive comparative study of the two diseases was thus launched along with renewed attempts to transmit kuru to laboratory animals and other species (GAJDUSEK et al. 1965). In earlier transmission attempts, inoculated animals had been kept for only several months; now observation for many years was planned. In 1965, after 1.5 years of incubation, chimpanzees inoculated intracerebrally with suspensions of human brain from kuru patients developed the disease (GAJDUSEK et al. 1966, 1967). They were found to have the same pathological process as that in the human kuru victims (BECK et al. 1966a). The disease can be serially transmitted from chimpanzee to chimpanzee using either suspensions of brain tissue from affected chimpanzees or visceral tissues containing no brain; it is inoculated by either intracerebral or peripheral (intravenous, intraperitoneal, intramuscular, or subcutaneous) routes (GAJDUSEK et al. 1967). The incubation period in the chimpanzee varies from 14 to 39 months after intracerebral inoculation. On second passage in the chimpanzee, it drops to about 10–12 months and remains so shortened after further serial passage (GAJDUSEK and GIBBS 1973).

For several years all work on the kuru virus was done with chimpanzees, the first species to which the disease was transmitted (GAJDUSEK 1977, 1978; GAJDUSEK and GIBBS 1975; GAJDUSEK et al. 1966, 1967). Eventually, other species of nonhuman primates developed the disease: first, several species of New World monkeys, the spider (*Ateles* sp.), squirrel (*Saimiri* sp.), capuchin (*Cebus* sp.), and wooley (*Lagothrix* sp.) monkeys, with longer incubation periods than in the chimpanzee (GAJDUSEK and GIBBS 1971, 1973; GIBBS et al. 1979), and later, several species of Old World monkeys with yet longer incubation periods (GIBBS et al. 1979). We have also transmitted kuru to the goat using both human kuru brain inocula and brain from experimentally infected monkeys. In the goat, the disease resembles scrapie. Kuru does not transmit to sheep (GIBBS et al. 1979).

The virus has been regularly isolated from the brain tissue of kuru patients. Incubation periods have ranged from 1 to 12 years in monkeys with intracerebral inoculation. It may attain high titers of more than 10^8 ID_{50}/g of brain tissue of patients or experimental animals. In peripheral tissues of man or animals (e.g., liver, spleen, kidney) it has been found only rarely at the time of death and in much lower titers. Blood, urine, saliva, leukocytes, CSF, milk, placenta, and embryonal membranes of patients with kuru or animals with experimental kuru have not yet yielded the virus (GAJDUSEK 1978; GAJDUSEK and GIBBS 1973).

3 Creutzfeldt–Jakob Disease

It is not an exaggeration to say that kuru was reponsible for the elevation of Creutzfeldt–Jakob disease (CJD) to star status in the firmament of neurology, and the stimulus to serious examination of its clinical spectrum and epidemiology. The neuropathology of kuru had early been recognized to resemble that of CJD (KLATZO et al. 1959), and when kuru was shown to be experimentally transmissible (GAJDUSEK et al. 1966), a similar success with CJD was accomplished shortly thereafter (GIBBS et al. 1968). If CJD was experimentally transmissible, it clearly deserved attention as a potentially contagious disease, and a natural consequence of this idea was an examination of its epidemiology. However, time was still needed before enough transmitted cases could be assembled for a reliable clinical definition of CJD (BROWN et al. 1986; MATTHEWS 1975; TRAUB et al. 1977) that would in turn permit a trustworthy identification of cases for epidemiological evaluation (BROWN et al. 1987; MASTERS et al. 1979).

3.1 Clinical Features

Prodromal symptoms of asthenia, weight loss, or disordered sleep patterns, beginning weeks to months before the onset of neurological signs, can be elicited in over one third of patients. A gradually progressive mental deterioration is the most common presentation, usually in the form of simple memory loss, errors in judgment, mood change (often depression), or uncharacteristic behavior. Less often, mental deterioration takes the form of higher cortical function deficits, such as aphasia or apraxia, and sometimes there are episodes of frank confusion. In about half the patients, cerebellar or visual signs dominate the clinical presentation, with only moderate to minimal mental deterioration. Often, this type of presentation is comparatively rapid, and the sudden onset of vertigo, diplopia, ataxia, and even paralysis or paresthesia that occurs in up to 20% of patients may lead to a mistaken initial diagnosis of cerebrovascular accident or multiple sclerosis.

The progression of disease is usually relentless. Mental deterioration and mood alteration evolve into a state of dementia, confusion, and mutism; visual deterioration continues to cortical blindness (often with hallucinations); and motor impairments progress to increasingly severe incoordination, marked oppositional rigidity, and abnormal movements (usually myoclonic, but often trembling and sometimes choreo-athetoid or complex movements).

The electroencephalogram may be normal early in the illness, but later begins to show abnormal slow wave activity that in most patients evolves into some form of periodicity, either as bursts of high voltage slow waves, or as a distinctive sharp, triphasic, periodic pattern occurring at a frequency of 1–2 cycles per second, often synchronous with myoclonic jerking.

The majority of patients die within 6 months, most often within 2–3 months of the onset of illness. There are no verified recoveries.

Although CJD almost always takes this consistent and characteristic clinical course, there exists a small proportion of neuropathologically confirmed, clinically atypical patients, including some cases of GSS (MASTERS et al. 1981; TATEISHI et al. 1984), a few cases with prominent amyotrophic signs (SALAZAR et al. 1983; WILL and MATTHEWS 1979), and some otherwise typical cases of unusually long duration (BROWN et al. 1984).

3.2 Epidemiology and Molecular Genetics

During the past 15 years, a number of careful regional and national studies in Europe, the Americas, Israel, and Japan have provided a generally consistent descriptive picture of the epidemiology of CJD around the world. Updates of most European studies were included in the recently published proceedings of an international congress held in 1986 in Paris (COURT et al. 1989). These and other studies were summarized in two subsequent reviews of CJD epidemiology (BROWN 1987, 1988), to which interested readers may refer for statistical details and citations.

The essential epidemiological facts about CJD are by now too well known to require more than a brief recapitulation. The disease has been identified in every country in which it has been sought and has a worldwide incidence of about 0.5 to 1.0 case per million population per year, with twice as many cases being diagnosed in urban areas as in the population as a whole. There has been no obvious increase or decrease in this frequency during the past 10 years, when standardized criteria have been used for case diagnosis. The age-specific incidence of disease approximates a normal distribution around the peak age group of 60–65 years, skewed somewhat to the younger side of the curve, with occasional cases occurring as early as the second decade and as late as the ninth decade of life. In most studies, the case sex ratio has not differed significantly from that of the general population.

Between 5% and 10% of the worldwide case total has occurred in families in which at least one other member has been strongly suspected or proven to have the disease, and it is clear from the study of several large pedigrees that a familial form of CJD exists in which the disease assumes an autosomal dominant pattern of inheritance. Familial CJD shows a somewhat younger age at onset than sporadic cases, but with no evidence of paternal or maternal blood line bias, or, with the possible exception of one Finnish kindred (HALTIA et al. 1979), of anticipation (a trend to earlier age at onset in succeeding generations).

Here and there, an unexpectedly large number of cases occurring over a comparatively short period of time in a particular geographic area has suggested the phenomenon of case clustering. This issue was recently addressed in a systematic way in France, and reported clusters elsewhere were critically evaluated (RAUBERTAS et al. 1989), with the conclusion that no basis exists for statistically significant clustering anywhere in the world, except in Israel and Czecho-Slovakia. In Israel, a very high incidence of CJD has been reported in Jews of Libyan origin (KAHANA et al. 1974; ZILBER et al. 1991), and in Slovakia, ongoing focal outbreaks of CJD are occurring in two rural areas (MITROVA 1989; GOLDFARB et al. 1990). These outbreaks have appeared suddenly in the past decade in persons born and raised in the regions of Orava and Lucenec, attaining in some villages a current annual incidence equivalent to 500 cases per million people.

The proportion of familial cases in these clusters is extraordinarily high (40%), and we have recently shown that an identical amino acid-altering point mutation (glutamic acid to lycine) in codon 200 of the amyloid precursor gene (PRIP) is linked to the occurrence of disease in all families from both the Slovak and Israeli clusters (GOLDFARB et al. 1990a, b, c). We have now found the same mutation in other families with Eastern European or Sephardic Jew ancestry who were born and lived in other countries: Tunisia, France, Poland, Greece, Chile, and the U.S.

Different mutations in the same gene have been identified in several other non-Slovak, non-Jewish families. Finnish, French, and U.S. families of Dutch and Hungarian ancestry, all have an identical point mutation in codon 178 that changes aspartic acid to asparagine (GOLDFARB et al. 1991; NIETO et al. 1991); and two kindreds of English origin (one living in the U.S. for the past 300 years) have been shown to have similar extra repeats of an octapeptide coding sequence inserted in the region between codons 51 and 91 (OWEN et al. 1989; OWEN et al. 1990; GOLDFARB et al. 1990a).

Still another mutation has been identified in the majority of patients with GSS (HSIAO et al. 1989; GOLDGABER et al. 1989; DOH-URA et al. 1989; GOLDFARB et al. 1990; BROWN et al. 1991). This rare subset of CJD has been described in only a few families, in which the inheritance pattern, like familial CJD, is that of autosomal dominance. However, certain distinctive characteristics set it apart from the usual picture of familial CJD: the age at onset is earlier (35–55 years), the illness evolves more slowly (3–5 years), with spinocerebellar signs as well as dementia, and the neuropathology features long tract degeneration and prominent multicentric amyloid plaques as well as spongiform change. The mutation (in codon 102, changing proline to leucine) may play a role in the distinctive clinical and neuropathological characteristics of GSS, but it cannot be the only factor, since three GSS families have been found not to have the mutation: a French family and an American family of German origin each have an identical mutation in codon 117 (TATEISHI et al. 1990; HSIAO et al. 1991), and an American family reportedly has no identifiable mutation (FARLOW et al. 1990).

The question of inherited versus environmentally acquired disease in familial cases has been argued for years, and it is now apparent that, at the least, mutations in the amyloid precursor gene vastly increase susceptibility to disease, if indeed they are not its proximate cause. Eventual resolution of the question will undoubtedly come from a combination of molecular genetic and epidemiological studies of case cluster areas.

In addition to the familial subsets of CJD (including GSS) which account for not more than 10% of all cases, there is another subset of CJD that is fortunately rarer still: iatrogenic disease (BROWN 1990). These cases can be grouped into 3 categories: transmission from contaminated grafts, surgical instruments, and biological products. The routes of inoculation, incubation periods, and clinical durations of all reported cases of iatrogenic disease are summarized in Table 2.

It is apparent that an infection originating in direct proximity to the brain evolves much faster than one initiated by a peripheral route. In the centrally infected group (for which the infecting event was precisely dated) the incubation period averaged about 18–24 months, whereas in the peripherally infected

Table 2. Summary of regional and national surveys of Creutzfeldt–Jakob disease (through 1990)

Country	Survey years	Total cases (L)	Male-female ratio	Familial cases (%)	National annual mortality rate (cases/million)	Cases in higher incidence regions	Regional annual mortality rate (cases/million)
Chile	1955–1972	19			0.10	Santiago (14)	0.25
	1973–1977	16	1.04[a]	27[a]	0.31	Santiago (11)	0.73
	1978–1983	46			0.69	Santiago (16)	0.77
	1980–1982					Rural (4)	8.33
Czechoslovakia	1972–1990	71	0.8	8[b]	0.80	Lacent (19)[c]	11.4
						Orava (28)[c]	24.4
England and Wales	1964–1973	46[a]	0.53	7	0.09	2 Rural (3,5)	
	1970–1979	158	0.60	4–6	0.31		
	1980–1984	120	0.69	6	0.47		
Finland	1974–1984	31	0.43	22	0.91[d]		
France	1968–1977	178			0.34	Paris (28)	1.22
	1978–1982	151	0.81	6–9	0.58	Paris (13)	1.19
Hungary	1960–1986	65[a]	0.55	11	0.39	Budapest (33)	0.64
						Rural (10)	
Israel: Libyan	1963–1987	49	0.82	40	42.87		
non-Libyan		65	0.44	5	0.91		
Italy	1958–1971	32	1.00	8	0.05	Rome (21)	0.50
	1974–1979					Genova (6)	0.93
	1975–1984					Parma rural (13)	0.90
	1972–1985	87	0.53	0	0.11		
Japan	1975–1977	75[e]	0.79	6	0.15 (0.45)[f]	Fukuoka (5)	1.10[g]
United States	1973–1977	265			0.26	Boston (22)	0.43
	1979	148[h]	0.95		0.66		
	1980	142[h]	0.84		0.63		

[a] Histopathologically confirmed cases only.
[b] Excluding cluster regions.
[c] 40% familial cases.
[d] 1979–1984
[e] Questionnaire survey of neurology services.
[f] Estimated period prevalence rate.
[g] Point prevalence rate (1978)
[h] U.S. mortality statistics (death certificates).

group (for which only minimum and maximum incubation intervals could be determined with certainty) the average incubation period was 10–15 years. The same phenomenon has been observed in central versus peripherally infected primates when low infective doses were used for inoculation (BROWN 1988).

The other interesting aspect of the iatrogenic hormone-treated cases is that the clinical picture strongly resembles that of kuru, with ataxia an invariable first symptom, progressive cerebellar and basal ganglia dysfunction dominating the course of illness, and dementia being a late and often minor component (BROWN 1988). The reasons for this similarity are unclear, but may have to do with the peripheral route of infection in both diseases, or the young age at which infection occurred.

Although the mode of transmission presents no problem in these cases, it is not known whether the small number of patients developing CJD from among several thousand unaffected hormone-treated patients is entirely dependent upon the conditions of infection (randomly distributed small doses given by an inefficient route of infection), or whether a genetic predisposition to infection might also play some role. None of the four affected patients from whom we have DNA has any of the known mutations, and we are presently sequencing their PRIP coding regions in a search for new mutations.

4 The Origins of Transmissible Spongiform Encephalopathy

It is virtually certain that naturally occurring scrapie in sheep has been responsible for outbreaks of transmissible spongiform encephalopathy in mink (HARTSOUGH and BURGER 1965) cats (unpublished data), exotic ungulates (cited in KIMBERLIN 1990) and cattle (WELLS et al. 1987; WILESMITH et al. 1988) in the form of scrapie-infected tissue products incorporated into the diets of those animals. It is likely that a random, sporadic case of naturally occurring CJD was similarly responsible for the outbreak of kuru in the Fore people of New Guinea (GAJDUSEK 1977, 1990; KLITZMAN et al. 1984). Juxtaposing these two observations, we may wonder how these two source diseases stand in relation to each other. (There were no sheep or cattle in New Guinea before European arrival, when kuru attained epidemic proportions among the Fore people). Is scrapie at the origin of some human CJD and thus the root cause of the entire group of spongiform encephalopathies? Could the opposite be true? Or could both diseases have arisen independently?

Since the occasions for people to eat sheep far outnumber those for sheep to eat people, it is tempting to argue for the primacy of scrapie. Yet after years of effort, no pursuasive evidence has been produced to implicate scrapie in the occurrence of human disease (BROWN 1980; BROWN et al. 1987). On the other hand, the infrequency of scrapie-infected tissue coming to the marketplace, the irregular distribution and small amounts of scrapie agent in all tissue but brain and lymph nodes (meat, for example, has never been shown to be infective), the inefficiency of the oral route of infection and, possibly, varying degrees of individual host susceptibility could all make it extremely difficult to recognize human infection with scrapie. We are now collaborating with Eva Mitrová and her colleagues in Slovakia to try to elucidate the focus of CJD in a few Orava villages of shepherds with extensive exposure to scrapie-infected sheep. The sudden outbreak in the past decade and the great discrepancy of ages in nearly simultaneously afflicted close relatives suggests common source infection, possibly from their infected sheep, yet all cases of CJD in Oravas have the 200 codon mutation (GOLDFARB et al. 1990b), and cases have also occurred in American-born members of one of the families.

We must nevertheless admit that CJD is the only spongiform encephalopathy that ordinarily occurs in a noncontagious, nonepidemic pattern. Thus, if scrapie and CJD are causally related, it is perhaps equally plausible to suppose that sometime in the distant past, contaminated tissue from a sporadic case of CJD found its way into a sheep that, by virtue of grazing and parturition habits of that species, led to its continuing dissemination as epidemic scrapie.

Whichever interpretation is correct, we are still left with the undeniable fact that CJD is a rare and, in the main, randomly distributed disease, with next to no chance for natural horizontal spread (Table 3). We must therefore give serious consideration to the possibility that most cases of CJD might arise de novo as a spontaneous stochastic event of conversion of normal full-length (35 Kd) precursor protein to an infectious configuration, which occurs in about one person per million per annum (the wide incidence of CJD). This infectious amyloid has the unexpected property of causing a homologous nucleation-like induction of the same configurational change in the precursor molecules with which it makes contact.

The conversion of host proteins into amyloid is of course not unique to the spongiform encephalopathies: amyloid deposition in the brain in normal aging, Alzheimer's disease, Down's syndrome, all share this same pathogenetic feature (GAJDUSEK 1978, 1987, 1988a, b, 1990; GOLDGABER et al. 1987). What is unique is the phenomenon of experimental transmissibility of the amyloidoses, and the explanation of this is the key issue in the further elucidation of these diseases. Our recognition a decade ago that we were dealing with transmissible

Table 3. Summary of proven of highly suspect cases of iatrogenic Creutzfeldt–Jakob disease

Mode of infection	Number of patients	Agent entry into brain	Mean incubation period (range)	Clinical presentation
Instrumentation				
Neurosurgery	4[a]	Intracerebral	20 months (18–28)	Demential
Stereotactic EEG	2	Intracerebral	18 months (16–20)	Demential
Tissue transfer				
Corneal transplant	1	Optic nerve	18 months	Demential
Dura mater implant	4	Cerebral surface	33 months (19–45)	Demential
Tissue extract transfer				
Growth hormone	18	Hematogenous	13 years (5–21)[b]	Cerebellar
Gonadotrophin	2	Hematogenous	13 years (13–13)	Cerebellar

[a]Several additional CJD patients with preceding neurosurgery but no proven link to CJD are not tabulated

[b]Calculated from the treatment mid-point to onset of symptomatic CJD in published and unpublished cases

amyloidoses of brain gave us the paradigm of the familial amyloidotic polyneuropathies (ARAKI 1989; FRANGIONI 1989; GAJDUSEK 1987, 1991) and other systemic amyloidoses. It made it possible for us to focus on the key issue: the transmissibility of there spongiform amyloids of brain as opposed to the non-transmissible amyloidoses of brain (normal aging, Alzheimer's and Down's).

We have for some time been proposing a mechanism of nucleation and subsequent autopatterned epitaxial growth of protein polymerization (just as may occur in mineral crystallization) as a paradigm for the pathogenesis of amyloid formation in the transmissible spongiform encephalopathies (GAJDUSEK 1987). Furthermore, we have recently completed a series of experiments in which preparations of purified scrapie-associated fibrils exposed for 1 h to 360°C still retained some infectivity (BROWN et al. 1990). If this finding should be confirmed, it will mean that a small number of infectious molecules can survive a temperature higher than the decomposition points of all naturally occurring proteins and nucleic acids, as well as their component amino acids and nucleotides, but not high enough to affect simple inorganic mineral structures. Although by no means the only possible candidates as nucleating agents (GUIROY and GAJDUSEK 1988), minerals are interesting for several reasons. It is known that when any one of a variety of minerals is exposed in vitro to any one of several different normal host proteins, epitaxial protein crystalization can occur (MCPHERSON and SCHLICHTA 1988). There is no a priori reason to reject the possibility of a similar process in vivo, since minerals have free access to the central nervous system, as shown by the codeposition of aluminium and silicon in the cores of senile plaques in patients with Alzheimer's disease (CANDY et al. 1986) and of aluminium, silicon, and calcium in neurons containing neurofibrillary tangles in Guamanian patients with the amyotrophic lateral scierosis-Parkinson's dementia complex (GARRUTO et al. 1985).

Several different amino acid-altering mutations in the PRIP gene have already been found in patients with familial CJD and GSS. Most are responsible for the diseases while several other mutations are silent polymorphisms also found in normal subjects. A remarkably similar situation is seen in familial amyloidotic polyneuropathy (ARAKI 1988), which also shares with the spongiform encephalopathies the fact that each disease can be reproduced in transgenic mice in which a mutated human gene has been introduced (WAKASUGI et al. 1988; YI et al. 1990; HSIAO et al. 1990). More mutations will undoubtedly be discovered, and it is not unreasonable to imagine that any number of relatively slight changes in the normal host precursor protein encoded by this gene could facilitate its spontaneous, de novo conversion to the infectious β-pleated sheet configuration (GAJDUSEK 1989, 1990).

In human disease, the presence of a mineral (or other) nucleating factor could, in conjunction with mutationally altered precursor protein, initiate the autopatterning of epitaxial crystalline protein growth; in experimentally transmitted disease, where mutations play no role, the threshold for configurational change in the precursor protein may be lowered by the artificial introduction of an already prepatterned nucleating seed complex.

References

Alpers MP (1968) Kuru: implications of its transmissibility for the interpretation of its changing epidemiologic pattern. In: The central nervous system. International Academy of Pathology Monograph No. 9, Baltimore, pp 234–251

Alpers MP, Gajdusek DC (1965) Changing patterns of kuru: epidemiological changes in the periods of increasing contact of the Fore people with Western civilization. Am J Trop Med Hyg 14: 852–879

Alpers MP, Gajdusek DC, Ono SG (1975) Bibliography of kuru, 3rd edn. National Institute of Neurological and Communicative Disorders and Stroke, National Institutes of Health, Bethesda

Araki S (1989) Familial amyloidotic polyneuropathy, Japanese type. Disc Neurosci 5: 73–79

Beck E, Daniel PM, Alpers M, Gajdusek DC, Gibbs CJ Jr (1966a) Experimental "kuru" in chimpanzees: a pathological report. Lancet 2: 1056–1059

Beck E, Daniel PM, Gajdusek DC (1966b) A comparison between the neuropathological changes in kuru and scrapie, a system degeneration. In: Proceedings of the fifth international congress of neuropathology. Excerpta Medica, Amsterdam (International congress series, no 100)

Beck E, Daniel PM, Gajdusek DC, Gibbs CJ Jr (1973) Experimental kuru in the chimpanzee: an neuropathological study. Brain 96: 441–442

Beck E, Bak IJ, Christ JF, Gajdusek DC, Gibbs CJ Jr, Hassler R (1975) Experimental kuru in the spider monkey. Histopathological and ultrastructural studies of the brain during early stages of incubation. Brain 98: 585–612

Beck E, Daniel PM, Davey A, Gajdusek DC, Gibbs CJ Jr (1982) The pathogenesis of spongiform encephalopathies: an ultrastructural study. Brain 104: 755–786

Brown P (1980) An epidemiologic critique of Creutzfeldt–Jakob disease. Epidemiol Rev 2: 113–135

Brown P, Cathala F, Raubertas R, Gajdusek DC, Castaigne P (1987) The epidemiology of Creutzfeldt–Jakob disease: conclusion of a 15-year investigation in France and review of the world literature. Neurology 37: 895–904

Brown P (1988) The clinical neurology and epidemiology of Creutzfeldt–Jakob disease, with special reference to iatrogenic cases. In: Bock G, Marsh J (eds) Novel infections agents and the central nervous system. Wiley, Chichester

Brown P (1990) Iatrogenic Creutzfeldt–Jakob disease. Aust NZ J Med 20: 633–635

Brown P, Rodgers-Johnson P, Cathala F, Gibbs CJ Jr, Gajdusek DC (1984) Creutzfeldt–Jakob disease of long duration: clinico-pathological charactieristics, transmissibility, and differential diagnosis. Ann Neurol 16: 295–304

Brown P, Cathala F, Castaigne P, Gajdusek DC (1986) Creutzfeldt–Jakob disease: clinical analysis of a consecutive series of 230 neuropathologically verified cases. Ann Neurol 20: 597–602

Brown P, Wolff A, Liberski PP, Gajdusek DC (1990) Resistance of scrapie infectivity to steam autoclaving after formaldehyde fixation, and limited survival after ashing at 360°C: practical and theoretical implications. J Infect Dis 161: 467–472

Brown P, Goldfarb LG, Gajdusek DC (1991) The neurobiology of spongiform encephalopathy: infectious amyloidoses with a genetic twist. Lancet 337: 1019–1022

Candy JM, Oakley AE, Klinowski J, Carpenter TA, Perry RH, Atack JR, Perry K, Blessed G, Fairbairn A, Edwardson JA (1986) Aluminosilicates and senile plaque formation in Alzheimer's disease. Lancet 1: 354–357

Court LA, Dormont D, Brown P, Kingsbury D (eds) (1989) Unconventional virus diseases of the central nervous system. Commissariat à l'Energie Atomique (CEA), Service de Documentation, Fontenay-aux-Roses, France

Diringer H, Braig HR, Pocchiari M, Bode L (1986) Scrapie-associated fibrils in the pathogenesis of diseases caused by unconventional slow viruses. Disc Neurosci 3: 95–100

Doh-ura K, Tateishi J, Sasaki H, Kitamoto T, Sakaki Y (1989) Pro-leu change at position 102 of prion protein is the most common but not the sole mutation related to Gerstmann–Sträussler syndrome. Biochem Biophys Res Commun 2: 974–979

Farlow MR, Yee RD, Delouhy SR, Conneally PM, Azzarelli B, Ghetti B (1989) Gerstmann–Sträussler–Scheinker disease. I. Extending the clinical spectrum. Neurology 39: 1446–1452

Farquhar J, Gajdusek DC (eds) (1980) Kuru: early letters and field notes from the collection of D. Carleton Gajdusek. Raven, New York

Frangione B (1989) Systemic amyloidosis and cerebral amyloid angiopathy. Disc Neurosci 5: 79–84

Gajdusek DC (1963) Kuru. Trans R Soc Trop Med Hyg 57: 151–169

Gajdusek DC (1972) Spongiform virus encephalopathies. J Clin Pathol (Suppl) 25: 78–83

Gajdusek DC (1973) Kuru in the New Guinea highlands. In: Spillane JD (ed) Tropical neurology. Oxford, New York

Gajdusek DC (ed) (1976) Correspondence on the discovery and original investigations of kuru. Smadel–Gajdusek correspondence 1956–1959. National Institutes of Health, Bethesda

Gajdusek DC (1977) Unconventional viruses and the origin and disappearance of kuru. Revised version. Science 197: 943–960

Gajdusek DC (1978) Slow infections with unconventional viruses. Harvey Lect 72: 283–353

Gajdusek DC (1985) Subacute spongiform virus encephalopathies caused by unconventional viruses. In: Maramorosch K (ed) Subviral pathogens of plants and animals: viroids and prions. Academic, New York

Gajdusek DC (1987) A newly recognized mechanism of pathogenesis in Alzheimer's disease, amyotrophic lateral sclerosis, and other degenerative neurological diseases: the β-fibrilloses of brain. In: Jariwalla RJ, Schwoebel SL (eds) Nutrition, health and peace. Linus Pauling Institute, Palo Alto

Gajdusek DC (1988a) Etiology versus pathogenesis: the cause of post-translational modifications of host specified brain proteins to amyloid configuration. In: Sinet PM, Lamour Y, Christen Y (eds) Genetics and Alzheimer's disease. Proceedings of a meeting held by the foundation IPSEN pour la Recherche Thérapeutique, Paris. Springer, Berlin Heidelberg New York

Gajdusek DC (1988b) Transmissible and non-transmissible amyloidoses: autocatalytic post-translational conversion of host precursor proteins to β-pleated configurations. J Neuroimmunol 20: 95–110

Gajdusek DC (1957–1989) Journals 1957–1989, 39 volumes. National Institutes of Health, Bethesda

Gajdusek DC (1989) Fantasy of a "virus" from the inorganic world: pathogenesis of cerebral amyloidoses by polymer nucleating agents and/or "viruses". In: Neth R, Gallo RC, Greaves MF, Gacdicke G, Ritter J (eds) Modern trends in human leukemia VIII. Springer, Berlin Heidelberg New York

Gajdusek DC (1990) Subacute spongiform encephalopathies: transmissible cerebral amyloidoses caused by unconventional viruses. In: Virology, 2nd edn. Raven, New York

Gajdusek DC (1991) Transthyretin amyloidoses of familial amyloidotic polyneuropathy as a paradigm for the genetic control of de novo generation of Creutzfeldt–Jakob disease infectious amyloid by a spontaneous change in configuration of host precursor protein. In: Marchant BA (ed) Proceedings of European Commission Seminar on Sub-Acute spongiform encephalopathies, Brussels, November 13, 1990

Gajdusek DC, Gibbs CJ Jr (1971) Transmission of the two subacute spongiform encephalopathies of man (kuru and Creutzfeldt–Jakob disease) to New World monkeys. Nature 230: 588–591

Gajdusek DC, Gibbs CJ Jr (1973) Subacute and chronic diseases caused by atypical infections with unconventional viruses in aberrant hosts. Perspect Virol 8: 279–311

Gajdusek DC, Gibbs CJ Jr (1975) Slow virus infections of the nervous system and the laboratories of slow, latent and temperate virus infections. In: Chase TN (ed) The nervous system, vol 2: The clinical neurosciences. Raven, New York

Gajdusek DC, Zigas V (1957) Degenerative disease of the central nervous system in New Guinea: the endemic occurrence of "kuru" in the native population. N Engl J Med 257: 974–978

Gajdusek DC, Zigas V (1959) Kuru: clinical, pathological and epidemiological study of an acute progressive degenerative disease of the central nervous system among natives of the Eastern Highlands of New Guinea. Am J Med 26: 442–469

Gajdusek DC, Gibbs CJ Jr, Alpers MP (eds) (1965) Slow, latent and temperate virus infections. NINDS monograph no 2, National Institutes of Health, PHS Publication no 1378, US Government Printing Office, Washington, DC

Gajdusek DC, Gibbs CJ Jr, Alpers MP (1966) Experimental transmission of a kuru-like syndrome in chimpanzees. Nature 209: 794–796

Gajdusek DC, Gibbs CJ Jr, Alpers M (1967) Transmission and passage of experimental 'kuru' to chimpanzees. Science 155: 212–214

Garruto RM, Swyt C, Yanagihara R, Fiori CE, Gajdusek DC (1985) Intraneural co-localization of silicon with calcium and aluminium in amyotrophic lateral sclerosis and Parkinsonism with dementia of Guam. New Engl J Med 315: 711–712

Gibbs CJ Jr, Gajdusek DC, Asher DM, Alpers MP, Beck E, Daniel PM, Matthews WB (1968) Creutzfeldt–Jakob disease (subacute spongiform encephalopathy): transmission to the chimpanzee. Science 161: 388–389

Gibbs CJ Jr, Gajdusek DC, Amyx HL (1979) Strain variation in the viruses of Creutzfeldt–Jakob disease (subacute spongiform encephalopathy): transmission to the chimpanzee. Science 161: 388–389

Gibbs CJ Jr, Amyx HL, Bacote A, Masters C, Gajdusek DC (1980) Oral transmission of kuru, Creutzfeldt–Jakob disease, and scrapie to nonhuman primates. J Infect Dis 142: 205–208

Glass RM (1963) Cannibalism in the kuru region. Department of Public Health, Papua New Guinea

Goldfarb LG, Brown P, Goldgaber D, Garruto RM, Yanagihara R, Asher DM, Gajdusek DC (1990a) Identical mutation in unrelated patients with Creutzfeldt–Jakob disease. Lancet 336: 174–175

Goldfarb LG, Mitrová E, Brown P, Toh BH, Gajdusek DC (1990b) Mutation in condon 200 of scrapie amyloid protein gene in two clusters of Creutzfeldt–Jakob disease in Slovakia. Lancet 336: 514–515

Goldfarb LG, Korczyn AD, Brown P, Chapman J, Gajdusek DC (1990c) Mutation in codon 200 of scrapie amyloid precursor gene linked to Creutzfeldt–Jakob disease in Sephardic Jews of Libyan and non-Libyan origin. Lancet 336: 637–638

Goldfarb LG, Brown P, Goldgaber D, Asher DM, Rubenstein R, Brown WT, Piccardo P, Kascsak, Boellaard, Gajdusek DC (1990d) Creutzfeldt–Jakob disease and kuru patients lack a mutation consistently found in the Gerstmann–Sträussler–Scheinker syndrome. Exp. Neurol 108: 247–250

Goldfarb LG, Haltia M, Brown P, Nieto A, Kovanen J, McCombie WR, Trapp S, Gajdusek DC (1991) New mutation in scrapie amyloid precursor gene in Finnish Creutzfeldt–Jakob disease kindred. Lancet 337: 425

Goldgaber D, Lerman MI, McBride OW, Saffiotti U, Gajdusek DC (1987) Characterization and chromosomal localization of cDNA encoding brain amyloid of Alzheimer's disease. Science 235: 877–880

Goldgaber D, Goldfarb LG, Brown P, Asher DM, Brown WT, Lin S, Teener JW, Feinstone SM, Rubenstein R, Kascsak RJ, Boellaard JW, Gajdusek DC (1989) Mutations in familial Creutzfeldt–Jakob disease and Gerstmann–Sträussler–Scheinker's syndrome. Exp Neurol 106: 204–206

Guiroy DC, Gajdusek DC (1989) Fibril-derived amyloid enhancing factors as nucleating agents in Alzheimer's disease and transmissible virus dementia. Disc Neurosci 5: 69–73

Hadlow WJ (1959) Scrapie and kuru. Lancet 2: 289–290

Haltia M, Kovanen J, van Crevel H, Bots GThAM, Stefanko S. Familial Creutzfeldt–Jakob disease (1979) J Neurol Sci 42: 381–389

Hartsough GR, Burger D (1965) Encephalopathy of mink. I. Epizootiologic and clinical observations. J Infect Dis 115: 387–392

Hornabrook RW, Wagner F (1975) Creutzfeldt–Jakob disease. P N G Med J 18: 226–228

Hsiao KK, Scott M, Foster D, Groth DF, DeArmond SJ, Prusiner SB (1990) Spontaneous neurodegeneration in transgenic mice with mutant prion protein. Science 250: 1587–1590

Hsiao K, Baker HF, Crow TJ, Poulter M, Owen F, Terwilliger JD, Westaway D, Ott J, Prusiner SB (1989) Linkage of a prion protein missense varient to Gerstmann–Sträussler syndrome. Nature 338: 342–345

Hsiao KK, Cass C, Schallenberg GD, Bird T, Devine-Gage E, Wisniewski H, Prusiner SB (1991) A prion protein variant in a family with telecephalic form of Gerstmann–Sträussler–Scheinker syndrome. Neurology 41: 681–684

Kahana E, Alter M, Braham J, Sofer D (1974) Creutzfeldt–Jakob disease: focus among Libyan Jews in Israel. Science 183: 90–91

Kimberlin RH (1990) Transmissible encephalopathies in animals. Can J Vet Res 54: 30–37

Klatzo I, Gajdusek DC, Zigas V (1959) Pathology of kuru. Lab Invest 8: 799–847

Klitzman RL, Alpers MP, Gajdusek DC (1984) The natural incubation period of kuru and the episodes of transmission in three clusters of patients. Neuroepidemiology 3: 13–20

Lampert PW, Gajdusek DC, Gibbs CJ Jr (1972) Subacutre spongiform virus encephalopathies: scrapie, kuru and Creutzfeldt–Jakob disease. Am J Pathol 68: 626–646

Landis DMD, Williams RS, Masters CL (1981) Golgi and electron microscopic studies of spongiform encephalopathy. Neurology 31: 538–549

Lindenbaum S (1979) Kuru sorcery. Mayfield, Palo Alto

Masters CL, Harris JO, Gajdusek DC, Gibbs CJ Jr, Bernoulli C, Asher DM (1979) Creutzfeldt–Jakob disease: patterns of worldwide occurrence and the significance of familial and sporadic clustering. Ann Neurol 5: 177–188

Masters CL, Gajdusek DC, Gibbs CJ Jr (1981) Creutzfeldt–Jakob disease virus isolations from the Gerstmann–Sträussler syndrome, with an analysis of the various forms of amyloid plaque deposition in the virus-induced spongiform encephalopathies. Brain 104: 559–588

Matthews WB (1975) The clinical aspects of slow virus infections of the human brain. In: Jllis LS (ed) Viral diseases of the central nervous system. Bailliere Tindall, London

McPherson A, Shlichta P (1988) Heterogeneous and epitaxial nucleation of protein crystals on mineral surfaces. Science 239: 385–387

Merz PA, Somerville RA, Wisniewski HM, Iqbal K (1981) Abnormal fibrils from scrapie infected brain. Acta Neuropathol (Berl) 54: 63–74

Merz PA, Rohwer RG, Somerville RA, Wisniewski HM, Gibbs CJ Jr, Gajdusek DC (1983a) Scrapie associated fibrils in human Creutzfeldt–Jakob disease. J Neuropathol Exp Neurol 42: 327

Merz PA, Somerville RA, Wisniewski HM (1983b) Abnormal fibrils in scrapie and senile dementia of Alzheimer's type. In: Court L, Cathala F (eds) Virus non conventionnels et affections du système nerveux central. Masson, Paris

Merz PA, Somerville RA, Wisniewski HM, Manuelidis L, Manuelidis EE (1983c) Scrapic associated fibrils in Creutzfeldt–Jakob disease. Nature 306: 474–476

Merz PA, Rohwer RG, Kascsak R, et al. (1984) An infection specific particle from the unconventional slow virus diseases. Science 225: 385–387

Mitrova E (1989) Analytical epidemiology and risk factors in CJD. In: Court LA, Dormont D, Brown P, Kingsbury DT (eds) Unconventional virus diseases of the central nervous system. Commissariat à 1'Energie Atomique (CEA), Service de Documentation, Fontenay-aux-Roses, France

Narang HK, Asher DM, Gajdusek DC (1987) Tubulofilaments in negatively stained scrapie-infected brains: relationship to scrapie associated fibrils. Proc Natl Acad Sci USA 84: 7730–7734

Narang HK, Asher DM, Gajdusek DC (1988) Evidence that DNA is present in abnormal tubulofilamentous structures found in scrapie. Proc Natl Acad Sci USA 85: 3575–3579

Nieto A, Goldfarb LG, Brown P, McCombie WR, Trapp S, Asher DM, Gajdusek DC (1991) Codon 178 mutation in ethnically diverse Creutzfeldt–Jakob disease families. Lancet 337: 622–623

Owen F, Poulter M, Shah T, Collinge J, Lofthouse R, Baker H, Ridley R, McVey J, Crow TJ (1990) An inframe insertion in the prion protein gene in familial Creutzfeldt–Jakob disease. Mol Brain Res 7: 273–276

Owen F, Poulter M, Lofthouse R, Collinge J, Crow TJ, Risby D, Baker HF, Ridley RM, Hsaio K, Prusiner SB (1989) Insertion in prion protein gene in familial Creutzfeldt–Jakob disease. Lancet 1: 51–52

Raubertas RF, Brown P, Cathala F, Brown I (1989) The question of clustering of Creutzfeldt–Jakob disease. Am J Epidemiol 129: 146–154

Salazar AM, Masters CL, Gajdusek DC, Gibbs CJ Jr (1983) Syndromes of amyotrophic lateral sclerosis and dementia: relation to transmissible Creutzfeldt–Jakob disease. Ann Neurol 14: 17–26

Tateishi J, Sato Y, Nagara H, Boellaard JW (1984) Experimental transmission of human subacute spongiform encephalopathy in small rodents. Acta Neuropathol 64: 55–58

Tateishi J, Kitamoto T, Doh-ura K, Sakaki Y, Steinmetz G, Tranchant C, Warter JM, Heldt N (1990) Immunochemical, molecular genetic, and transmission studies on a case of Gerstmann–Sträussler–Scheinker syndrome. Neurology 40: 1578–1581

Trabattoni G, Lechi A, Bettoni L, Macchi G, Brown P (1990) Considerations on a group of 13 patients with Creutzfeldt–Jakob disease in the region of Parma (Italy). Eur J Epidemiol 6: 239–243

Traub R, Gajdusek DC, Gibbs CJ Jr (1977) Transmissible virus dementia: the relation of transmissible spongiform encephalopathy to Creutzfeldt–Jakob disease. In: Kinsbourne M, Smith L (eds) Aging and dementia. Spectrum, New York

Wakasugi S, Inomoto T, Yi S et al. (1988) Potential animal model for familial amyloidotic polyneuropathy through introduction of human mutant transthyretin gene into mice. In: Takashi I, Shakuro A, Fumike O, Shumuko K, Eiro T (eds) Amyloid and Amyloidoses. Plenum Press, New York, pp 393–398

Wells GA, Scott AC, Johnson CT, Gunning RF, Hancock RD, Jeffrey M, Dawson M, Bradley R (1987) A novel progressive spongiform encephalopathy in cattle. Vet Rec 121: 419–420

Will RG, Matthews WB (1979) A retrospective study of Creutzfeldt–Jakob disease in England and Wales 1970–1979. J Neurol Neurosurg Psychiatry 47: 134–140

Wilesmith JW, Wells GAH, Cranwell MP, Ryan JBM (1988) Bovine spongiform encephalopathy: epidemiological studies. Vet Rec 123: 638–644

Zigas V, Gajdusek DC (1957) Kuru: clinical study of a new syndrome resembling paralysis agitans in natives of the Eastern Highlands of Australian New Guinea. Med J Aust 2: 745–754

Yi S, Takahashi K, Tashiro F, Wakasugi S, Yamamura K, Araki S (1990) Pathological similarity to human familial amyloidotic polyneuropathy (FAP) type I in transgenic mice carrying the human mutant transthyretin gene. Abstract no. 07/5 in Programme and Abstracts of the VIth International Symposium on Amyloidosis, Oslo, Norway, August 5–8, p 58

Zilber N, Kahana E, Abraham M (1991) Creutzfeldt–Jakob disease in Israel: an epidemiological evaluation. Neurology (in press)

Bovine Spongiform Encephalopathy

J. W. WILESMITH[1] and G. A. H. WELLS[2]

1 Introduction

The emergence in domestic cattle of a spongiform encephalopathy homologous with scrapie of sheep and goats (WELLS et al. 1987; HOPE et al. 1988; FRASER et al. 1988) has renewed interest in many aspects of the unconventional slow virus diseases of man and other mammals, none more so than their methods of transmission. From early clinical descriptions it is clear that scrapie, the exemplar of this disease group, occurred in Europe from 1700 onwards and has remained, or since become, endemic in many countries, including the UK (PARRY 1983a). Although "calamitous epidemics" featured in the scrapie occurrences in Northwest Europe during the latter part of the 18th century (PARRY 1983a), the magnitude of the present epidemic of bovine spongiform encephalopathy (BSE) is, among animal populations, without precedent in the recorded history of this group of neurodegenerative diseases. The origin of BSE is therefore fundamental to our knowledge of the properties of their unusual, if not unique, causal agents. The chronology of recognition of the naturally occurring transmissible spongiform encephalopathies and their geographic distribution is summarised in Table 1. Prior to the occurrence of BSE, the only ruminant species, apart from domestic sheep and goats, known to be affected by such diseases

[1]Epidemiology Unit and [2]Pathology Department, Central Veterinary Laboratory, New Haw, Weybridge, Surrey KT15 3NB, UK

Current Topics in Microbiology and Immunology, Vol. 172

Table 1. Chronology of recognition of naturally occurring transmissible spongiform encephalopathies

Host	Disease	Geographic distribution	Original clinical observation	Initial report
Sheep	Scrapie	Widely distributed except Australia, New Zealand, some European countries	ca. 1730, possibly much earlier	Comber (1772) (cited by Parry 1983a)
Goat	Scrapie	?	?	Eisenmayer and Göbel (1872) (cited by Parry 1983a)
Man	Kuru	Papua New Guinea	ca. 1900	Zigas and Gajdusek (1957) Gajdusek and Zigas (1957)
	Creutzfeldt–Jakob disease	Worldwide	1920	Jakob (1921a–c)
	Gerstmann–Sträussler–Scheinker syndrome	Worldwide	1926	Gerstmann, Sträussler, Scheinker (1936)
Farmed mink (*Mustela vison*)	Transmissible mink encephalopathy	North America, Europe	1947	Burger and Hartsough (1964)
Mule deer (*Odocoileus hemionus hemionus*)	Chronic wasting disease	North America	1967	Williams and Young (1980)
Elk (*Cervus elaphus nelsoni*)				Williams and Young (1982)
Cattle	Bovine spongiform encephalopathy	UK, Republic of Ireland	1985	Wells et al. (1987) Bassett and Sheridan (1989)
Nyala (*Tragelaphus angasi*)	[a]	UK	1986	Jeffrey and Wells (1988) (1986, 1987)
Gemsbok (*Oryx gazella*)	[a]	UK	1987	
Eland (*Taurotragus oryx*)	[a]	UK	1989	Fleetwood and Furley (1990)
Arabian oryx (*Oryx leucoryx*)	[a]	UK	1989	Kirkwood et al. (1990) (1989, 1989)
Greater kudu (*Tragelaphus strepsiceros*)	[a]	UK	1989	
Cat (*Felis domesticus*)	[a]	UK	1990	Wyatt et al. (1990)

[a] Transmissibility not established, but occurrence geographically and temporally associated with the BSE epidemic in the UK

were captive groups of the North American cervids: mule deer (*Odocoileus hemionus hemionus*) and elk (*Cervus elaphus nelsoni*) (WILLIAMS and YOUNG 1980; WILLIAMS and YOUNG 1982).

In this chapter BSE is reviewed with particular emphasis on case definition and the epidemiology and probable origin.

2 Case Definition

The present nosological grouping of the scrapie-like diseases has evolved, in the absence of conventional etiological criteria, principally from close similarities between their neuropathology, their transmissibility and the biology of their causal agents. The properties of the agents have so far precluded the development of antemortem diagnostic methods. Case finding has depended upon clinical and neuropathological data and transmission to laboratory animals. The impracticality of establishing transmissibility routinely in individual cases of naturally occurring spongiform encephalopathies in animals has necessitated reliance upon neuropathological criteria for diagnosis and case definition.

2.1 Clinical Case Definition

Clinical neurological examinations in cattle rely, of practical necessity, on observation of the animals' interaction with the environment, supplemented with limited clinical neurological testing. Descriptions of the clinical features of BSE (WELLS et al. 1987; CRANWELL et al. 1988; GILMOUR et al. 1988; WILESMITH et al. 1988; WELLS 1989; WINTER et al. 1989) indicate a close semblance to natural scrapie in sheep with some contrasts of emphasis in the expression of individual clinical signs. The progressive and often prolonged clinical course, of 1–6 months, noted in the first recorded cases of BSE provided a fortuitous contrast with the mainly acute metabolic and inflammatory neurological diseases of adult domestic cattle common in the British Isles. The predominant neurological signs, apprehensive behavior, hyperesthesia and ataxia of gait, reflect a diverse disturbance of central nervous system functions but, together with clinical progression, characterise the clinical diagnosis. The relative constancy of these features is borne out by the unchanging primacy of their reported frequency through the course of the epidemic. Initially described in the first six cases to be recorded (WELLS et al. 1987), they were the most frequently observed signs in a set of 192 neuropathologically confirmed cases with clinical onset before April 1988 (WILESMITH et al. 1988) and have remained so to the end of December 1990 (WILESMITH, unpublished observations). Loss of general bodily condition, reduced liveweight or diminishing milk yield accompany neurological signs in almost 79% of cases. The constellation of clinical signs comprising apprehensiveness, hyperesthesia, locomotor signs and wasting are present in the majority of cases from 6–8 weeks after onset. Truncation of the clinical course by euthanasia, initially for humane and economic reasons, but later also because of regulatory control measures (ORDER 1988), has inevitably limited information on features of the advanced disease. The distribution of the duration of illness from onset to slaughter or death, in the set of 192 cases with clinical onset before the mandatory slaughter policy, is given in Fig. 1. Clinical disease duration ranged from 7 days to 14 months. Muscular weakness, affecting primarily the pelvic limbs, frequently accompanied the ataxia of gait, and may be among the presenting signs. With clinical progress and in the advanced phase of illness, paresis may be the predominant locomotor problem, resulting in prolonged recumbency. Clinical illness of less than 15 days, characterised by acute onset of neurological signs including gait deficits and usually culminating in recumbency with marked muscular weakness, occurs in a small proportion of cases. BSE must therefore be considered in the differential diagnosis of the relatively common occurrence of pathological decubitus in adult cattle. Remarkably short clinical histories have also been recorded in cases of scrapie in sheep (JOUBERT et al. 1972) and Creutzfeldt–Jakob disease (CJD) (BERNOULLI et al. 1979). Should BSE occur in an area where rabies is endemic, a similar timescale of presenting signs may give rise to difficulties in their clinical differentiation (WELLS 1988). In the UK, where rabies is absent, there is usually a close but variable agreement

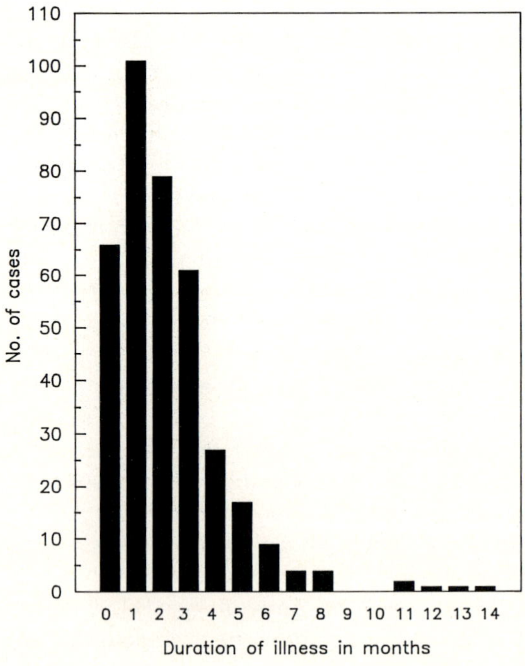

Fig. 1. Distribution of the duration of illness in 192 confirmed cases of BSE

between clinical diagnosis and neuropathological confirmation of BSE (WELLS et al. 1989; WILESMITH, unpublished observations). A similar correlation pertains in scrapie (MCDANIEL and MOREHOUSE 1964; PARRY 1983c).

In natural sheep scrapie there are two main clinical signs, pruritus and incoordination of gait, one of which usually emerges to dominate the clinical course (PALMER 1976; KIMBERLIN 1981; PALSSON and SIGURDSSON 1958; ZLOTNIK and KATIYAR 1961). PALSSON (1979) observed that Icelandic sheep with *rida* (synonymous with scrapie), did not show signs of pruritus prior to the occurrence of the disease outside the original endemic area, but in newly affected areas pruritus was the first and most prominent sign. In BSE, pruritus, though present in some cases, does not seem to become predominant. Nevertheless, affected cattle have altered sensory perception which manifests particularly as exaggerated responses to a variety of tactile and auditory stimuli (WILESMITH et al. 1988; WELLS 1989).

In Creutzfeldt–Jakob disease (CJD), involuntary movements, noteably myoclonus, are regarded as important diagnostic components (BROWN et al. 1986). Similar adventitial movement disorders have been recorded in scrapie (PALMER 1957) and in BSE (WELLS 1989). Tremor, reported in 62% of 192 cases of BSE (WILESMITH et al. 1988) was not differentiated further and almost certainly includes muscle fasciculation and myoclonus. In the series of six cases examined by SCOTT et al. (1988) myoclonus was a constant feature.

Autonomic dysfunctions are described in scrapie, including altered thermoregulation and tachycardia (PARRY 1983b). Detailed sequential clinical studies of the disease course in BSE are not reported, but bradycardia is recorded in one series of 14 cases from a single dairy herd (WINTER et al. 1989).

Variations of host genotype and/or agent strains are known to influence profoundly the clinicopathology in mouse models of scrapie (DICKINSON 1976; FRASER 1979). Similarity of the clinical signs among cases of sheep scrapie within the same flock (MITCHELL and STAMP 1983), the changes in clinical signs with time within sheep populations (PALSSON 1979), the differences in the clinical syndrome between sheep breeds, and the emergence in goats of two distinct syndromes, "drowsy" and "scratching", from passage of the SSBP/1 sheep scrapie source probably all reflect variations of these factors in the ruminant disease (DICKINSON 1976). No differences in the clinical signs between cases of BSE have as yet been attributed to differences in breed, herd or geographic location.

Abnormalities of the electroencephalogram using non-invasive surface electrodes in cases of BSE (SCOTT et al. 1988; CLARKE et al. 1989) are not yet adequately characterised and validated to use in case definition, and the method may prove impractical for routine diagnostic application. No consistent abnormalities of blood biochemistry (JOHNSON and WHITAKER 1988) or cerebrospinal fluid (CSF) biochemistry, cytology and electrophoretic pattern (SCOTT et al. 1990) have been detected in BSE. The latter authors have, however, suggested that such examinations of CSF allow the differentiation of BSE from certain viral and bacterial infections of the CNS.

2.2 Neuropathological Case Definition

The transmissible encephalopathies ascribed to novel infectious agents have pathomorphological generalities in common with many other neurodegenerative diseases. They present progressive, symmetrically distributed, selective degenerations of neuronal "systems". They also have clinicopathological variability with common pathological changes which exhibit overlapping topographies. More specifically, the pathology of the transmissible neurodegenerations includes vacuolar degeneration of neurons, astrocyte reaction and amyloidogenic accumulations of an altered neuronal membrane glycoprotein. Variations in the relative representation of these changes among different diseases within the group and between individual cases provide inconsistencies which are problematic to the histopathological case definition.

The light microscopic pathology of BSE is, qualitatively, essentially the same as that of natural scrapie, differing only in the relative prominence of the individual component features. Neuronal vacuolation in BSE, like scrapie, comprises two separate features, a microcystic vacuolation of grey matter neuropil (spongiosis or spongiform change) and characteristic large single, or multiple, vacuoles in the perikarya of neurons (WELLS et al. 1987). Vacuoles in clearly discernible neurites of affected neurons appear identical to those of spongiosis which are ultrastructurally localised mainly to postsynaptic processes of dendrites (LIBERSKI et al., to be published). In BSE, spongiosis is the more conspicuous change in many affected neuroanatomic nuclei, but there are predilection sites for the occurrence of each feature. The distribution pattern of vacuolation throughout the brain in BSE presents remarkable uniformity (WELLS and WILESMITH 1989). In the medulla oblongata, where the nuclei of the solitary tract and the spinal tract of the trigeminal nerve are affected in 99.6% of cases, spongiosis predominates, often in the absence of vacuolated perikarya (WELLS et al. 1989). Confirmation of vacuolar changes in these areas is the basis of the statutory diagnosis (ORDER 1988). The predominance of spongiosis in BSE is in contrast to natural scrapie of sheep, in which, historically, greater diagnostic emphasis has been placed on the vacuolated perikaryon (MCDANIEL and MOREHOUSE 1964; FRASER 1976).

The diagnostic specificity of vacuolar changes in unconventional virus encephalopathies is in general, however, subject to several qualifications. For each of the diseases within the group, case definition must be based on the precise morphology of the vacuolar changes, their topographic distribution and other associated pathological features.

The term spongiform change has been widely used in the human and comparative neuropathology literature to describe vacuolar changes in grey or white matter attributable to a number of viral infections, toxins and metabolic disturbances, and similar changes can result from certain postmortem artefacts (MASTERS and RICHARDSON 1978; MASTERS and GAJDUSEK 1982; WELLS and WELLS 1989).

MASTERS and RICHARDSON (1978) made an important distinction between the appearances of spongiform change and status spongiosus in CJD. The former, defined as small round or ovoid vacuoles in the grey matter neuropil, they considered as characteristic of the subacute spongiform encephalopathies, whereas the latter, comprising irregular cavitation of the neuropil, they regarded as non-specific and associated with an end stage gliosis. Status spongiosus as defined by these authors, does not feature in natural scrapie (FRASER 1976, 1979) nor is it observed in BSE (WELLS, personal observation). This may relate to differences between survival times of the human patients and the ruminants relative to clinical course, but perhaps also differences between the species in the tempo of the degenerative process.

Vacuolation of perikarya is also not an absolute criterion in the diagnosis of scrapie, since vacuoles in small numbers and in otherwise unremarkable neurons are reported in clinically healthy sheep (ZLOTNIK and RENNIE 1957), including those from flocks considered free of scrapie (ZLOTNIK and RENNIE 1958). Rarely, vacuolated neurons occur in clinically normal goats (PATTISON et al. 1959).

In domestic cattle large intraneuronal vacuoles identical to those observed in scrapie and BSE were reported in the brains of clinically healthy cows sampled from an abattoir and in cattle of widely differing ages, irrespective of accompanying pathologies (WELLS et al. 1987). Such reports indicate that the occurrence of this apparently incidental vacuolation in cattle is mainly confined to the red nucleus, but further studies are required in healthy cattle to examine the possibility of the occurrence of incidental neuronal vacuolation at a low frequency in other neuroanatomic locations.

There are also instances of scrapie (FRASER 1976), CJD (MANUELIDIS and MANUELIDIS 1979; BARINGER et al. 1983), kuru (KAKULAS et al. 1967) and experimental transmissible mink encephalopathy (TME) (MARSH et al. 1976) where advanced clinical disease is associated with minimal or insufficient vacuolar changes to confirm a diagnosis.

Minimal changes necessitating an inconclusive diagnosis are also a feature of a small proportion of clinically suspect cases of BSE (WELLS et al. 1989).

Pathological homologies between BSE and scrapie include characteristic scrapie associated fibrils (SAF), demonstrated by electron microscopy, in proteinase-K treated, detergent extracts of brain (WELLS et al. 1987; HOPE et al. 1988) and the corresponding molecular marker, the major fibril protein PrP (HOPE et al. 1988; FARQUHAR et al. 1989; SCOTT et al. 1990).

The specificity of fibril detection in BSE has been confirmed in a large case sample of cattle brains, but the sensitivity of detection is apparently dependent on the brain region examined, with the highest sensitivity confined to the brain stem areas (SCOTT et al. 1990). Although requiring validation for routine use, preliminary work indicates that fibril demonstration in BSE, even in the presence of postmortem decomposition, where a histopathological diagnosis is precluded, provides an invaluable supplement or substitute for conventional neuropathological case definition (SCOTT AC, personal communication).

3 Epidemiology

3.1 Initial and Descriptive Epidemiological Findings

Following the recognition of BSE in November 1986 it became apparent that the disease was geographically widespread, and an epidemiological study was started in June 1987. The results of the initial phase of this study have been described previously (WILESMITH et al. 1988). Briefly, they indicated that the form of the epidemic was typical of an extended common source in which all affected animals were index cases. Clinical histories suggested also that the earliest cases associated with the epidemic occurred in April 1985. The use of therapeutic or agricultural chemicals and biological products were not common factors and there was no evidence that BSE had been introduced into Great Britain by imported cattle, semen or animal products. The transmission of the scrapie agent from sheep to cattle via direct or indirect contact on affected farms was also precluded. These initial epidemiological findings together with the compelling clinicopathological similarity of BSE with scrapie were consistent with exposure of cattle to a scrapie-like agent via cattle feedstuffs containing ruminant-derived protein in the form of meat and bone meal. Further investigations of the epidemiological data by computer simulation modelling suggested that such exposure, sufficient to result in a detectable clinical incidence, began in 1981/82 and, although both adults and calves had been exposed, the majority of affected animals became infected in calfhood. This simulation also indicated that the incubation period has a log normal distribution and that the incidence in animals 5 years of age and older would necessarily increase in the period 1988 to 1990, because, at the time of the initial analyses and simulation, insufficient time had elapsed to witness the full extent of the incubation period distribution.

The collection of details on all affected animals and their herds has continued since the start of the epidemic. No unexpected changes have been revealed, but since reporting of the disease became mandatory in Great Britain in June 1988 and indemnity payments made for all clinically suspect cases slaughtered from August 1988 (ORDER 1988), a better estimate of the true incidence of the disease has been realised. The epidemic curve (Fig. 2) depicts data on both histopathologically confirmed cases and clinically suspect cases during the period April 1985 until December 1989. The interpretation of the epidemic curve is confounded by an inconstant case ascertainment rate consequent on the change from voluntary, to compulsory notification. This is further influenced by an improvement in awareness of the clinical signs. However, data from those herds for which early voluntary notifications were provided do not indicate any major change in incidence until July 1989, when a stepwise increase occurred. This increase would appear to be consistent with the recycling of significant amounts of infected cattle material, via meat and bone meal, from the beginning of the epidemic in 1984/85 (WILESMITH and RYAN, unpublished observations).

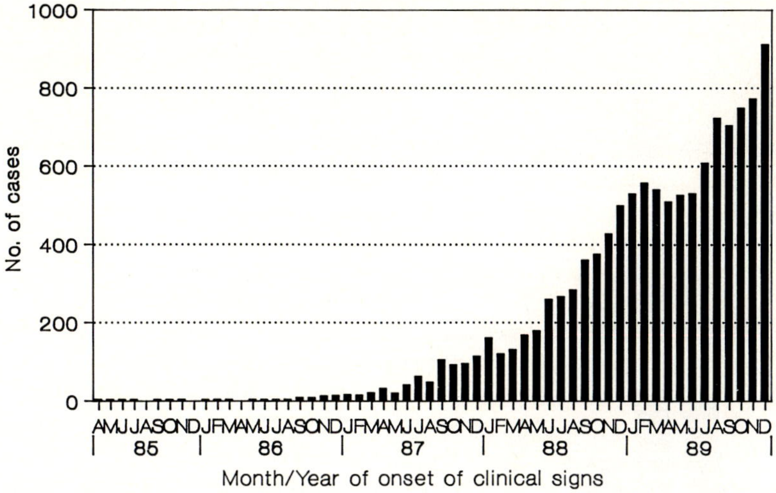

Fig. 2. Distribution of suspected and confirmed cases of BSE by month and year of clinical onset from April 1985 to December 1989

Because there is an inconstant period between the date of clinical onset and the official reporting there will inevitably be changes to the number of cases with a clinical onset in 1989 (Fig. 2). This inconstancy is due to the fact that the clinical onset is often determined in retrospect on the basis of the subsequent clinical course. This aside, the incidence of BSE in Great Britain in the latter half of 1989 represented an annual incidence of two cases per 1000 adult animals. A parallel can be drawn between the BSE epidemic and the kuru epidemic in Papua New Guinea, where the annual incidence of kuru in 1957, when surveillance commenced, was of a similar order of magnitude at 8 cases per 1000 total population in the North and South Fore regions (ALPERS 1987).

The age specific incidences in herds with at least one confirmed case in 1987, 1988 and 1989 (Fig. 3) indicate a peak incidence in 4- and 5-year-old animals in 1989. The youngest animal affected up to December 1989 was 22 months of age at the onset of clinical signs. The predicted increase in incidence in animals 5 years old or greater in 1988 and 1989 compared with 1987 is evident and, as discussed, is a result of a better estimate of the incubation period distribution and consequently also of the contribution from adulthood exposure. This change supports the estimate of the time of the initial exposure in 1981/82 and therefore the commencement of the epidemic in 1985. The age specific incidences of BSE in 1987 were not dissimilar to those estimated for sheep scrapie (DICKINSON 1976), which undoubtedly reflects the fact that the majority of sheep acquire infection early in life. While, from the results of initial and subsequent simulation studies (WILESMITH et al. 1988), this is also the case for BSE, the degree of adulthood exposure in cattle is apparently greater than that for sheep. Simulation modelling indicates that the age specific incidences are essentially the product of two overlapping incubation period distributions. In

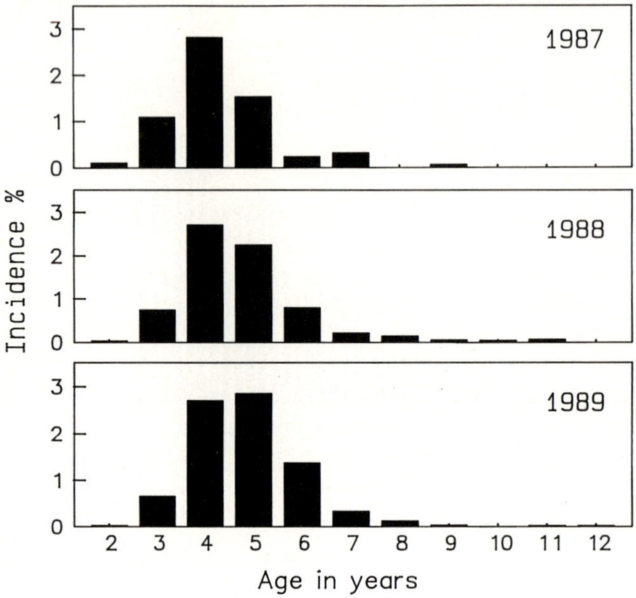

Fig. 3. Age specific incidence (%) of BSE in herds with at least one confirmed case of BSE in 1987, 1988 and 1989

kuru, there was a more obvious bimodality of the age specific incidence pattern as a result of the exposure periods in childhood and adulthood (ALPERS 1987). The full contribution of adulthood exposure in cattle will become evident in future years. It is unlikely, however, that the incubation period distribution resulting from adult exposure will be different from that resulting from calfhood exposure, as is suggested for childhood and adulthood exposure in kuru (ALPERS 1987), since the potential for exposure of adult cattle is less than for calves given the inclusion rates of meat and bone meal in feedstuffs.

The incidence of affected dairy herds is ten times that for beef suckler herds. This can be explain by the difference between the two types of herd in the feeding of concentrate rations. The risk of a herd experiencing BSE increases with increasing herd size irrespective of the type of herd (WILESMITH et al. 1988). Taking this factor into account reduces the crude difference in risk for beef suckler herds and dairy herds to fourfold. There is no evidence of any breed or sex predisposition.

The mean within herd incidence is low and was 2.16% of adult animals in herds with at least one confirmed case of BSE in 1989. In general the attack rate is low within putatively exposed animal populations, and the majority of herds (64%) have experienced only one case. This apparently low risk of infection is a function of the exposure dose, the route of infection or the susceptibility of cattle, or combinations of these. Possible variation in susceptibility is the subject of current research to investigate the possible presence of an allele analogous to *sinc* and *sip*, which control the incubation period in mice and sheep respectively (DICKINSON and FRASER 1979). At present, the epidemiological evidence in cattle does not suggest any major variation in susceptibility, as attack rates approaching 100% have occurred in a small number of birth cohorts whose members are not highly related. Similarly the proportion of affected herds in some counties in the south of England is not indicative of a significant variation in susceptibility. In addition, experimental transmission of the BSE agent to cattle (DAWSON et al. 1990) has resulted in these parenterally exposed cattle all succumbing to BSE and exhibiting little variation in incubation period (DAWSON and WELLS, unpublished observations). The exposure dose in the field is impossible to estimate directly. The use and inclusion of meat and bone meal in cattle feedstuffs, together with the proportion of animals fed such rations, provide an indirect estimate, and the data on this suggest that a considerable proportion of the dairy cattle population could consume meat and bone meal, and therefore be exposed, at some time during their life. Not all meat and bone meal will have been derived from ruminant material, and there is some opportunity for a non-uniform distribution of a scrapie-like agent even in ruminant-derived meat and bone meal. However, the known repeated dietary exposure argues that dose is not the limiting factor responsible for the low attack rate. The low attack rate is more consistent with an oral exposure of cattle to the BSE agent rather than an effect of susceptibility or exposure dose. Similarly, original epidemiological data on kuru (ALPERS 1979) and TME (HARTSOUGH and BURGER 1965) support an oral exposure, but subsequently parenteral routes have been suggested in these latter two diseases. In kuru, cutaneous scarification and conjunctival contamination have been considered of possible importance (GAJDUSEK 1985), while in TME, biting during feeding has been thought the likely natural route of infection (MARSH and HANSON 1979). The oral route is less efficient than parenteral routes of infection for the scrapie agent in mice (KIMBERLIN and WALKER 1989) and the TME agent in mink (BURGER and HARTSOUGH 1965). Some caution is required in interpreting the likely route of transmission in naturally occurring disease from such comparisons, as experimental oral exposure has usually involved a single oral dosing rather than repeated or continuous dietary exposure.

Associated temporally with the BSE epidemic and consistent also with oral exposure from consumption of contaminated feedstuffs is the occurrence of spongiform encephalopathies in several species of exotic ruminant captive in the UK (JEFFREY and WELLS 1988; WILESMITH et al. 1988; FLEETWOOD and FURLEY 1990; KIRKWOOD et al. 1990). The contemporaneous occurrence of a scrapie-like encephalopathy in a domestic cat (WYATT et al. 1990) is of interest with respect to the BSE epidemic, but further surveillance and investigation will be required to determine its significance.

The route of infection in natural scrapie in sheep and goats has not been determined, but experimental studies have indicated that animals can be infected by oral dosing (PATTISON and MILLSON 1961; GORDON 1966). Furthermore, the only source of the scrapie agent so far identified for environmental contamination is the placenta, and this has also proved to be an efficient means of transmitting scrapie (PATTISON et al. 1974; HOURRIGAN et al. 1979).

Age specific incidence data, as discussed previously, suggest that the majority of sheep become infected in early life, and in the light of the epidemiological findings on BSE the oral route of exposure may well maintain natural sheep scrapie.

The geographical variation in incidence of BSE, as described originally (WILESMITH et al. 1988), has persisted, with a considerably greater incidence in the south of England. This is a reflection of variation in exposure for which two possible reasons have been identified. The first is the variation in the use of meat and bone meal between commercial compounders of cattle feedstuffs and in their geographical market share of sales of cattle feedstuffs. The second reason is an apparent geographical variation in the supply of meat and bone meal which, in the course of its production, has been subjected to additional heat treatment as a result of reprocessing the partially processed product called greaves. Meat and bone meal produced in this way could have a differentially reduced titre of the scrapie-like agent (WILESMITH et al. 1991).

The low incidence in Scotland and the northern part of England can also be accounted for by a potential decontaminating affect of the hydrocarbon solvent extraction of fat from meat and bone meal in the rendering process which is only in use in Scotland. This is discussed in Sect. 3.2.

No evidence of cattle-to-cattle transmission had emerged by the time of writing (January 1991). However, the epidemiological findings indicate that exposure of cattle via feedstuffs continued until at least 1986, and given that the use of meat and bone meal in commercial cattle rations did not change until July 1988 the exposure would have continued until this latter time.

3.2 The Source of Infection and Reasons for Occurrence of BSE in Great Britain

The results of the continued detailed surveillance of the epidemic are consistent with the original hypothesis of exposure of cattle to a scrapie-like agent via cattle feedstuffs containing ruminant-derived protein (WILESMITH et al. 1988). This is the subject of a case-control study of feeding practices in the first year of life in BSE-affected and BSE-unaffected herds. The preliminary results of this study provide further supporting evidence for meat and bone meal as the source of exposure (WILESMITH and RYAN, unpublished observations).

The inclusion of meat and bone meal in domestic animal rations is not a recent event; such meal has been used as a protein source for several decades (MMC 1985). The occurrence of scrapie in sheep and the inclusion of sheep tissues in the production of meat and bone meal are also long established. A survey of the rendering processes used in the UK (WILESMITH et al. 1991) has also established that, irrespective of the type of process employed, the maximum temperatures achieved and the method of heat treatment were insufficient to inactivate the scrapie agent given the recommendations for disinfection of the most resistant strain (22A) (TAYLOR and MCCONNELL 1988) and of the CJD agent (DHSS 1984). Therefore the mere inclusion of meat and bone meal in cattle rations could not alone account for an abrupt change in exposure of cattle in 1981/82 sufficient to result in clinical disease. The survey of rendering processes suggested two major changes which could have been important in altering the exposure: first, a move away from batch rendering of material to continuous processes, and second, the reduction in the use of hydrocarbon solvent extraction of fat from meat and bone meal. The former change commenced in 1972, and from then until 1988 there was a gradual transition from batch rendering to continuous rendering processes. The reduction in use, in all but two rendering plants (both in Scotland), of hydrocarbon solvent extraction of fat from meat and bone meal was, however, coincident with the estimated time at which exposure of cattle changed. The proportion of meat and bone meal produced using solvent extraction went down by 46% in the period 1980–1983. This change is currently considered to be the single most important factor in the appearance of BSE in the UK in 1985 (WILESMITH et al. 1991).

The apparent absence of BSE in countries other than the British Isles is likely to be attributable to differences more fundamental than this change in the rendering process. The summation of three basic risk factors in the British Isles underlies the potential for exposure to the scrapie-like agent via meat and bone meal: a large sheep population, endemic scrapie in the sheep population and the feeding of diets containing meat and bone meal. Although a formal study is required to confirm the observation, general knowledge of these three factors in some other countries, particularly in Europe and North America, suggests that they are not and have not been present in combination. However, the use of hydrocarbon solvent for fat extraction from meat and bone meal in a majority of plants may provide a protective effect in countries where the factors are present in combination.

The epidemiological findings in the UK are therefore consistent with two possibilities regarding the origin of a scrapie-like agent. The first is that cattle have experienced an increase in exposure to a pool of sheep scrapie agents, infection with one or more of which has resulted in clinical disease. The second possibility is that BSE resulted from a similar increase in exposure, but to a strain of a scrapie-like agent which had been previously selected by cattle and passaged in the species, presumably via meat and bone meal. Prior to the increase in exposure, subclinical and perhaps a low incidence of clinical disease may have occurred. Such cases would probably have involved relatively old animals, consistent with low exposure dose and resultant extended incubation periods, and may therefore have been assigned alternative clinical diagnoses. A precedent for this latter hypothesis can be drawn from mouse models of scrapie (BRUCE and DICKINSON 1987). If this hypothesis is correct, there is also the possibility that such subclinical infection of cattle with a scrapie-like agent or, indeed, a very low incidence of clinical BSE is similarly present in other countries. There has been a suggestion that this may be the case in the United States (MARSH and HARTSOUGH 1988).

3.3 The Future Course of the Epidemic

Although some aspects of the epidemiology of BSE, particularly those relating to direct cattle-to-cattle transmission, remain uncertain, it is possible to make broad predictions on the possible future courses of the disease. Three scenarios are considered (Fig. 4). The simplest and most certain of these, postulating successful prohibition of the feedstuffs and no direct cattle-to-cattle transmission of the BSE agent, is analogous to the decline in incidence of kuru in Papua New Guinea (ALPERS 1987) and the self-limiting nature of TME outbreaks (HADLOW and KARSTAD 1968). Following the cessation of exposure the subsequent cases are those animals which were already in the incubation period and live long enough to develop the clinical disease. With this, and the other two scenarios, no attempt has been made to depict predictions of the incidence until 1992, hence no fluctuations are illustrated. As indicated in Sect. 3.1 a stepwise increase in incidence occurred in 1989. However, given that exposure ceased after July 1988 such changes will have no effect on the course of the epidemic and the incidence will decline. Further cases will essentially represent the "tail" of the incubation period distribution, with virtual extinction by the end of the century.

In the second scenario, despite cessation of exposure from feedstuffs in 1988 direct transmission occurs, but only in utero. However, the current breeding regime in dairy herds in Great Britain and the annual herd replacement rate of 20% would not facilitate sustenance of infection in the cattle population by this method of transmission. A contact rate of 1:1, necessary to sustain infection, would not be achieved in the national cattle population for a prolonged period of time. The decay in incidence according to this scenario is slower than in the first scenario, as in a small proportion of herds BSE-affected animals could produce more than one female offspring, which would themselves produce calves.

The third scenario is unavoidably more speculative. It assumes a high component of maternal transmission and a certain degree of lateral transmission to animals in contact, as a result of infection acquired from the

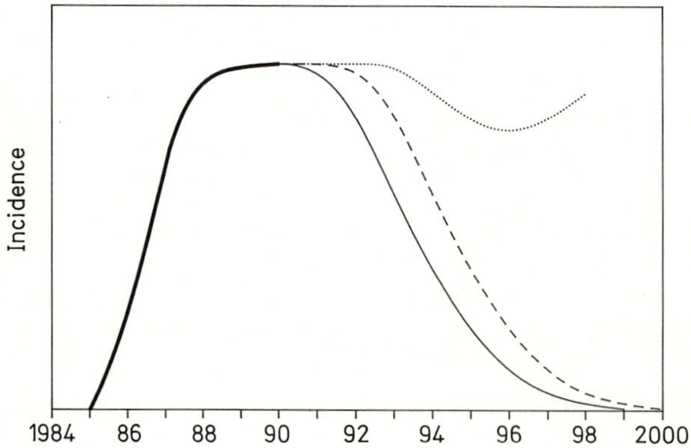

Fig. 4. Three examples of possible courses of the BSE epidemic in the UK ..... feed ban effective but cattle-to-cattle transmission occurs; ---- feed ban effective but in utero transmission occurs; —— feed ban effective and no cattle-to-cattle transmission

placenta—as would appear to be the case in sheep scrapie (PATTISON et al. 1974). Even with such transmission there is a possibility that the incidence could decline from 1992 onwards as a result of the relatively long incubation period and its distribution. The duration of this decline is impossible to estimate, as it is dependent on the rate of transmission, should the latter occur. If no disease control measures were introduced, sooner or later the incidence would rise at an unknown rate to an as yet unpredictable level and certain endemic status. The question of the possible maternal transmission of BSE and the incidence under farm conditions is addressed by a specific study currently in progress.

4 Conclusion

The recent sudden occurrence of a new or previously undetected scrapie-like disease of domestic cattle is of considerable interest in relation to the epidemiology of the entire group of neurodegenerations attributable to unconventional slow viruses. The degree of clinicopathological homology between BSE and scrapie of sheep and goats and the relative uniformity of both the clinical presentation and the neuropathology of BSE has facilitated case definition for accurate monitoring of the epidemic in the UK.

The nature of the disease, the extended common source form of the epidemic and the elimination of other environmental factors, have provided substantial evidence for dietary exposure to a scrapie-like agent via cattle feedstuffs containing ruminant-derived protein as meat and bone meal. Computer simulations of the epidemic further indicate that exposure, sufficient to result in clinical disease, began abruptly in 1981/82. Specific changes in the manufacturing of meat and bone meal probably account for such an alteration in exposure.

The dietary origin of the BSE agent is consistent also with the contemporaneous occurrence of scrapie-like spongiform encephalopathies in several exotic ruminant species in the UK and in a domestic cat; all of the animals concerned had exposure to ruminant-derived meat and bone meal.

Though consistent with an oral route of exposure to infection the epidemiologies of natural sheep scrapie, kuru and TME have been open to alternative interpretations which remain speculative. In BSE, the findings indicate an unequivocal and relatively efficient dietary exposure.

Prediction of the future course of the disease is constrained by the uncertainty of the occurrence of direct cattle-to-cattle transmission. Should such transmission indeed take place, it could, depending upon specific mode and efficiency and in the absence of any intervention, result in BSE becoming endemic in the UK, similar to scrapie of sheep. If, however, the food-borne route, now curtailed by legislation, proves the sole means of transmission of BSE the disease will decline in incidence and will cease to exist by the end of the 20th century.

References

Alpers MP (1979) Epidemiology and ecology of kuru. In: Prusiner SB, Hadlow WJ (eds) Slow transmissible diseases of the nervous system. Academic, New York, pp 67–90

Alpers MP (1987) Epidemiology and clinical aspects of kuru In: Prusiner SB, McKinley MP (eds) Prions: novel infectious pathogens causing scrapie and Creutzfeldt–Jakob disease. Academic, San Diego, pp 451–464

Baringer JP, Bowman KA, Prusiner SB (1983) Replication of the scrapie agent in hamster brain precedes neuronal vacuolation. J Neuropathol Exp Neurol 42: 539–547

Bassett HF, Sheridan C (1989) Case of BSE in the Irish Republic. Vet Rec 124: 151

Bernoulli CC, Masters CL, Gajdusek DC, Gibbs CJ Jr, Harris JO (1979) Early clinical features of Creutzfeldt–Jakob disease (subacute spongiform encephalopathy). In: Prusiner SB, Hadlow WJ (eds) Slow transmissible diseases of the nervous system, vol 1. Academic, New York, pp 229–251

Brown P, Cathala F, Castaigne P, Gajdusek DC (1986) Creutzfeldt–Jakob disease: clinical analysis of a consecutive series of 230 neuropathologically verified cases. Ann Neurol 20: 597–602

Bruce ME, Dickinson AG (1987) Biological evidence that scrapie has an independent genome. J Gen Virol 68: 79–89

Burger D, Hartsough GR (1964) A 'scrapie'-like disease of mink. In: Report of a scrapie seminar. ARS 91–53. United States Department of Agriculture, Paper no 27, pp 225–227

Burger D, Hartsough GR (1965) Encephalopathy of mink. II Experimental and natural transmission. J Infect Dis 115: 393–399

Clarke M, Aldridge BM, Scott PR, McInnes A, Will R (1989) EEG studies in bovine spongiform encephalopathy. Electroencephalogr Clin Neurophysiol 23: 61

Comber T (1772) Real improvements in agriculture (on the principles of A Young Esq.). In: Letters to Reade Peacock, Esq. and to Dr Hunter, Physician in York, concerning the rickets in sheep. Nicoll, London

Cranwell MP, Hancock RD, Hudson JR, Hall SA, Daniel NJ, Hopkins AR, Wonnacott B, Vivian M, Hunt P (1988) Bovine spongiform encephalopathy. Vet Rec 122: 190

Dawson M, Wells GAH, Parker BNJ (1990) Preliminary evidence of the experimental transmissibility of bovine spongiform encephalopathy to cattle. Vet Rec 126: 112–113

Dickinson AG (1976) Scrapie in sheep and goats. In: Kimberlin RH (ed) Slow virus diseases of animals and man. North-Holland, Amsterdam, pp 209–241

Dickinson AG, Fraser H (1979) An assessment of the genetics of scrapie in sheep and mice. In: Prusiner SB, Hadlow WJ (eds) Slow transmissible diseases of the nervous system, vol 1. Academic, New York, pp 367–385

DHSS (1984) Management of patients with spongiform encephalopathy (Creutzfeldt–Jakob disease). DHSS Circular DA (84): 16

Eisenmayer S, Göbel C (1872) In: The healing of animals, etc. Pest

Farquhar CF, Somerville RA, Ritchie LA (1989) Post mortem immunodiagnosis of scrapie and bovine spongiform encephalopathy. J Virol Methods 24: 215–222

Fleetwood AJ, Furley CW (1990) Spongiform encephalopathy in an eland. Vet Rec 126: 408–409

Fraser H (1976) The pathology of natural and experimental scrapie. In: Kimbelin RH (ed) Slow virus diseases of animals and man. North-Holland, New York, pp 267–305

Fraser H (1979) Neuropathology of scrapie: the precision of the lesions and their diversity. In: Prusiner SB, Hadlow WJ (eds) Slow transmissible diseases of the nervous system, vol 1. Academic, New York, pp 387–406

Fraser H, McConnell I, Wells GAH, Dawson M (1988) Transmission of bovine spongiforrm encephalopathy to mice. Vet Rec 123: 472

Gajdusek DC (1985) Subacute spongiform virus encephalopathies caused by unconventional virusues. In: Maramorosch K, McKelvey J Jr (eds) Subviral pathogens of plants and animals: viroids and prions. Academic, Orlando, pp 483–544

Gajdusek DC, Zigas V (1957) Degenerative disease of the central nervous system in New Guinea. N Engl J Med 257: 974–978

Gerstmann J, Sträussler E, Scheinker I (1936) Über eine eigenartige hereditär-familiäre Erkrankung des Zentralnervensystems. Zugleich ein Beitrag zur Frage des vorzeitigen lokalen Alterns. Z Gesamte Neurol Psychiatr 154: 736–762

Gilmour JS, Buxton D, Macleod NSM, Brodie TA, More JB (1988) Bovine spongiform encephalopathy. Vet Rec 122: 142

Gordon WS (1966) Transmission of scrapie and evidence of spread of infection in sheep at pasture. In: Report of scrapie seminar, Washington, DC, Jan 27–30 1964. US Department of Agriculture, Washington, pp 8–18
Hadlow WJ, Karstad L (1968) Transmissible encephalopathy of mink in Ontario. Can Vet J 9: 193–196
Hartsough GR, Burger D (1965) Encephalopathy of mink. I Epizootiologic and clinical observations. J Infect Dis 115: 387–392
Hope J, Reekie LJD, Hunter N, Multhaup G, Beyreuther K, White H, Scott AC, Stack MJ, Dawson M, Wells GAH (1988) Brain fibrils of novel British cattle disease contain scrapie-associated protein. Nature 336: 390–392
Hourrigan J, Klingsporn A, Clark WW, de Camp M (1979) Epidemiology of scrapie in the United States. In: Prusiner SB, Hadlow WJ (eds) Slow transmissible diseases of the nervous system, vol 1. Academic, New York, pp 331–356
Jakob A (1921a) Uber eine eigenartige Erkrankung des Zentral-nervensystems mit bemerkenswertem anatomischem Befunde (spstische pseudosklerotische Encephalomyelopathie mit disseminierten Degenerationsherden) Dtsch Z Nervenheilk 70: 132–146
Jakob A (1921b) Uber eine eigenartige Erkrankung des Zentral-nervensystems mit bemerkenswertem anatomischem Befunde (spastische pseudosklerotische Encephalomyelopathie mit disseminierten Degenerationsherden) Z Gesamte Neurol Psychiatr 64: 147–228
Jakob A (1921c) Uber eine der multiplen Sklerose klinisch nahestehende Erkrankung des Zentralnervensystems (spastische Pseudosklerose) mit bemerkenswertem anatomischem Befunde. Med Klin 13: 372–376
Jeffrey M, Wells GAH (1988) Spongiform encephalopathy in a Nyala (*Tragelaphus angasi*). Vet Pathol 25: 398–399
Johnson CT, Whitaker CJ (1988) Bovine spongiform encephalopathy. Vet Rec 122: 142
Joubert L, Lapras M, Gastellu J, Prave M, Laurent D (1972) Un foyer de tremblante du mouton en Provence. Sci Vet 74: 165
Kakulas BA, Lecours AR, Gajdusek DC (1967) Further observations on the pathology of Kuru. (A study of the two cerebra in serial section). J Neuropathol Exp Neurol 26: 85–97
Kimberlin RH (1981) Scrapie. Br Vet J 137: 105–112
Kimberlin RH, Walker CA (1989) Pathogenesis of scrapie in mice after intra-gastric infection. Virus Res 12: 213–200
Kirkwood JK, Wells GAH, Wilesmith JW, Cunningham AA, Jackson SI (1990) Spongiform encephalopathy in an Arabian oryx (*Oryx leucoryx*) and a greater kudu (*Tragelaphus strepsiceros*). Vet Rec 127: 418–420
Liberski PP, Yanagihara R, Wells GAH, Gibbs CJ Jr, Gajdusek DC (to be published) Comparative ultrastructural neuropathology of bovine spongiform encephalopathy, scrapie and Creutzfeldt–Jakob disease
Manuelidis EE, Manuelidis L (1979) Clinical and morphological aspects of transmissible Creutzfeldt–Jakob disease. In: Zimmerman HM (ed) Progress in neuropathology, vol 4. Raven, New York, pp 1–26
Marsh RF, Hanson RP (1979) On the origin of transmissible mink encephalopathy. In: Prusiner SB, Hadlow WJ (eds) Slow transmissible diseases of the nervous system, vol 1. Academic, New York, pp 451–460
Marsh RF, Hartsough GR (1988) Evidence that transmissible mink encephalopathy results from feeding infected cattle. In: Proceedings of the Fourth International Scientific Congress in Fur Animal Production. Toronto, Canada, pp 204–207
Marsh RF, Sipe JC, Morse SS, Hanson RP (1976) Transmissible mink encephalopathy. Reduced spongiform degeneration in aged mink of the Chediak–Higashi genotype. Lab Invest 34: 381–386
Masters CL, Gajdusek DC (1982) The spectrum of Creutzfeldt–Jakob disease and the virus induced subacute spongiform encephalopathies. In: Smith WT, Cavanagh JB (eds) Recent advances in neuropathology, vol 2. Churchill Livingstone, Edinburgh, pp 139–163
Masters CL, Richardson EP (1978) Subacute spongiform encephalopathy (Creutzfeldt–Jakob disease)—the nature and progression of spongiform change. Brain 101: 333–334
McDaniel HA, Morehouse LG (1964) The diagnosis of scrapie. In: Proc. 67th Ann. Mtng. U.S. Lives. San. Ass., pp 550–564
Mitchell B, Stamp JT (1983) Scrapie. In: Martin WB (ed) Diseases of sheep. Blackwell, Oxford, pp 71–75
MMC (1985) Animal waste. A report on the supply of animal waste in Great Britain. HMSO, London
Order (1988) The bovine spongiform encephalopathy order 1988. Statutory Instrument no 1039, HMSO, London

Palmer AC (1957) Studies in scrapie. Vet Rec 69: 1318–1328

Palmer AC (1976) Introduction to animal neurology, 2nd edn. Blackwell, Oxford, pp 30, 60

Palsson PA (1979) Rida (scrapie) in Iceland and its epidemiology. In: Prusiner SB, Hadlow WJ (eds) Slow transmissible diseases of the nervous system, vol 1. Academic, New York, pp 357–366

Palsson PA, Sigurdsson B (1958) Rida: a slow progressive disease affecting the central nervous system of sheep. In: Proc 8th Nord Vet Congress, Helsinki, A8, pp 179–191

Parry HB (1983a) Recorded occurrences of scrapie from 1750. In: Oppenheimer DR (ed) Scrapie disease in sheep. Academic, London, pp 31–59

Parry HB (1983b) Clinical features in natural scrapie. In: Openheimer DR (ed) Scrapie disease in sheep. Academic, London, pp 60–72

Parry HB (1983c) Epidemiological studies. In: Oppenheimer DR (ed) Scrapie disease in sheep. Academic, London, pp 60–72

Pattison IH, Millson GC (1961) Experimental transmission of scrapie to goats and sheep by the oral route. J Comp Pathol 71: 171–176

Pattison IH, Gordon WS, Millson GC (1959) Experimental production of scrapie in goats. J Comp Pathol 69: 300–312

Pattison IH, Hoare MN, Jebbett JN, Watson WA (1974) Spread of scrapie to sheep and goats by oral dosing with foetal membranes from scrapie affected sheep. Br Vet J 130: lxv–lxvii

Scott AC, Wells GAH, Stack MJ, White H, Dawson M (1990) Bovine spongiform encephalopathy: detection and quantitation of fibrils, fibril protein (PrP) and vacuolation in brain. Vet Microbiol 23: 295–304

Scott PR, Aldridge BM, Clark M, Will R, McInnes A (1988) Bovine spongiform encephalopathy-electroencephalographic studies. In: Proc 15th World Buiatrics Congress, Palma, p 1530

Scott PR, Aldridge BM, Clarke M, Will RG (1990) Cerebrospinal fluid studies in normal cows and cases of bovine spongiform encephalopathy. Br Vet J 146: 88–90

Taylor DM, McConnell I (1988) Autoclaving does not decontaminate formol-fixed scrapie tissues. Lancet i: 1463–1464

Wells GAH (1988) Bovine spongiform encephalopathy. Foreign animal disease report, US Department of Agriculture 16-4: 4–7

Wells GAH (1989) Bovine spongiform encephalopathy. In: Grunsell CSG, Raw M–E, Hill FWG (eds) The veterinary annual, 29th edn. Wright, London, pp 59–65

Wells GAH, Wells M (1989) Neuropil vacuolation in brain: a reproducible histological processing artefact. J Comp Pathol 101: 355–362

Wells GAH, Wilesmith JW (1989) The distribution pattern of neuronal vacuolation in bovine spongiform encephalopathy (BSE) is constant. Neuropathol Appl Neurobiol 15: 591

Wells GAH, Scott AC, Johnson CT, Gunning RF, Hancock RD, Jeffrey M, Dawson M, Bradley R (1987) A novel progressive spongiform encephalopathy in cattle. Vet Rec 121: 419–420

Wells GAH, Hancock RD, Cooley WA, Richards MS, Higgins RJ, David GP (1989) Bovine spongiform encephalopathy: diagnostic significance of vacuolar change in selected nuclei of the medulla oblongata. Vet Rec 125: 521–524

Wilesmith JW, Wells GAH, Cranwell MP, Ryan JBM (1988) Bovine spongiform encephalopathy: epidemiological studies. Vet Rec 123: 638–644

Wilesmith JW, Ryan JBM, Atkinson MJ (1991) Bovine spongiform encephalopathy: epidemiological studies on the origin. Vet Rec 128: 199–203

Williams ES, Young S (1980) Chronic wasting disease of captive mule deer: a spongiform encephalopathy. J Wildl Dis 16: 89–98

Williams ES, Young S (1982) Spongiform encephalopathy in a Rocky Mountain Elk. J Wildl Dis 18: 465–471

Winter MH, Aldridge BM, Clarke M (1989) Occurrence of 14 cases of bovine spongiform encephalopathy in a closed dairy herd. Br Vet J 145: 191–194

Wyatt JM, Pearson GR, Smerdon T, Gruffydd-Jones TJ, Wells GAH (1990) Spongiform encephalopathy in a cat. Vet Rec 126: 513

Zigas V, Gajdusek DC (1957) Kuru: clinical study of a new syndrome resembling paralysis agitans in natives of the Eastern Highlands of Australian New Guinea. Med J Aust 2: 745–754

Zlotnik I, Katiyar RD (1961) The occurrence of scrapie disease in sheep of the remote Himalayan foothills. Vet Rec 73: 543–544

Zlotnik I, Rennie JC (1957) The occurrence of vacuolated neurones and vascular lesions in the medullas of apparently healthy sheep. J Comp Pathol 67: 30–37

Zlotnik I, Rennie JC (1958) A comparative study of the incidence of vacuolated neurones in the medulla from apparently healthy sheep of various breeds. J Comp Pathol 68: 411–415

Purification of Scrapie Agents: How Far Have We Come?*

D. C. BOLTON and P. E. BENDHEIM

1 Introduction

It might seem, after a careful reading of the literature on the etiologic agents of scrapie and Creutzfeldt–Jakob disease (CJD), that at least three or four distinct pathogens cause these diseases. Independent investigators have assigned different names for the agents (agent, virus, virino, prion) and different names for the disease-specific protein (PrP^{Sc}, Sp33–37, Gp34) and its partial digestion product (PrP-27–30, SAF protein, p26). Substantial evidence from many studies indicates that this disease-specific protein is an essential component of the agent particle (BOLTON et al. 1982, 1984, 1987; MCKINLEY et al. 1983; PRUSINER et al. 1984a; BENDHEIM et al. 1985; BOCKMAN et al. 1985; GABIZON et al. 1988a, 1988b), but this conclusion is contradicted by other reports that appear to show that it is not (MANUELIDIS et al. 1987; AIKEN et al. 1989; SKLAVIADIS et al. 1989). The agents are heterogeneous in size, with reported sizes from less than 55000 daltons

Department of Molecular Biology, New York State Institute for Basic Research in Developmental Disabilities, 1050 Forest Hill Road, Staten Island, NY 10314, USA

* This work was supported by grants NS-23948 (DCB) and NS-24720 (PEB) from the National Institute of Neurological and Communicative Disorders and Stroke, and by the New York State Office of Mental Retardation and Developmental Disabilities

(PRUSINER 1982; GABIZON et al. 1987; BELLINGER-KAWAHARA et al. 1988) to over 10000 *S* (PRUSINER et al. 1978a, 1978b, 1980a; PRUSINER 1982; DIRINGER et al. 1983a; SKLAVIADIS et al. 1989). Reported densities for these agents also cover a broad spectrum (BROWN et al. 1978; MARSH et al. 1984a, b; SKLAVIADIS et al. 1989). Despite these apparent contradictions, there are no compelling reasons to reject the hypothesis that a single group of closely related pathogens causes scrapie, CJD and the related animal and human diseases (bovine spongiform encephalopathy, chronic wasting disease of mule deer, transmissible mink encephalopathy, kuru and the Gerstmann–Sträussler syndrome). Like the story of the blind men attempting to describe an elephant, the view one gets of the scrapie and CJD agents depends on how, and where, one looks.

One reason for the discrepant views of the scrapie and CJD agents is that the agents have proved difficult to purify and characterize. Some properties of the scrapie agent are advantageous for designing purification protocols (e.g., the fact that it is not easily inactivated is an asset from the standpoint of recovery of activity), but these are offset by the association of the agent with membranes, its propensity for autoaggregation, and the resultant heterogeneous physical behavior (PRUSINER et al. 1978b, 1982; PRUSINER 1982). The biggest obstacle to purification of these pathogens, however, is the lack of a convenient assay for them. The animal bioassay widely in use has two major drawbacks: (1) incubation periods of at least 60–150 days, and (2) a requirement for large numbers of animals. These prerequisites put significant practical limits on the number samples that can be assayed during the course of a year, and thus, limit the number of fractions from purification trials that can be tested.

Purification is a prerequisite for determining the chemical composition and molecular structure of the scrapie and CJD agents. At the heart of this approach is the expectation that knowledge of the molecular structure of the etiologic agents will lead to a better understanding of the pathogenesis of the disease. Additionally, purification of the etiologic agents of scrapie and similar diseases may be important for developing diagnostic methods and for designing treatment and prevention strategies.

Some would argue that essentially nothing is known of the molecular structure of these agents, but this view is unnecessarily pessimistic. The above difficulties have slowed progress, but have not prevented it. In our opinion, substantial progress has been made toward purification and characterization of the scrapie agent, but our knowledge of the chemical composition and structure of these agents remains incomplete.

The purpose of this review is to discuss progress in purification and characterization of the etiologic agents of the transmissible slow neurodegenerative diseases. We will focus on the scrapie agent, under the assumption (justified by past experience) that the conclusions will be relevant to studies of the related pathogens. Purification methods, biochemical analyses of enriched fractions, and biological characterization of the agent in these fractions are discussed. Theoretical considerations such as criteria for assessing purity and the interpretation of data also are included in this discussion. Our intent is to highlight areas where progress has been made, but also to discuss areas where gaps in our knowledge exist and controversy remains.

2 Terminology

One cannot proceed very far toward purifying the scrapie agent without establishing acceptable criteria for the endpoint of this search. These criteria provide the descriptive, and perhaps quantitative, definition for the scrapie agent. The scrapie agent is operationally defined by its ability to induce a well-defined neurological condition following a predictable, prolonged asymptomatic incubation period. The agent cannot be grown in pure culture and is filterable (under some conditions), and therefore has been categorized as a "virus" by some. However, the modern molecular definition of viruses has progressed far beyond these simple criteria and at a minimum includes virus-encoded coat proteins and a virus-specific nucleic acid. At this time, classification of the scrapie agent and related pathogens as "viruses" is not warranted based on the available data.

The requirement for reproducing and assaying the agent in laboratory animals leads to an operational definition for it based on two fundamental biological attributes: (1) the scrapie agent induces a slow neurodegenerative disease with characteristic neuropathology in a susceptible host; and (2) it replicates or accumulates in the host to many times the amount of the original inoculum. For this definition to be useful, however, it must be more precise. To formally distinguish between a crude homogenate and a discrete particle, the definition must exclude any molecules that are not essential to inducing the disease or agent replication. Also, the definition must exclude normal host components that, as in the case of viruses, are essential to replication of the agent but are not part of the physical particle that initiates the process (e.g., ribosomes, tRNA, glycolytic enzymes, etc.).

The above criteria define the biological, but not the chemical, features of the particle that induces scrapie. We must also consider that there may be particles that are indistinguishable from the scrapie agent by many criteria, but that are not capable of inducing disease. During the purification of viruses, for example, it is often observed that some loss of biological activity, due to modest denaturation of viral proteins or inactivation of the viral nucleic acid, does not compromise the overall structure and composition of the virion. A preparation containing these particles may appear homogeneous by EM or analytical methods such as SDS-PAGE, but the activity of the preparation (usually measured by plaque count or ID_{50}) is lower than expected based on a count of virions by EM. An examination of the virus titer at different dilutions can be used to differentiate between two possibilities: (1) that some of the virions are inactive and (2) that more than one virion per cell is required to initiate infection. In most cases the former explanation applies and it is common for purified virus preparations assayed in cell culture to have particle-to-infectivity ratios (or particle-to-plaque ratios) much greater than one. Values of the particle-to-infectivity ratio (particle/ID_{50} or particle/LD_{50}) of 100 or more are not unexpected for purified viruses assayed in animals, where the efficiency of infection depends on the route of inoculation, access to the target organ and the rate of clearance of the virus by the host reticuloendothelial and immune systems.

Thus, there is a broader class of scrapie-specific particles that have the physical and chemical attributes of the biologically active units; some of these particles are biologically active, others are not. In an ideal assay system each biologically active particle would yield a positive result (cytopathic effect, plaque, disease, or death), but in practice this is rarely true. In animal bioassay systems, it is unlikely that each biologically active particle would be able to cause disease for the reasons noted above. After inoculation into a susceptible animal host there is a statistical probability that any one of the biologically active physical particles will survive to actually initiate disease.

We think that the following definitions provide useful criteria to guide studies of the purification and characterization of the scrapie agent and similar agents:

—*Scrapie agent:* The biologically active molecular complex containing only those molecules essential for (1) inducing scrapie disease in an appropriate host and (2) initiating reproduction of itself in substantial amounts as a result of either de novo biosynthesis or conversion of existing host components.

—*Scrapie agent physical particle:* Macromolecular complexes having the physical properties and biochemical composition of the scrapie agent. Some, but not all, physical particles are biologically active.

3 Theoretical Considerations

3.1 Criteria for Establishing Purity

The appropriate measure of enrichment is the specific activity, i.e., the biological activity per unit of total mass. Scrapie specific activity is usually measured as LD_{50} units per milligram of protein, but other measures of total mass (milligrams of nucleic acid or carbohydrate for example) could be employed. Since scrapie is invariably fatal, LD_{50} and ID_{50} units are equivalent. Purification of the agent to homogeneity will theoretically yield the highest values of specific activity, regardless of which measure of total mass is used. However, it should be noted that empirical values of specific activity will depend on both the method of purification and the relative standard used for estimating total mass.

Theoretical maximum specific activities can be calculated based on the mass and chemical composition of the physical particle (if known), assuming a particle/infectivity ratio of 1. For example, the theoretical specific activity for one of the smallest RNA viruses, poliovirus, is 1×10^{14} pfu/mg protein or 2.2×10^{14} pfu/mg RNA (FIELDS et al. 1990). For parvovirus, the smallest DNA virus, these values are about 1.3×10^{14} pfu/mg protein and 5×10^{14} pfu/mg DNA (FIELDS et al. 1990). Larger viruses have a higher mass per particle and therefore lower values of specific activity. Additionally, larger viruses have relatively less nucleic acid than protein per physical particle, and thus the value of the specific activity relative to nucleic acid is much higher than the value relative to protein. For example, retroviruses have maximum values in the range of 1×10^{13} pfu/mg RNA, but only 2×10^{11} pfu/mg protein (DUESBERG 1970; BADER 1975). Empirical values of specific activity would undoubtedly be lower since some loss of activity is likely to occur during purification.

In any case, the theoretical values are useful because they provide a range of hypothetical specific activities for the scrapie agent based on different assumptions about its structure. Two predictions can be made based on the assumption that the scrapie agent is a virus. First, the values of specific activity relative to nucleic acid mass and protein mass should increase together during purification. Ultimately, the specific activity relative to nucleic acid would be approximately equal to, or significantly greater than, the value relative to protein for the reasons discussed above. The relative ratio of these values should be consistent with the theoretical values calculated from the known composition of other members of the putative virus family. Secondly, the specific activity of a purified scrapie agent preparation cannot exceed the theoretical value for a proposed structure (within experimental error). As the specific activity approaches this maximum value, the preparation must approach homogeneity, and thus virus particles would be easily visualized by EM, viral proteins would be readily detected by SDS-PAGE, and the virus nucleic acid would be readily detected by agarose gel electrophoresis and staining with ethidium bromide. Failure to detect virus-specific components under these conditions, or attainment of specific activities exceeding the theoretical maximum, would exclude that structure for the agent.

If the scrapie agent is very small, as some have suggested (ALPER et al. 1966; PRUSINER 1982; ALPER 1985; BELLINGER-KAWAHARA et al. 1988), its theoretical specific activity would be much higher than for any virus. For example, if the scrapie agent is pure protein with a particle mass of 100 kDa and a particle to LD_{50} ratio of 1.0, the theoretical specific activity is 6×10^{15} LD_{50}/mg protein. Also, if it is pure protein, its theoretical specific activity relative to nucleic acid is infinite. Empirically, we might expect values for the particle to LD_{50} ratio of 100 to 1000 in an animal bioassay system, thus reducing the specific activity to 6×10^{12} LD_{50}/mg protein. A maximum expected value for specific activity of 2×10^{14} LD_{50}/mg nucleic acid can be calculated based on certain assumptions: (1) The total amount of scrapie agent is $10^{9.5}$ LD_{50}/g of hamster brain (KIMBERLIN and WALKER 1977, 1986); (2) no biological activity is lost during purification, and (3) contamination of the purified fraction with host (non-agent) nucleic acids is about 0.0001% of the level found in the tissue homogenate (i.e., a 1 000 000-fold enrichment with respect to nucleic acid).

3.2 Yield

The amount of scrapie agent obtained in the purified fraction is termed the yield. Yield measures the recovery of activity without respect to purity and therefore it is independent of enrichment. An acceptable yield is important from the standpoint of efficiency, but also may be critical in interpreting analytical results. For example, if a highly purified fraction contains only a very small percentage of the original activity (e.g., $< 1\%$), it is reasonable to question whether that fraction is representative of the 99% which remains unaccounted.

It is desirable for this reason to document by bioassay the amount of the agent in each fraction produced throughout a purification procedure. This analysis will document the amount of enrichment and yield at each step and thus will guide the investigator toward a more efficient procedure. The tendency to assay only a few fractions is hard to resist, however, because of the expense and logistical problems associated with animal bioassay of large numbers of samples.

The accuracy of the scrapie agent bioassay is limited; standard deviations for replicate samples are almost never less than 0.5 $\log_{10}$ LD_{50} units. This means that if the bioassay results show that a purified fraction contains 10% of the total activity of the starting homogenate (for example 10^9 LD_{50} units), there is a high probability that the actual yield is somewhere between 1% (10^7 LD_{50} units) and 100% (10^9 LD_{50} units). Of course, the estimate for the homogenate carries essentially the same margin of error and could be as high as 10^{10} or as low as 10^8 LD_{50} units. It is evident that large differences are required between the measured values in one fraction versus another for one to have confidence that a separation method produced significant enrichment.

Two additional complications of the bioassay need some discussion: the influence of solvent on the bioassay and observer bias in diagnosis of disease. During purification, fractions are often produced that differ substantially in their solvent composition. Detergent concentration, pH, salt concentration, protease inhibitors, organic solvents, among others, may influence the bioassay in unpredictable ways. The uncertainty introduced by the presence of these factors can be reduced by empirical determinations of the effects of any suspect compounds on fractions of known activity. Again, a careful accounting of the distribution of agent at each step of purification may reveal anomalies in the bioassay of one or more fractions.

The problem of observer bias is more difficult to remedy. Each bioassay depends to a significant extent on the ability to accurately diagnose disease. In a titration using the LD_{50} endpoint, the animals must be diagnosed with clinical scrapie before death to distinguish between scrapie and nonscrapie deaths. The incubation period assay depends heavily on diagnosing disease, because the day of diagnosis (i.e., end of the incubation period) is used to calculate the titer. High titers are expected to cause disease after relatively short incubation times, and thus the natural tendency is to attempt to diagnose the disease as early as possible. Since animals inoculated with the same sample are routinely caged together, there also may be a tendency to diagnose all animals as "scrapie positive" once the first animal is diagnosed. Since scrapie is invariably fatal, the observer is almost always correct, eventually.

Several precautions can be taken to reduce the possibility of systematic error from these sources. Obviously, each cage of animals should carry only coded identification numbers (experiment number, sample number, etc.), so that different fractions or treatments cannot be distinguished from each other or from controls. In addition, the practice of diagnosing animals at "early stages" should be avoided and animals should be diagnosed as "positive" only after absolute and unmistakable evidence of clinical disease is present. Diagnosis by two independent observers (at separate times) would be desirable, but this may not be possible in many smaller research groups. Finally, individual caging of animals and neuropathologic confirmation of disease in each assay animal would be ideal, but is impractical in almost all instances.

3.3 Characterization of Purified Fractions

Producing a fraction having an increased specific activity compared to the starting material completes part of the purification task, but it is not an endpoint. The next step is to document the components that are present in this fraction. If a preparation is substantially enriched, agent-specific components will be identifiable; if the method is reproducible, these components should be seen with regularity. Proteins and nucleic acids are obvious choices for analysis, but carbohydrates, lipids, and metals may also be relevant. Ultrastructural analysis by electron microscopy can be used to document the presence of unique structures. Also, at this stage it is advisable to prepare an identical fraction using normal brain as the starting material. A comparison of the components present in the fractions from diseased and normal brains will give some indication of the degree of enrichment and may provide some clue as to the identity of any contaminants.

Since it is known that the scrapie agent contains an essential protein, a measure of the complexity of the sample with respect to protein components is probably the best starting point. SDS-PAGE combined with sensitive silver staining techniques or chemical radiolabeling provide the most sensitive and highest resolution techniques for analysis of protein constituents. Two-dimensional gel electrophoresis provides additional resolution and is useful for addressing specific questions regarding charge isomers. Ultimately, isolation and quantitation of individual proteins from an enriched fraction will document the relative contribution of each to the total mass of protein in the sample. Sequencing of the proteins, or of peptides produced by enzymatic or chemical cleavage, is a natural extension of this analysis. In a highly purified fraction, few proteins or peptides should be identified and these should account for a very large percentage of the protein mass of the fraction.

Analysis of nucleic acids can be more difficult than for proteins. Estimates of total amounts of nucleic acid can be obtained by measuring the absorbance at 254 nm of a phenol-chloroform extract or by ethidium bromide binding assays. More accurate estimates of RNA concentration can be obtained using the orcinol assay; DNA can be measured using the DAPI assay. Several different cloning strategies have been used in an attempt to detect agent-specific nucleic acids, but none have been successful to date (WIETGREFE et al. 1985; OESCH et al. 1988).

4 Purification Methods and Results

Several methods for purification of the scrapie and CJD agents will be discussed below. Some of the studies that provided a foundation for these methods have been reviewed elsewhere (MILLSON et al. 1976; PRUSINER et al. 1984b, 1987; CARP et al. 1989); additional references can be found in the papers describing each purification method. The purification methods and analytical techniques employed vary, thus direct comparisons of the results from one report with those of another are not always possible. None of the reports provide a complete chemical and structural analysis of the purified fraction. Interpretations of the results from the purification studies have been controversial and polarized; scientific opinion regarding the structure of the scrapie agent remains divided.

4.1 Is PrP-27–30 an Agent Component?

A discontinuous sucrose gradient method (PRUSINER et al. 1982) (reviewed in PRUSINER et al. 1984b, 1987) provided a significant step forward in the effort to purify the scrapie agent. Based upon a series of earlier studies using the mouse and hamster models (PRUSINER et al. 1977, 1978a, b, 1980a, b, c, d), this method provided the first substantial purification of the agent from any source. The method incorporates detergent extraction followed by precipitation with polyethylene glycol (PEG), nuclease and proteinase K digestion, ammonium sulfate precipitation and sedimentation to the interface of a discontinuous sucrose gradient. Purification relative to protein was reported at 100- to 1000-fold, with a yield of about 6% (PRUSINER et al. 1982). Concentrations of RNA and DNA were below the level of detection at less than $10\,\mu g/10^9$ ID_{50} units. The specific activities were 1.4×10^{10} ID_{50}/mg protein and over 10^{11} ID_{50}/mg nucleic acid. Subsequently, a large scale version of this procedure was reported to give purification of 3600-fold with 59% yield (PRUSINER et al. 1984b). The specific activity obtained by this procedure was 5×10^{10} ID_{50}/mg protein, but quantitation with respect to nucleic acid was not reported.

The degree of purification achieved in these preparations coupled with sensitive radiochemical labeling procedures led to the identification of PrP-

27–30 (BOLTON et al. 1982). Fibrillar structures termed "prion rods" were also found in these fractions. The prion rods appear as shortened (and perhaps slightly modified) versions of the SAF described previously (MERZ et al. 1981, 1983, 1984). PrP-27–30 is found only in the purified fractions obtained when scrapie diseased brain is the starting material; the protein has not been found in the equivalent fractions from aged normal brain (BOLTON et al. 1982, 1984). These and other data prompted the suggestion that PrP-27–30 is a component of the scrapie agent (BOLTON et al. 1982; PRUSINER et al. 1982, 1983; MCKINLEY et al. 1983; BENDHEIM et al. 1984; BOLTON et al. 1984, 1985).

A note about purification and characterization of scrapie-associated fibrils (SAF) is warranted. SAF were first identified in homogenates and various subcellular fractions of scrapie-diseased mouse brain (MERZ et al. 1981). Additional characterization of SAF isolated from scrapie- and CJD-affected brains from several species (including human CJD and hamsters with CJD or scrapie) was reported (MERZ et al. 1983, 1984) and the fibrils were demonstrated to be disease specific, but none of the SAF-containing fractions were bioassayed to determine the scrapie agent concentration. Copurification of the scrapie agent and SAF was suggested in a subsequent report (SOMERVILLE et al. 1986), but the degree of purification obtained was less than 100-fold. Other studies have shown that PrP-27–30 is the major protein component of SAF and prion rods (BARRY et al. 1985; DEARMOND et al. 1985; MERZ et al. 1987; WILEY et al. 1987).

Shortly after the report of the discovery of PrP-27–30, Diringer and colleagues described several methods for purifying the scrapie agent (DIRINGER et al. 1983a, 1983b; HILMERT and DIRINGER 1984). In the first of these reports, the authors describe purification of the scrapie agent from CLAC hamsters using proteinase K digestion, micrococcal nuclease digestion, detergent and high salt extractions and differential ultracentrifugation (DIRINGER et al. 1983b). Purification of 100- to 1000-fold was achieved with respect to protein, 20-fold with respect to RNA, and 200-fold with respect to DNA. The yield was estimated to be about 20%. Specific activities calculated from their data are approximately 1.3×10^{10} LD_{50}/mg protein, 1.3×10^{10} LD_{50}/mg RNA and 1×10^{11} LD_{50}/mg DNA.

An agent-specific protein component was not detected in the Diringer group's preparations by SDS-PAGE, but SAF were found in the final fraction. Failure to detect PrP-27–30 could be due to several factors. The agent titer in brain homogenates from the CLAC hamsters was reported to be about 10-fold less than in the LVG hamsters used by Prusiner and others, and thus, the agent concentration was probably lower in the final fraction. The staining method used to detect proteins in the gels was not described and may not have been sensitive enough to detect low concentrations of PrP-27–30. Finally, the preparations may not have been sufficiently pure to reliably detect PrP-27–30 in the presence of other proteins of similar M_r.

Two subsequent papers by this group reported improvements on the purification method and identified SAF and PrP-27–30 (called p26 in one report and SAF-protein in the other) as specific components (DIRINGER et al. 1983a; HILMERT and DIRINGER 1984). These papers lack the detailed analysis provided in the first report, but confirm the apparent association of PrP-27–30, SAF and the scrapie agent reported previously (MERZ et al. 1981; BOLTON et al. 1982; PRUSINER et al. 1982; MERZ et al. 1983). The specific activity obtained using the latter procedure (HILMERT and DIRINGER 1984) appears to be higher than those from the previous methods but numerical values were not provided.

Protein sequencing and molecular cloning studies demonstrated that PrP-27–30 is the product of a normal host gene and that it was cleaved from a larger precursor (OESCH et al. 1985; KRETZSCHMAR et al. 1986; LOCHT et al. 1986; ROBAKIS et al. 1986). Subsequent studies demonstrated that the precursor protein has an apparent mass in the range of about 33–37 kDa (OESCH et al. 1985; BARRY et al. 1986; BASLER et al. 1986; BENDHEIM and BOLTON 1986; BOLTON et al. 1986; MEYER et al. 1986). Using antibodies specific to PrP-27–30 to monitor purification of the precursor protein, we demonstrated that the precursor (termed Sp33–37 by us and PrP^{Sc} by others), and not PrP-27–30, purified with the scrapie agent when proteinase K digestion was eliminated (BOLTON et al. 1986, 1987). The purification method used detergent extraction, differential centrifugation, nuclease digestion, high salt extraction, and ultracentrifugation through sucrose cushions to produce specific activities as high as 3×10^{11} LD_{50}/mg protein and yields of 20%–30%. Bioassay of fractions from the purification showed that over 90% of the agent recovered at each step was distributed in a single fraction (BOLTON et al. 1987). Concentrations of nucleic acid were not reported, but subsequent studies demonstrated that the specific activity with respect to total nucleic acid (RNA plus DNA) is over 1×10^{14} LD_{50}/mg (D. BOLTON, S. WIETGREFE and A. HAASE, unpublished data).

Sp33–37 was essentially the only protein identified by silver staining of SDS-PAGE gels and by amino acid sequencing; a minor sequence signal accounting for 10% of the Sp33–37 signal identified a partial cleavage product larger than PrP-27–30 (BOLTON et al. 1987). As in a previous study of PrP-27–30 (PRUSINER et al. 1984a), Sp33–37 in the scrapie agent fraction was sufficiently pure to be sequenced from the amino terminus after denaturation, but without further purification (BOLTON et al. 1986, 1987). Proteinase K digestion of the purified scrapie agent fraction removed 67 residues from the amino terminus to produce PrP-27–30 (BOLTON et al. 1987; BENDHEIM et al. 1988), and trypsin digestion removed at least 35 residues from the amino terminus to produce a 29- to 32-kDa fragment (BOLTON et al. 1987). Neither of these protease digestions reduced scrapie agent titer (BOLTON et al. 1987; BENDHEIM et al. 1988). To date, no other scrapie-specific macromolecules have been identified in these purified fractions.

Several other studies have identified the abnormal proteins PrP-27–30 or Sp33–37 in human CJD-affected brain (BENDHEIM et al. 1985; BOCKMAN et al. 1985; SKLAVIADIS et al. 1986), scrapie-diseased hamster (LIBERSKI et al. 1989), mouse (TAKAHASHI et al. 1986) or sheep brain (TAKAHASHI et al. 1986; RUBENSTEIN et al. 1987) and BSE-diseased bovine brain (HOPE et al. 1988). In addition, copurification of PrP-27–30 or Sp33–37 with the scrapie agent has demonstrated by immunoaffinity chromatography (GABIZON et al. 1988a) and copartitioning into liposomes (GABIZON et al. 1987).

4.2 Contradictory Evidence

Reports from two laboratories appear to refute the hypothesis that Sp33–37 is the major protein component of the scrapie agent, in that the investigators fail to identify the protein in enriched fractions. A discussion of these results follows.

A series of papers from Marsh et al. describe purification of the scrapie agent by density gradient centrifugation in a nonionic iodinated medium called Nycodenz, sometimes in combination with isopycnic sedimentation in CsCl (MARSH et al. 1984a, b; DEES et al. 1985). The authors report enrichment of more than 500-fold and specific activities of 7×10^9 to 6×10^{10} LD_{50}/mg protein. The yields from these preparations were not reported. In some cases a 30-kDa protein was identified in preparations from both healthy and scrapie-diseased brains (MARSH et al. 1984b), while in other fractions many different proteins were found, none of which was specific to the disease or agent (MARSH et al. 1984a; DEES et al. 1985). A recent paper purports to show that the scrapie agent is separable from Sp33–37 in scrapie agent-enriched mitoplast fractions (AIKEN et al. 1989). However, no data are provided regarding the degree of purification in these fractions, and the preparations are clearly contaminated by numerous normal host proteins. (A fraction called "Gp" is reported to have specific activities of 2×10^9 LD_{50}/mg protein and 1×10^{10} LD_{50}/mg nucleic acid, but this fraction apparently bears no relationship to those used in the critical experiments.) By immunoblotting, Sp33–37 was detected in synaptosomal and SAF fractions (each containing the scrapie agent), but was not detected in an agent-enriched mitoplast fraction (AIKEN et al. 1989). A critical piece of data, i.e., the actual number of LD_{50} units of scrapie agent analyzed for each sample, is not provided; the gels were loaded based solely on total protein. Scrapie-specific components were not detected in any of the enriched fractions.

In contrast to those studies, Sklaviadis et al. documented the presence of Sp33–37 (termed Gp34 in their report) in CJD agent-enriched fractions using antibodies to PrP-27–30 (SKLAVIADIS et al. 1986). The protein (seen as three bands by SDS-PAGE) was present in fractions enriched from CJD-diseased brain but not normal brain, and cosedimented with the infectious agent. The degree of enrichment or purity of the preparations was not documented, however. The authors concluded that Sp33–37 is not a component of the CJD agent, based in part on assumptions about the glycosylation and structure of the protein that have subsequently been shown to be incorrect (ENDO et al. 1989; HARAGUCHI et al. 1989). In a subsequent paper, these investigators attempt to demonstrate that Sp33–37 is not "the infectious agent" in CJD (MANUELIDIS et al. 1987). They showed that glycosidase digestions did not inactivate the CJD agent and that some Sp33–37 was recovered in a fraction having a substantially reduced concentration of the CJD agent following lectin affinity chromatography. The authors interpretation of the data seems forced, however. Assuming first of all that Sp33–37 is a component of the agent, one can not assume that removal of sugars from the protein would necessarily have a negative effect on biological activity. Secondly, the lectin chromatography data can easily be interpreted to demonstrate copurification of Sp33–37 with the CJD agent. In their experiment, approximately half of the Sp33–37 was recovered from the top of the column as a flocculent residue having significantly reduced activity; this observation is consistent with protein denaturation and inactivation of the agent. The remainder of the Sp33–37 (about 50%) was eluted in the void volume with "practical quantitative recovery" of the CJD agent (MANUELIDIS et al. 1987). The actual amount of CJD agent recovered in the void volume was 25% and 65% in two experiments, which roughly corresponds to the recovery of Sp33–37 in this fraction (about 50%) (MANUELIDIS et al. 1987). Thus, these data can alternatively be interpreted to confirm copurification of Sp33–37 and the CJD agent. The fact that some of the protein and agent may be denatured and inactivated during a separation step does not provide evidence that the protein is not a component of the agent.

In a study of the physical properties of the CJD agent by this group the authors argue that it "has both a viruslike size and density" (SKLAVIADIS et al. 1989). Interpretation of these experiments is complicated by the low titers obtained ($< 10^7$ LD_{50}/g of brain tissue, except in one crude fraction) and the fact that Sp33–37 concentrations were estimated by immunoblotting, at best a semiquantitative procedure. Specific activity relative to protein and degree of purification were not determined for any fractions. An approximately 1000-fold reduction in nucleic acids relative to total brain homogenate was reported for isopycnic gradient fractions, but these values were not related to actual CJD agent concentrations in these fractions, i.e., LD_{50}/mg nucleic acid in these fractions.

In any case, Sp33–37 and the CJD agent clearly copurified in most of the separations shown. The apparent size of the CJD agent was heterogeneous and dependent on the fraction used for sedimentation studies. In general, sizes of over 60 *S* were obtained, but in at least one experiment with the purified CJD agent a significant fraction (approximately 10% of the largest peak) had a sedimentation coefficient of less than 6.7 *S* (Fig. 4B in SKLAVIADIS et al. 1989). The density of the CJD agent was between 1.12 and 1.30 g/cm^3 and a peak at 1.27–1.29 g/cm^3 appeared to have a reduced amount of Sp33–37. The authors use these data to argue that the CJD agent resembles viruses in its size and density. They fail to point out that the known aggregation of the scrapie and CJD agents makes determination of the minimum size of these agents difficult and that many proteins have densities within this range (Table 1). Even the smallest viruses have sedimentation coefficients of over 100 *S*, so that some of the scrapie and CJD agent particles (with apparent sizes of 6–60 *S*) would clearly fall outside the range of viruses. Aggregation of very small particles to form large structures would readily explain the heterogeneous size of these agents, but explaining how functional particles of very small apparent size are obtained from a large unit structure is more difficult. Flotation in the rate zonal gradient is one possibility, but is unlikely in the case noted above (SKLAVIADIS et al. 1989). Another possible explanation would be that sedimentation of a large fibrillar particle is significantly slowed due to the increased frictional force on this type of structure

Table 1. Physical properties of proteins and viruses[a]

Example	ρ(g/cm^3)	$S_{20,w} \times 10^{13}$	Mass (daltons)	References
Ribonuclease	1.374	1.64	12400	TANFORD (1961)
Lysozyme	1.453	1.87	14100	TANFORD (1961)
β-Lactoglobin	1.332	2.83	35000	TANFORD (1961)
Catalase	1.37	11.3	250000	TANFORD (1961)
Urease	1.37	18.6	480000	TANFORD (1961)
Myosin	1.374	6.4	570000	TANFORD (1961)
Parvovirus	1.39–1.42	~122	6200000	BERNS (1990)
Poliovirus	1.460	156	8430000	COMPANS and CHOPPIN (1975)
Bushy stunt virus	1.35	132	10700000	TANFORD (1961)
Tobacco mosaic virus	1.37	170	50000000	TANFORD (1961)
Influenza A virus	~1.18	—	~300000000	RUECKERT (1990)
Retroviridae	1.16–1.18	—	~1000000000	BADER (1975)

[a] Adapted and modified from Table 22-1 in TANFORD (1961)

(TANFORD 1961; VAN HOLDE 1971). Particle density cannot be used to infer structure; doing so one could suggest that ribonuclease, β-lactoglobin, urease, or myosin are viruses because they have densities in the virus range.

5 Conclusions

A consensus view of the structure of the scrapie agent and similar agents is unlikely to occur without definitive biochemical analyses of highly purified preparations. Several of the purification methods discussed above could lead to preparations worthy of extensive analysis by adding or altering a few steps and increasing the amount of starting material. Specific activities of over 10^{12} LD_{50}/mg protein and over 10^{14} LD_{50}/mg nucleic acid should be possible with the 263 K scrapie agent produced and assayed in LVG hamsters. At these levels of purity, meaningful chemical compositions can be obtained with reasonable amounts of the endproduct using microscale techniques for protein and nucleic acid analysis. Other systems for studying these agents, such as murine scrapie or hamster CJD, will require more starting material and a greater degree of purification because the initial yield per gram of brain is lower by a factor of 10–100.

At this time there is compelling evidence indicating that Sp33–37 is a component of the scrapie agent. Copurification of the agent and Sp33–37 (BOLTON et al. 1982, 1987; PRUSINER et al. 1982, 1984a; DIRINGER et al. 1983a), proteolytic degradation of the protein and coincidental inactivation of the agent (MCKINLEY et al. 1983), neutralization of activity by specific antibodies (GABIZON et al. 1988a) as well as other studies (GABIZON et al. 1987; BELLINGER-KAWAHARA et al. 1988; BOLTON and BENDHEIM 1988; CARLSON et al. 1988; CARP et al. 1989; DEARMOND et al. 1989; PRUSINER 1989) provide strong arguments in support of this thesis. The studies questioning this hypothesis cannot be ignored, however. Those studies beg the question, if not Sp33–37 then what protein is a scrapie agent component?

Three hypotheses for the structure of the scrapie and CJD agents were recently reviewed (CARP et al. 1989). The structures proposed under the three hypotheses are mutually exclusive: a conventional virus structure (virus hypothesis); a small nucleic acid coated by a host protein (virino hypothesis); and an abnormally modified host protein (modified host protein hypothesis). The virino and modified host protein models accept Sp33–37 as an essential component of the etiologic agent, but differ in that the virino model requires an essential nucleic acid molecule. Purification of the scrapie agent and its chemical analysis will determine which, if any, of these models is correct.

For larger viruses, the theoretical maximum specific activity with respect to nucleic acid has been approached or exceeded in some scrapie agent preparations. Thus, these structures can essentially be discounted. The smallest viruses cannot yet be ruled out on those grounds, but become improbable if virus particles and proteins are not readily evident once the empirical specific activity with respect to nucleic acid increases beyond 10^{14}. Assuming the agent is composed only of protein, the maximum theoretical specific activity with respect to protein will depend on the average number of molecules necessary to induce disease (N_{avg}). If N_{avg} is small, we are far from achieving a pure preparation, but if N_{avg} is on the order of 10000 to 100000 (i.e., about 6×10^{-15}g of Sp33–37 per LD_{50}), then some of the preparations described above are nearly homogeneous.

References

Aiken JM, Williamson JL, Marsh RF (1989) Evidence of mitochondrial involvement in scrapie infection. J Virol 63: 1686–1694

Alper T (1985) Scrapie agent unlike viruses in size and susceptibility to inactivation by ionizing or ultraviolet radiation [letter]. Nature 317: 750

Alper T, Haig DA, Clarke MC (1966) The exceptionally small size of the scrapie agent. Biochem Biophys Res Commun 22: 278–284

Bader JP (1975) Reproduction of RNA tumor viruses. In: Fraenkel–Conrat H, Wagner RR (eds) Comprehensive virology, vol 4. Plenum, New York, pp 253–332

Barry RA, McKinley MP, Bendheim PE, Lewis GK, DeArmond SJ, Prusiner SB (1985) Antibodies to the scrapie protein decorate prion rods. J Immunol 135: 603–613

Barry RA, Kent SB, McKinley MP, Meyer RK, DeArmond SJ, Hood LE, Prusiner SB (1986) Scrapie and cellular prion proteins share polypeptide epitopes. J Infect Dis 153: 848–854

Basler K, Oesch B, Scott M, Westaway D, Walchli M, Groth DF, McKinley MP, Prusiner SB, Weissmann C (1986) Scrapie and cellular PrP isoforms are encoded by the same chromosomal gene. Cell 46: 417–428

Bellinger-Kawahara CG, Kempner E, Groth D, Gabizon R, Prusiner SB (1988) Scrapie prion liposomes and rods exhibit target sizes of 55,000 Da. Virology 164: 537–541

Bendheim PE, Bolton DC (1986) A 54-kDa normal cellular protein may be the precursor of the scrapie agent protease-resistant protein. Proc Natl Acad Sci USA 83: 2214–2218

Bendheim PE, Barry RA, DeArmond SJ, Stites DP, Prusiner SB (1984) Antibodies to a scrapie prion protein. Nature 310: 418–421

Bendheim PE, Bockman JM, McKinley MP, Kingsbury DT, Prusiner SB (1985) Scrapie and Creutzfeldt–Jakob disease prion proteins share physical properties and antigenic determinants. Proc Natl Acad Sci USA 82: 997–1001

Bendheim PE, Potempska A, Kascsak RJ, Bolton DC (1988) Purification and partial characterization of the normal cellular homologue of the scrapie agent protein. J Infect Dis 158: 1198–1208

Berns KI (1990) Parvoviridae and their replication. In: Fields BN, Knipe DM, Chanock RM, et al. (eds) Virology, 2nd edn. Raven, New York, pp 1743–1763

Bockman JM, Kingsbury DT, McKinley MP, Bendheim PE, Prusiner SB (1985) Creutzfeldt–Jakob disease prion proteins in human brains. N Engl J Med 312: 73–78

Bolton DC, Bendheim PE (1988) A modified host protein model of scrapie. Ciba Found Symp 135: 164–181

Bolton DC, McKinley MP, Prusiner SB (1982) Identification of a protein that purifies with the scrapie prion. Science 218: 1309–1311

Bolton DC, McKinley MP, Prusiner SB (1984) Molecular characteristics of the major scrapie prion protein. Biochemistry 23: 5898–5906

Bolton DC, Meyer RK, Prusiner SB (1985) Scrapie PrP-27–30 is sialoglycoprotein. J Virol 53: 596–606

Bolton DC, Bendheim PE, Marmorstein AD, Angus-Rooss T, Potempska A (1986) The scrapie agent contains a 33–37 kDa protein: a brief review of the evidence. In: Court LA, Dormont D, Brown P, Kingsbury DT (eds) Unconventional virus diseases of the central nervous system. Commissariat a l'Energie Atomique, Paris, pp 627–640

Bolton DC, Bendheim PE, Marmorstein AD, Potempska A (1987) Isolation and structural studies of the intact scrapie agent protein. Arch Biochem Biophys 258: 579–590

Brown P, Green EM, Gajdusek DC (1978) Effect of different gradient solutions on the buoyant density of scrapie infectivity. Proc Soc Exp Biol Med 158: 513–516

Carlson GA, Westaway D, Goodman PA, Peterson M, Marshall ST, Prusiner SB (1988) Genetic control of prion incubation period in mice. Ciba Found Symp 135: 84–99

Carp RI, Kascsak RJ, Wisniewski HM, Merz PA, Rubenstein R, Bendheim PE, Bolton DC (1989) The nature of the unconventional slow infection agents remains a puzzle. Alzheimer Dis Assoc Disord 3: 79–99

Compans RW, Choppin PW (1975) Reproduction of myxoviruses. In: Fraenkel–Conrat H, Wagner RR (eds) Comprehensive virology, vol 4. Plenum, New York, pp 179–252

DeArmond SJ, McKinley MP, Barry RA, Braunfeld MB, McColloch JR, Prusiner SB (1985) Identification of prion amyloid filaments in scrapie-infected brain. Cell 41: 221–235

DeArmond SJ, Gonzales M, Mobley WC, Kon AA, Stern A, Prusiner H, Prusiner SB (1989) PrPSc in scrapie-infected hamster brain is spatially and temporally related to histopathology and infectivity titer. Prog Clin Biol Res 317: 601–618

Dees C, German TL, Wade WF, Marsh RF (1985) Characterization of proteins in membrane vesicles from scrapie-infected hamster brain. J Gen Virol 66: 851–859

Diringer H, Gelderblom H, Hilmert H, Ozel M, Edelbluth C, Kimberlin RH (1983a) Scrapie infectivity, fibrils and low molecular weight protein. Nature 306: 476–478

Diringer H, Hilmert H, Simon D, Werner E, Ehlers B (1983b) Towards purification of the scrapie agent. Eur J Biochem 134: 555–560

Duesberg PH (1970) On the structure of RNA tumor viruses. Curr Top Microbiol Immunol 1: 79–104

Endo T, Groth D, Prusiner SB, Kobata A (1989) Diversity of oligosaccharide structures linked to asparagines of the scrapie prion protein. Biochemistry 28: 8380–8388

Fields BN, Knipe DM, Chanock RM, Hirsch MS, Melnick JL, Monath TP, Roizman B (1990) Virology, 2nd edn. Raven, New York, pp 1–2236

Gabizon R, McKinley MP, Prusiner SB (1987) Purified prion proteins and scrapie infectivity copartition into liposomes. Proc Natl Acad Sci USA 84: 4017–4021

Gabizon R, Mckinley MP, Groth D, Prusiner SB (1988a) Immunoaffinity purification and neutralization of scrapie prion infectivity. Proc Natl Acad. Sci USA 85: 6617–6621

Gabizon R, McKinley MP, Groth DF, Kenaga L, Prusiner SB (1988b) Properties of scrapie prion protein liposomes. J Biol Chem 263: 4950–4955

Haraguchi T, Fisher S, Olofsson S, Endo T, Groth D, Tarentino A, Borchelt DR, Teplow D, Hood L, Burlingame A, et al. (1989) Asparagine-linked glycosylation of the scrapie and cellular prion proteins. Arch Biochem Biophys 274: 1–13

Hilmert H, Diringer H (1984) A rapid and efficient method to enrich SAF-protein from scrapie brains of hamsters. Biosci Rep 4: 165–170

Hope J, Reekie LJD, Hunter N, Multhaup G, Beyreuther K, White H, Scott AC, Stack MJ, Dawson M, Wells GAH (1988) Fibrils from brains of cows with new cattle disease contain scrapie-associated protein. Nature 336: 390–392

Kimberlin RH, Walker CA (1977) Characteristics of a short incubation model of scrapie in the golden hamster. J Gen Virol 34: 295–304

Kimberlin RH, Walker CA (1986) Pathogenesis of scrapie (strain 263K) in hamsters infected intracerebrally, intraperitoneally or intraocularly. J Gen Virol 67: 255–263

Kretzschmar HA, Stowring LE, Westaway D, Stubblebine WH, Prusiner SB, DeArmond SJ (1986) Molecular cloning of a human prion protein cDNA. DNA 5: 315–324

Liberski PP, Plucienniczak A, Hrabec E, Bogucki A (1989) Isolation and purification of scrapie-associated fibrils and prion protein from scrapie-infected hamster brain. J Comp Pathol 100: 177–185

Locht C, Chesebro B, Race R, Keith JM (1986) Molecular cloning and complete sequence of prion protein cDNA from mouse brain infected with the scrapie agent. Proc Natl Acad Sci USA 83: 6372–6376

Manuelidis L, Sklaviadis T, Manuelidis EE (1987) Evidence suggesting that PrP is not the infectious agent in Creutzfeldt–Jakob desease. EMBO J 6: 341–347

Marsh RF, Castle BE, Dees C, Wade WF (1984a) Equilibrium density gradient centrifugation of the scrapie agent in Nycodenz. J Gen Virol 65: 1963–1968

Marsh RF, Dees C, Castle BE, Wade WF, German TL (1984b) Purification of the scrapie agent by density gradient centrifugation. J Gen Virol 65: 415–421

McKinley MP, Bolton DC, Prusiner SB (1983) A protease-resistant protein is a structural component of the scrapie prion. Cell 35: 57–62

Merz PA, Somerville RA, Wisniewski HM, Iqbal K (1981) Abnormal fibrils from scrapie-infected brain. Acta Neuropathol (Berl) 54: 63–74

Merz PA, Somerville RA, Wisniewski HM, Manuelidis L, Manuelidis EE (1983) Scrapie-associated fibrils in Creutzfeldt–Jakob disease. Nature 306: 474–476

Merz PA, Rohwer RG, Kascsak RJ, Wisniewski HM, Somerville RA, Gibbs CJ Jr, Gajdusek DC (1984) Infection-specific particle from the unconventional slow virus diseases. Science 225: 437–440

Merz PA, Kascsak RJ, Rubenstein R, Carp RI, Wisniewski HM (1987) Antisera to scrapie-associated fibril protein and prion protein decorate scrapie-associated fibrils. J Virol 61: 42–49

Meyer RK, McKinley MP, Bowman KA, Braunfeld MB, Barry RA, Prusiner SB (1986) Separation and properties of cellular and scrapie prion proteins. Proc Natl Acad Sci USA 83: 2310–2314

Millson GC, Hunter GD, Kimberlin RH (1976) The physico-chemical nature of the scrapie agent. Front Biol 44: 243–266

Oesch B, Westaway D, Walchli M, McKinley MP, Kent SB, Aebersold R, Barry RA, Tempst P, Teplow DB, Hood LE, Prusiner SB, Weissmann C (1985) A cellular gene encodes scrapie PrP-27–30 protein. Cell 40: 735–746

Oesch B, Groth DF, Prusiner SB, Weissmann C (1988) Search for a scrapie-specific nucleic acid: a progress report. Ciba Found Symp 135: 209–223

Prusiner SB (1982) Novel proteinaceous infectious particles cause scrapie. Science 216: 136–144

Prusiner SB (1989) Scrapie prions. Annu Rev Microbiol 43: 345–374

Prusiner SB, Hadlow WJ, Eklund CM, Race RE (1977) Sedimentation properties of the scrapie agent. Proc Natl Acad Sci USA 74: 4656–4660

Prusiner SB, Hadlow WJ, Eklund CM, Race RE, Cochran SP (1978a) Sedimentation characteristics of the scrapie agent from murine spleen and brain. Biochemistry 17: 4987–4992

Prusiner SB, Hadlow WJ, Garfin DE, Cochran SP, Baringer JR, Race RE, Eklund CM (1978b) Partial purification and evidence for multiple molecular forms of the scrapie agent. Biochemistry 17: 4993–4999

Prusiner SB, Garfin DE, Cochran SP, McKinley MP, Groth DF, Hadlow WJ, Race RE, Eklund CM (1980a) Experimental scrapie in the mouse: electrophoretic and sedimentation properties of the partially purified agent. J Neurochem 35: 574–582

Prusiner SB, Groth DF, Bildstein C, Masiarz FR, McKinley MP, Cochran SP (1980b) Electrophoretic properties of the scrapie agent in agarose gels. Proc Natl Acad Sci USA 77: 2984–2988

Prusiner SB, Groth DF, Cochran SP, Masiarz FR, McKinley MP, Martinez HM (1980c) Molecular properties, partial purification, and assay by incubation period measurements of the hamster scrapie agent. Biochemistry 19: 4883–4891

Prusiner SB, Groth DF, Cochran SP, McKinley MP, Masiarz FR (1980d) Gel electrophoresis and glass permeation chromatography of the hamster scrapie agent after enzymatic digestion and detergent extraction. Biochemistry 19: 4892–4898

Prusiner SB, Bolton DC, Groth DF, Bowman KA, Cochran SP, McKinley MP (1982) Further purification and characterization of scrapie prions. Biochemistry 21: 6942–6950

Prusiner SB, McKinley MP, Bowman KA, Bolton DC, Bendheim PE, Groth DF, Glenner GG (1983) Scrapie prions aggregate to form amyloid-like birefrigent rods. Cell 35: 349–358

Prusiner SB, Groth DF, Bolton DC, Kent SB, Hood LE (1984a) Purification and structural studies of a major scrapie prion protein. Cell 38: 127–134

Prusiner SB, McKinley MP, Bolton DC, Bowman KA, Groth DF, Cochran SP, Hennessey EM, Braunfeld MB, Baringer JR, Chatigny MA (1984b) Prions: Methods for assay, purification, and characterization. In: Maramorosch K, Koprowski H (eds) Methods in virology, vol 8. Academic, Orlando, pp 293–345

Prusiner SB, Bowman KA, Groth DF (1987) Purification of scrapie prions. In: Prusiner SB, McKinley MP (eds) Prions: novel infectious pathogens causing scrapie and Creutzfeldt–Jakob disease. Academic, San Diego, pp 149–171

Robakis NK, Sawh PR, Wolfe GC, Rubenstein R, Carp RI, Innis MA (1986) Isolation of a cDNA clone encoding the leader peptide of prion protein and expression of the homologous gene in various tissues. Proc Natl Acad Sci USA 83: 6377–6381

Rubenstein R, Merz PA, Kascsak RJ, Carp RI, Scalici CL, Fama CL, Wisniewski HM (1987) Detection of scrapie-associated fibrils (SAF) and SAF proteins from scrapie-affected sheep. J Infect Dis 156: 36–42

Rueckert RR (1990) Picornaviridae, their replication. In: Fields BN, Knipe DM, Chanock RM et al. (eds) Virology, 2nd edn. Raven, New York, pp 507–548

Sklaviadis T, Manuelidis L, Manuelidis EE (1986) Characterization of major peptides in Creutzfeldt–Jakob disease and scrapie. Proc Natl Acad Sci USA 83: 6146–6150

Sklaviadis TK, Manuelidis L, Manuelidis EE (1989) Physical properties of the Creutzfeldt–Jakob disease agent. J Virol 63: 1212–1222

Somerville RA, Merz PA, Carp RI (1986) Partial copurification of scrapie-associated fibrils and scrapie infectivity. Intervirology 25: 48–55

Takahashi K, Shinagawa M, Doi S, Sasaki S, Goto H, Sato G (1986) Purification of scrapie agent from infected animal brains and raising of antibodies to the purified fraction. Microbiol Immunol 30: 123–131

Tanford C (1961) Physical chemistry of macromolecules. Wiley, New York, pp 1–710

Van Holde KE (1971) Physical biochemistry. Prentice-Hall, Englewood Cliffs, pp 1–246

Wietgrefe S, Zupancic M, Haase AT, Chesebro B, Race R, Frey W 2nd, Rustan T, Friedman RL (1985) Cloning of a gene whose expression is increased in scrapie and in senile plaques in human brain. Science 230: 1177–1179

Wiley CA, Burrola PG, Buchmeier MJ, Wooddell MK, Barry RA, Prusiner SB, Lampert PW (1987) Immuno-gold localization of prion filaments in scrapie-infected hamster brains. Lab Invest 57: 646–656

The Scrapie Fibril Protein and Its Cellular Isoform

J. HOPE and J. MANSON

1 Introduction

Over 20 years ago the association of the transmissible agents of scrapie with the surface membranes and endoplasmic reticulum of infected cells was widely recognised (HUNTER et al. 1971; MILLSON et al. 1971; KIMBERLIN et al. 1971). Membranes of scrapie-affected brain cells are less fluid than those found in healthy brain tissue (VIRET et al. 1981), and show characteristic changes in structure when viewed by the electron microscope (DUBOIS-DALCQ et al. 1979). Indeed, intraneuronal membrane-bound vacuoles appear as disease-specific structures in the brains of scrapie-affected animals. Today, the question of what goes wrong with membrane biogenesis during scrapie infection remains, although it has become sharply focused on the gene structure and expression of a single membrane component, the PrP protein. This article reviews recent

Institute for Animal Health, AFRC and MRC Neuropathogenesis Unit, Ogston Building, West Mains Road, Edinburgh, EH9 3JF, UK

Current Topics in Microbiology and Immunology, Vol. 172

progress in our understanding of the biosynthesis and function of PrP and of its key role in the development of the transmissible degenerative encephalopathies (TDEs).

2 Membranes, Infectivity and PrP

Early attempts to purify the scrapie agent showed its association with membranes and the relative resistance of scrapie infectivity to detergents (HUNTER et al. 1964; MILLSON et al. 1971; MILLSON and MANNING 1979). This latter property was exploited during the subcellular fractionation of scrapie-affected brain which yielded highly infectious fractions containing little else but fibrillar aggregates (scrapie-associated fibrils or SAF) of the neuronal membrane protein, PrP (MERZ et al. 1981, 1984). Both these clumps of membrane protein and infectivity appeared relatively resistant to proteases and other hydrolases included in the purification procedure (BOLTON et al. 1982; MCKINLEY et al. 1983; DIRINGER et al. 1983). The protease resistant protein of molecular weight 27000–30000 Da (PrP27–30) was partially sequenced and complimentary oligonucleotides were synthesised and used to isolate cDNA clones of PrP mRNA from scrapie-infected hamster brain (OESCH et al. 1985). These cDNAs were used to show that PrP is coded for by a highly conserved host gene (OESCH et al. 1985; ROBAKIS et al. 1986; LOCHT et al. 1986; BASLER et al. 1986; WESTAWAY and PRUSINER 1986). Immunoblotting using antisera to PrP27–30 detected PrP in detergent extracts of uninfected as well as scrapie-affected brain and its apparent molecular weight was 33000–35000 Da (OESCH et al. 1985; MANUELIDIS et al. 1985; KASCSAK et al. 1986; HOPE et al. 1986, 1988). It was soon clear that PrP27–30 was an artefact of the methods used to purify scrapie infectivity and that the molecular species in infectious fractions and healthy brain differed in their physicochemical properties but both were of molecular weight 33000–35000 Da and had similar, if not identical, covalent structures (HOPE et al. 1986).

Some groups have provided evidence for a dissociation between PrP and infectivity (BRAIG and DIRINGER 1985; MANUELIDIS et al. 1987), while others have shown infectivity to remain in a 30000-Da protein fraction following sodium dodecyl sulphate–polyacrylamide gel electrophoresis (SDS-PAGE) or gel filtration chromatography (SAFAR et al. 1990; BROWN et al. 1990). Our recent comparision of the titres of infectivity in scrapie-infected brain homogenate and purified SAF fractions has confirmed the partial copurification of infectivity with abnormal forms of PrP. The ratio of protein to infectious unit was in the region of 100000 PrP molecules to one infectious particle, similar to that reported by others. Figure 1 shows a two-dimensional size analysis of components in the SAF fractions which can be detected by silver staining. Both monomeric and oligomeric forms of PrP are seen, as well as minor amounts of other components. Further attempts are needed to resolve these non-PrP molecules in infectious

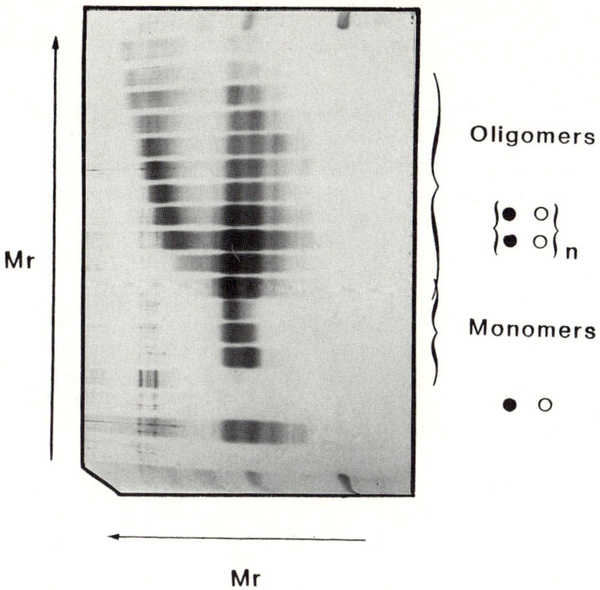

Fig. 1. Two-dimensional PAGE analysis and silver staining of SAF components in ME7 scrapie-affected mouse brain (HOPE et al.1986, 1988: Hope and Chong, unpublished). Mouse SAF purified from ME7-infected VM (*Sinc*^{P7}) mice contain 10^9 infectious particles per microgram of protein. The major protein component(s) of these fractions are various molecular variants of PrP (Mr 33000-Da, 26000 Da and 25000 Da; see also HOPE et al.1988). In SAF, PrP is in an aggregated form: these aggregates can be analysed by partial dissociation and electrophoresis in low concentrations of SDS. Re-electrophoresis in high SDS concentrations allows the protein composition of these aggregates to be visualised

fractions in order to determine whether PrP is strictly required for expression of infectivity.

There is a dramatic change in the amount and physicochemical properties of the PrP protein during the development of scrapie and related diseases. The effect of disease on the metabolic cycle of PrP biosynthesis and degradation is both devastating and specific. To put these processes into perspective, it is first necessary to describe what is known of the structure and function of the PrP gene and the cell types in which it is normally transcribed and translated into protein.

3 The Cellular Isoform

3.1 PrP Is Encoded by a Host Chromosomal Gene

The structure of the hamster gene was the first to be studied and this gene remains the one about which most is known (BASLER et al. 1986). The hamster PrP gene has two exons: a 5′ exon of 56–82 nucleotides preceded by a G + C rich element (which may serve as a transcription promoter), and a 3′ exon of 1960 nucleotides which contains the entire protein coding region of the gene. These exons are separated by a 10-kb intron. This gene organisation is conserved in mouse and sheep, although the exact size of the intron remains to be determined in these species (WESTAWAY et al. 1987; GOLDMANN et al. 1990). The gene has been localised to homologous chromosomes in mouse (chromosome 2) and man (chromosome 20) (SPARKES et al. 1986).

3.2 PrP mRNA of Different Host Species Has Variable Length and Structure

The length of the hamster mRNA, not including the poly(A) tail, was estimated at 2000–2042 nucleotides. In the 5′ non-coding region of the sequence, 56–82 nucleotides are derived from the 5′ exon and 11 nucleotides from the 3′ exon. The coding region of the mRNA is 762 nucleotides, and there is a 1198-nucleotide non-coding 3′ region before the addition signal for the poly(A) + tails. In sheep, goats and cattle, the size of the PrP mRNA is 4.6 kb and the distance from the 3′ intron/exon border to the polyadenylation site of the ovine PrP mRNA is about 4 kb (GOLDMANN et al. 1990). By comparison with the structure of the hamster PrP cDNA (OESCH et al. 1985; BASLER et al. 1986), the remaining 0.6 kb may be made up of sequence from the 5′ noncoding exon and a poly (A) tail. There is 80%–90% homology of nucleotide sequence in the protein coding region of all known PrP mRNAs, and the 3′ untranslated region of these mRNAs is similarly conserved. The larger size of ovine (and probably caprine and bovine) PrP mRNA is mainly due to the insertion of interspersed repetitive elements in this 3′ untranslated region. These repeat sequences show greater than 80% homology with the art2, Alu-like repetitive sequence family of the cow and are also found in the intron region of the ovine PrP gene (GOLDMANN et al. 1990b). These sequences are believed to have originated by reverse transcription, integration and amplification of species of RNA (7SL RNA, tRNA) involved in the biosynthesis of proteins, but their significance, if any, in the PrP mRNAs of ruminants remains to be explained (GOLDMANN et al. 1990b). In contrast, other RNA sequences (notably that near the site(s) of polyadenylation) in the 3′ untranslated region of PrP mRNA are highly conserved, and, speculatively, these RNA motifs may act as part of a control system to regulate the translation of PrP mRNA into protein.

No evidence of alternative splicing, gene rearrangement or elevated gene transcription was found during the development of experimental hamster scrapie (OESCH et al. 1985), implicating post-transcriptional processes in the aberrant metabolism of PrP in TDE diseases. However, nothing is known about somatic mutation, rearrangement or transcription of the PrP gene in brain or other tissues during the course of natural disease in man or sheep.

3.3 The Structure of PrP Is Highly Conserved

The primary structure of the PrP protein is highly conserved in man, mouse, hamster, rat, sheep and cow. From genomic DNA, the ovine PrP is a 256-amino-acid protein with an overall homology of 88%–90% with PrPs of other species. It has a predicted 24-amino-acid signal peptide, two possible sites of asparagine-linked glycosylation (at residues 184 and 200) and an extremely hydrophobic C-terminal sequence. This C-terminal region is cleaved from the nascent polypeptide and a phosphoinositol (PI)-linked glycolipid covalently attached to the new C-terminal amino acid (STAHL et al. 1987). Stahl and colleagues have recently confirmed the prediction of Ser215 as the C-terminal residue of hamster PrP (STAHL et al. 1990b). Surprisingly, the sequence around the putative site of PI-glycolipid attachment is the most variable part of the whole molecule (GOLDMANN et al. 1990).

The highly basic N-terminal region of the protein precedes a series of keratin-like glycine-rich segments that are conserved in PrPs across species (OESCH et al. 1985), although the number of these octapeptide domains can vary within species. For example, different forms of the cow PrP gene encode five or six copies of this sequence (GOLDMANN et al. 1991) and, although genes from human and all other species sequenced to date encode PrP proteins containing five copies off this repetitive element, rare alleles in two pedigrees of CJD contain 11 and 14 copies respectively (OWEN et al. 1990).

Interestingly, other amino acid polymorphisms in the PrP protein are linked to the alleles of *Sip* in sheep (HUNTER, this volume), *Sinc* in mice (BRUCE and FRASER, this volume) and equivalent host survival time genes in hamster (LOWENSTEIN et al. 1990) and man (HSIAO et al. 1989; GOLDGABER et al. 1989; GOLDFARB et al. 1990a, b; DOH-URA et al. 1989; COLLINGE et al. 1989). Several, but significantly not all, of these mutations seem to map to one domain of the protein (a putative stop-transfer signal domain, YOST et al. 1990; LOPEZ et al. 1990; see below), and much current research is aimed at understanding the molecular mechanisms by which these amino acid sequence changes might affect the survival time of man and animals.

3.4 PrP mRNA Is Expressed in Many Cells and Cell Lines

Early work by several groups established the ranking of rodent tissues and organs in respect of their PrP mRNA content by northern blot analysis (OESCH et al. 1985; CHESEBORO et al. 1985, ROBAKIS et al. 1986). Hybridisation stringency and the sensitivity of detection are crucial in these studies, and some disagreement in levels of expression in different tissues (notably spleen) was initially observed. These discrepancies have now been largely resolved. For example, Westaway and colleagues were able to elegantly demonstrate PrP mRNA in mouse spleen by oligonucleotide primer extension and S1 nuclease analysis. They estimated that the spleen contained about 1% of the level of PrP mRNA found in the same weight of brain (WESTAWAY et al. 1987).

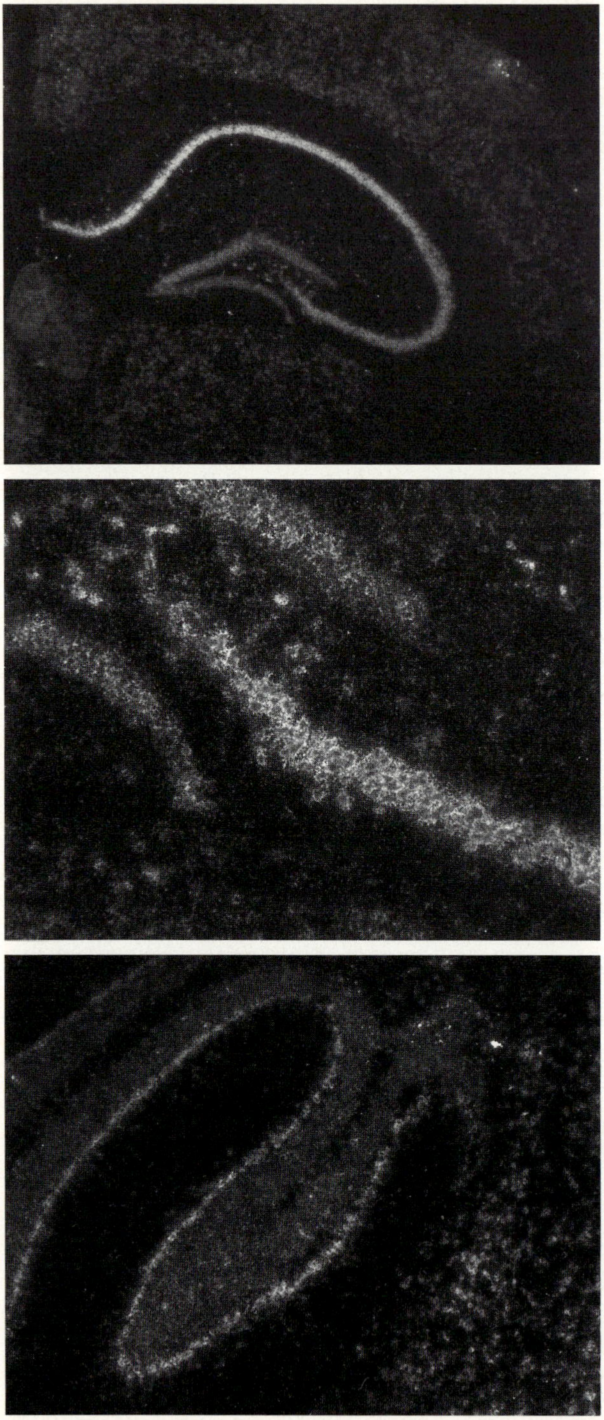

Fig. 2 a–c. Localisation of PrP mRNA in uninfected mouse brain using ^{35}S-labelled single-strand RNA probes complementary to PrP mRNA. Hybridisation was detected with nuclear track emulsion and dark-field microscopy (MANSON et al. 1991). **a** Low-power view of the hippocampus, thalamus and neurocortex showing intense hybridisation in the pyramidal cells of the hippocampus and the neurons of the dentate gyrus. Hybridisation is also found in neurons within the cortex and in the large neurons of the thalamus. **b** High-power view showing the pyramidal cells of the hippocampus and the neurons of the dentate gyrus. **c** View of the cerebellum showing high levels of PrP mRNA within the Purkinje cells and lower levels in the granular layer. Hybridisation is also detected in cells of the medulla

In situ hybridisation using various PrP gene probes has been used to find out in which cells of a tissue the PrP gene is transcribed. PrP mRNA has been located within neurones in the brains of hamsters (KRETSCHMAR et al. 1986; BROWN et al. 1990) and mice (BROWN et al. 1990; MANSON et al. 1991). A high concentration of PrP mRNA is found in the pyramidal cells of the hippocampus, the large neurones of the thalamus and neocortex and the Purkinje cells of the cerebellum. Smaller neurones, such as those of the granular layer of the cerebellum, contain lower levels of PrP mRNA (Fig. 2). These data may indicate a higher rate of PrP RNA transcription (and/or lower rate of degradation) in large neurones than in small ones (MANSON et al. 1991). In contrast to earlier studies, which found PrP mRNA to be located almost exclusively in neurones (KRETSCHMAR et al. 1986), Brown and colleagues found equivalent grain densities (indicating equivalent PrP mRNA levels) in neuronal and non-neuronal cells in hamster and mouse brain (BROWN et al. 1990). They found PrP mRNA in ependymal and choroid plexus epithelial cells, astrocytes, meninges and endothelial cells. Lung and heart contain almost the same levels of PrP mRNA as brain (OESCH et al. 1985), and this mRNA has been localised to muscle cells in heart and the alveolar lining and septal interstitial cells of lung (BROWN et al. 1990). Low levels of PrP mRNA were also found in liver, spleen and kidney, but individual cell types were not reported (KRETSCHMAR et al. 1986).

Studies on the synthesis of PrP in cells and cell lines have underlined the wide range of cell types in which the PrP gene is expressed (CAUGHEY et al. 1988, 1989; RACE et al. 1988). This wider cellular distribution of PrP may indicate a pluripotent function for the protein which is dependent on the tissue or cell type in which it is expressed. However, the ranking of rodent tissues and organs in respect of their PrP mRNA content does not correlate with terminal titres of infectivity in these tissues (EKLUND et al. 1967; KITAMOTO et al. 1989; BROWN et al. 1990), and factors other than PrP mRNA levels may control the amount of PrP and infectivity in different tissues.

3.5 Is There Post-Transcriptional Regulation of PrP mRNA?

It is not known to what extent the production of the PrP protein is regulated at the mRNA level. However the data on mRNA levels obtained by northern blotting and in situ hybridisation appear in agreement with the anatomical localisation of the normal cellular protein as seen by immunohistochemistry, tissue extraction and immunoblotting studies (DEARMOND et al. 1987). Early attempts to do PrP immunohistochemistry in brain from healthy animals were unsuccessful, until DeArmond and colleagues used a periodate-lysine-paraformaldehyde fixative to enhance the antigenicity of the PrP protein. They found that many nerve cell bodies of the dentate gyrus, Ammon's horn (CA4 region) and thalamus were strongly immunoreactive, but few stained in the caudate nucleus and neocortex. In the cerebellum, most of the Purkinje cells stained whereas little staining was seen in the neurones of the granular cell layer. At a single cell level, the PrP immunoreactivity appeared localised to the non-nuclear areas of the cell body and to the proximal parts of the dendrites of many neurones. Diffuse, weak staining was seen in the spaces between nerve cell bodies occupied by the axon-synapse-dendrite complex. Using this methodology, BRUCE and her colleagues (1989) have described similar findings in mouse brain. Few data have been published on the cellular location of PrP synthesised in cells of non-neural tissues, although their involvement in the replication of the scrapie micro-organism and the accumulation of protease-resistant isoforms of PrP in these tissues is well documented (KITAMOTO et al. 1989; FARQUHAR et al. 1991).

3.6 PrP is a Plasma Membrane Protein

Several groups have used immunohistofluorescence techniques to locate PrP on the outer surface of the plasma membrane of primary cell cultures and cell lines. The release of PrP from the surface of nerve cell-derived tumour cell lines (N2A neuroblastoma cells, PC 12 phaeochromocytoma cells) and primary cultured cells from neonatal hamster brain is catalysed by a bacterial, PI-specific phospholipase C. This provided evidence that the cellular isoform of PrP is attached to the plasma membrane through a PI glycolipid (STAHL et al. 1987, 1990a; CAUGHEY et al. 1989, 1990). Stahl and colleagues observed a patchy distribution of fluorescence on the surface of unfixed cells after incubation with anti-PrP antisera, similar to that of acetylcholinesterase on cultured sympathetic neurones. Whether this is an immunoglobulin-induced effect akin to receptor "capping" or reflects an in vivo punctate location of the protein, perhaps at cell–extracellular matrix junctions, remains to be elucidated. Ultrastructural studies on the normal cellular distribution of PrP in brain cells and other tissues are lacking.

3.7 Purification of the Cellular Isoform of PrP

The cellular isoform of PrP has been purified from hamster brain membranes by detergent extraction and cycles of affinity chromatography using anti-PrP monoclonal antibodies (MAbs) cross-linked to support media. In two studies, 0.8 μg (BENDHEIM et al. 1988) and 0.7 μg (TURK et al. 1988) of PrP per gram of hamster brain were purified and characterised by N-terminal sequencing, confirming the predicted cleavage site of signal peptidase during the biosynthesis of PrP (HOPE et al. 1986; BASLER et al. 1986). The total PrP content of hamster brain was estimated at 20–80 μg. Turk and colleagues emphasised the need for caution in interpreting data for the quantitive and qualitative analysis of PrP; an unrelated protein of molecular weight 33 000 Da (p33) was found to copurify with the cellular isoform of PrP through two cycles of immunoabsorption (TURK et al. 1988). These data have confirmed the earlier conclusion that the covalent structures of major forms of hamster PrP differ little, if at all, in normal and scrapie-affected brain (HOPE et al. 1986).

3.8 The Kinetics of PrP Biosynthesis in Neuroblastoma Cells

The biosynthesis of PrP and the kinetics of its post-translational modification have been investigated by pulse-chase experiments in cell cultures. These studies have provided insights into the private life cycle of this membrane protein.

CAUGHEY and colleagues (1989) analysed the forms of 35S-Met-PrP produced in mouse neuroblastoma cells by SDS-PAGE fluorography following their immunoprecipitation from detergent-lysates of cells using anti-PrP peptide antisera. Two early products of the PrP gene were of relative molecular weights 28000 and 33000 Da. By following a 2 h pulse of radioactive methionine with non-radioactive amino acid for varying times, they showed the processing of these PrP precursors to 30000-Da and 35000- to 41000-Da proteins. Using glycosidases, lipases and inhibitors of *N*-linked glycosylation, it was found that the 28000-Da and 33000-Da precursors resulted from the addition of high-mannose glycans to a 25000-Da polypeptide containing a PI moiety and that maturation of the precursors involved the conversion of the high manose glycans to complex glycans. These 28000-Da and 33000-Da proteins were expressed solely on the cell surface (half-life 3–5 h), where they were attached by PI glycolipid. Borchelt and colleagues produced similar data, confirming the plasma membrane topography of PrP by selectively tagging the protein with sulfo-NHS-biotin and estimating its half-life in N2A neuroblastoma cells to be about 5 h. Multiple forms of PrP were also seen in normal mouse brain extracts, and their ratio appeared to be dependent on the brain area sampled (SOMERVILLE and RITCHIE 1990). Hence the cellular fate of PrP appears similar to that of other plasma membrane glycoproteins, but may vary according to the subgroup of nerve cell in which it is synthesised.

3.9 Topographic Models of PrP Structure

Interestingly, the membrane orientation of PrP inferred from its biosynthesis in cultures of primary cells and cell lines differs from that originally predicted from modelling of protein structure (BAZAN et al. 1987) and in vitro topography experiments (HAY et al. 1987).

BAZAN and colleagues (1987) inferred a secondary structure of PrP from the amino acid sequence of the human, hamster and mouse proteins. Several hydrophobic domains, a proline- and glycine-rich repeat region and putative glycosylation sites were incorporated into a model in which PrP spanned the membrane lipid bilayer twice and had a membrane-associated C-terminal segment. The flexible, coiling structure at the N-terminus (residues 1–87) was recognised as a biologically important domain, located on the luminal side of the membrane. Residues 90–113 and 119–147 were modelled as transmembrane domains (the latter possibly as part of a channel pore), leaving the cysteine-flanked *N*-glycan sites in an extracytoplasmic orientation. This model was in agreement with studies on the biogenesis of PrP in a SP6-transcription-linked, translation-coupled, wheat germ cell-free system (HAY et al. 1987). Translation of full-length SP6 transcripts or of native hamster brain mRNA in this system resulted in a single protein (Mr 28 000), which could be precipitated with peptide antisera specific to the N- or the C-terminal domains of PrP. This precursor PrP (PrPo) was processed to two additional forms (PrP1 and PrP2) by the co-translational addition of dog pancreas membranes to the wheat germ system, Using proteases, glycosidases and anti-PrP peptide antisera, PrP2 (Mr 33 000) was shown to be glycosylated and to span the bilayer twice with both N- and C-terminal regions in the in vitro equivalent of the extracellular space. By this analysis, PrP1 (Mr 25 000) was an unglycosylated protein of similar orientation, produced from PrPo by cleavage of a hydrophobic leader sequence by signal peptidase. Paradoxically, while this membrane topography was consistent with our studies of the masking and proteolysis of PrP resulting from scrapie infection in mouse brain (HOPE et al. 1988), this transmembrane orientation did not fit with the release of PrP from cell membranes by PI-PLC (STAHL et al. 1987).

Recent studies comparing the translation/ membrane translocation of PrP in the reticulocyte lysate and wheat germ cell-free systems have indicated that these systems differ in their ability to use the PrP-90–114 segment as a membrane-spanning domain. A transmembrane form predominates if native or recombinant PrP mRNA is translated in cytosolic extracts of wheat germ-containing membranes. In the membrane/reticulocyte lysate system, the newly synthesised protein appears to be fully translocated through the membrane and this gives an "outside" topology, equivalent to that observed for PrP in cultured cells. Yost, Lopez and colleagues were able to show that the topogenic sequence responsible for this difference in behaviour was encoded between residues 74 and 114 of PrP (YOST et al. 1990; LOPEZ et al. 1990). They concluded that a heat-labile factor in cytosolic extracts of reticulocyte lysates, but not in wheat germ, might interact with the hydrophilic region (residues 74–89) of this topogenic sequence and facilitate the secretion of PrP through the membrane. In wheat germ extracts, this hydrophilic extracytoplasmic domain was actually shown to promote the integration of adjacent hydrophobic sequences of chimaeric proteins into the bilayer. Interestingly, YOST and colleagues (1990) noted that although glycosylation was not "a factor in determining transmembrane or secretory topology of PrP or its deletion mutants" the proportion of chains glycosylated varied between the native protein and each mutant. Similar differences in glycosylation have been observed in the scrapie isoforms of PrP produced in the same host animal by different strains of scrapie (HOPE et al. 1988; SOMERVILLE and RITCHIE 1990).

Intriguingly, some natural mutations in or near the stop-transfer effector sequence have been linked to the survival time of a number of host animals (and man) exposed to the subviral pathogens of scrapie. This domain may act as a binding site for a molecule or molecules (chaperones) involved in the folding and assembly of nascent polypeptides, and since misfolding of PrP may underlie the molecular pathology of the TDEs (see below), this process may be a primary target of the TDE pathogen (Ellis and Hemmingsen 1989).

3.10 Does PrP Function as a Monomer or as Part of an Oligomeric Complex?

To explain the biological properties of scrapie infectivity and its interaction with the host (Bruce and Fraser, this volume), Dickinson suggested that at least one protein product of the *Sinc* locus in mice formed part of an oligomeric complex which acted as a binding and replication site for the infectious agent (Dickinson 1975). Two sets of evidence support the notion that the PrP protein is a product of the *Sinc* (or host equivalent) locus and so, by inference, might require oligomerisation to be physiologically (or pathologically) active. Firstly, it has become clear that mutations in and around the PrP gene are linked to the relative survival time of mice, sheep and hamsters experimentally exposed to scrapie, and to the incidence of related diseases in man, Creutzfeldt–Jakob disease (CJD) and Gerstmann–Sträussler syndrome (GSS) (see above and Hunter, this volume). Secondly, transgenic experiments have shown that *Sinc* effects, such as susceptibility and survival times following scrapie exposure, can be drastically altered by the expression (and overexpression) of foreign PrP genes in mice (Scott et al. 1989; Prusiner et al. 1990). Indeed, overexpression of a Leu(101) mutant of mouse PrP in transgenic mice leads to the development of spontaneous neurodegeneration (Hsiao et al. 1990), similar to that seen in experimental TDEs. These data are consistent with PrP (as a product of the *Sinc* gene) acting as a functional subunit in an oligomeric complex (Hope et al. 1986; 1988): mutant subunits can affect the physiological function of their wild-type isoforms and so produce disease (Herskowitz 1987; Stacey et al. 1988).

Although there is no biochemical evidence that PrP functions as part of an oligomeric structure, Oesch and colleagues have documented several PrP ligands in hamster brain and other tissues which appear to specifically bind PrP27–30 and its cellular isoform (Oesch et al. 1990). These ligands all have an acidic pI (~ 5) and their molecular weights depend on their tissue of origin: brain, Mr 45 000 and Mr 110 000 (and, intriguingly, a minor form of Mr 32 000 of unknown pI); lung, Mr 56 000, Mr 110 000 and Mr 170 000; heart and skeletal muscle, Mr 52 000; spleen, Mr 56 000 and 110 000. Their identity and their significance, if any, in the development of TDEs remain to be determined.

3.11 Physiological Function of PrP

The structure of PrP is virtually constant in mammalian species, and nucleic acid hybridisation data show there may be a homologous gene in lower organisms

(WESTAWAY and PRUSINER 1986). This conservation of structure indicates an important role for this protein in the metabolic processes of the cell, but so far its physiological function has not been discovered. Levels of PrP mRNA in the developing rodent brain vary, and can be experimentally elevated using nerve growth factor (MCKINLEY et al. 1987; MOBLEY et al. 1988). The PrP mRNA is barely detectable in 1-day-old hamster brain, remains at a low level until postnatal day 10 and then increases to adult levels by day 20. Interestingly, this ontogeny may be related to differences in the development of rodent susceptibility to peripheral inoculation of the infectious agent (OUTRAM et al. 1973; MCKINLEY et al. 1989), but the timing of expression of the PrP gene has not yet been determined in extraneural tissues. Its dynamics of synthesis and its location at the cell surface have provoked speculation that, like the N-CAMs and cadhedrins, the protein is involved in the expression of a "morphogenetic" code, directing and maintaining the complex architecture of the nervous system (EDELMAN 1989). Such a physiological function might influence the targetting of lesions in the central nervous system which is a feature of the various strain isolates of the infectious agent. Key studies on the disruption of the PrP gene in vivo and in vitro are in progress to provide insight into these processes.

Recently, Cashman and colleagues have suggested that the cellular isoform of PrP participates in the activation of lymphocytes. Both PrP mRNA and protein were found to be expressed in normal human lymphocytes and lymphoid cell lines, and the surface abundance of PrP was increased 3.5-fold by the action of concanavalin A (Con A), phytohaemagglutinin and anti-CD3 MAbs (CASHMAN et al. 1990). Con A-induced mitogenesis of lymphocytes was found to be specifically blocked by polyclonal antisera to PrP. Paradoxically, Con A activation appeared to decrease the level of PrP mRNA in human lymphocytes. Cashman and coworkers suggested that PrP might be an activation-released soluble signal for the activation of other subsets of lymphoid cells.

Other groups have suggested a role for neural PrP in the control of proliferation of astroglial cells based on the production of a GFAP-stimulating factor in cultured cell lines from CJD-infected brain (OLESCIK et al. 1989) or the gliosis which develops in the brain of victims of the TDEs (DEARMOND et al. 1987). Whatever its normal function, there is a dramatic change in the amount and physicochemical properties of the PrP protein during the development of scrapie and related diseases.

4 The Scrapie Isoform

The typical life cycle of a membrane glycoprotein involves biosynthesis on the ribosome-endoplasmic reticulum, vesicular transport to the sites of *N*-glycan-glycolipid addition and processing in the cis-Golgi-trans-Golgi system. For cell surface proteins, this is followed by targetted fusion into the plasma membrane where a variable functional lifetime ensues before the molecule is engulfed in a coated vesicle for dispatch to lysosomes for degradation. Cell biological studies show that PrP biosynthesis follows a similar course of events in normal cells and tissues. In TDE-affected brain and other tissues, PrP accumulates in a protease-resistant isoform (ScPrP), which copurifies with high levels of infectivity during biochemical fractionation.

The distribution of PrP in the central nervous system alters dramatically during the course of disease, and the pathological end-point depends primarily on the genetic make-up of the host animal and pathogen. DeArmond and colleagues (1987) showed a redistribution of PrP immunoreactivity from nerve cell bodies to the neuropil of hamsters infected with the 263K strain of scrapie. This was widespread throughout the brain but particularly intense in areas of gliosis and vacuolation. Similarly, Bruce and colleagues (1989) using the 87V and ME7 strains of scrapie in mice, were able to reproduce contrasting patterns of PrP immunopathology which reflected the differences in targetting of vacuolation produced by the two strains. With ME7 in *Sinc* s7 or p7 mice, severe vacuolation occurs in the grey matter in most parts of the brain, with a few focal deposits of amyloid; in these models diffuse immunostaining with antisera to PrP was seen in the neuropil in areas of vacuolation and neuronal loss. In the 87V/*Sinc* p7 mouse combination, the major lesions are a severe vacuolation of the neuropil mainly confined to parts of the brain stem, neuronal loss from the thalamus and numerous plaques in the cerebral cortex; in this model, diffuse staining was seen in the neuropil and in areas of neuronal loss, with intense labelling of the central structures of amyloid plaques (Bruce et al. 1989; Lowe et al. 1990). The amyloid plaques of hamster scrapie, and of GSS and kuru victims also avidly cross-react with anti-PrP immunoglobulins. It is evident from this type of study that the amount of PrP accumulating in different parts of the brain (or in different tissues) is not simply related to its concentration (or the level of PrP mRNA) in unaffected tissues.

Various groups have documented the increase in ScPrP isoform in brain and other organs during the development of scrapie. Some have concluded that the kinetics of accumulation of ScPrP correlate precisely with the change of infectivity titre measured by bioassay (McKinley et al. 1983), while others maintain these events (Braig and Diringer 1985), and ScPrP and infectivity (Manuelidis et al. 1987), can be dissociated. Recently, Kitamoto and colleagues (1989) mapped the organ distribution of murine PrP in brain and tissues affected by the Fukuoka-1 strain of CJD. Tissues ranked for ScPrP content as follows: brain > spleen > spinal cord, lymph node > thymus > intestine, with no ScPrP detected in liver, lung or kidney. While no direct measurement of infectivity titre was made in these studies, this ranking was similar to that derived many years ago by Eklund and colleagues (1967) for the distribution of infectivity in mice following subcutaneous inoculation of scrapie. The authors concluded that 0.3 ng of protein could be detected in organs previously estimated to contain a titre higher than $10^{-4.5}$/0.03 ml i.c. LD^{50} units but not in tissues with a titre below $10^{-3.8}$/0.03 ml i.c. LD^{50} units, and that ScPrP correlated well with infectivity estimates. They also observed that ScPrP was much less easy to find in human peripheral tissues from victims of CJD. What is clear from these studies is that expression of PrP in a tissue may be a necessary but not sufficient condition for replication of the pathogen in that tissue.

Opinions differ as to the need for PrP in the replication of the scrapie agent. Current strategies are aimed at elucidation of the kinetics of ScPrP production in infected cells.

4.1 The Kinetics and Localisation of ScPrP Accumulation in Infected Neuroblastoma Cells

In contrast to its cellular isoform, ScPrP cannot be released from the plasma membrane of infected neuroblastoma cells, and, using enzymes and fluorescence-activated cell sorting, several groups have provided evidence for the intracellular accumulation of this protease-resistant form of PrP (CAUGHEY et al. 1990; STAHL et al. 1990a). The half-life of ScPrP has been estimated at > 15 and differs considerably from that of its normal isoform (BUTLER et al. 1988; BORCHELT et al. 1990). Hence in cell culture, as well as in the central nervous system, infection can specifically interfere with the metabolic turnover of this protein. Recent experiments have focused on how and where in the cell this pathological process takes place. Taraboulos and colleagues have shown that the acquisition of protease resistance by PrP is not dependent on asparagine-linked glycosylation (TARABOULOS et al. 1990b) and that ScPrP colocalises in infected cells with the Golgi marker, wheat germ agglutinin, and with MC-160, a protein of the medial-Golgi stack (TARABOULOS et al. 1990a). High-resolution immunogold electron microscopy has revealed discrete foci of immunoreactive ScPrP within cytoplamic vesicles of scrapie-infected cells (Taraboulos, personal communication). These vesicles, which show acid phosphatase cytochemistry as well as guanidium thiocyanate-enhanced PrP (ScPrP) immunohistochemistry, resemble secondary lysosomes (MCKINLEY et al. 1990). Their involvement in the pathogenesis of TDEs remains to be elucidated but could indicate a pathological process shared by other neurodegenerative diseases.

5 Summary

Proteins need help to fold and attain their functional conformation (ELLIS and HEMMINGSEN 1989), and mechanisms have evolved to prevent the accumulation of misfolded protein aggregates within cells (PELHAM 1988). These mechanisms fail to prevent the formation of protease-resistant, misfolded forms of PrP (ScPrP) during the development of scrapie and other transmissible spongiform encephalopathies, and ScPrP is a biochemical marker of these diseases. Much is now known about the structure and expression of the PrP gene, but the physiological function of the PrP protein and the mechanism by which the TDE pathogen replicates and specifically interferes with PrP metabolism remain a mystery—a mystery which will entertain prion-ophiliacs for some time yet.

References

Basler K, Oesch B, Scott M, Westaway D, Walchli M, Groth DF, McKinley MP, Prusiner SB, Weissmann C (1986) Scrapie and cellular PrP isoforms are encoded by the same chromosomal gene. Cell 46: 417–428

Bazan JF, Fletterick RJ, McKinley MP, Prusiner SB (1987) Predicted secondary structure and membrane topology of the scrapie prion protein. Protein Engineering 1: 125–135

Bendheim PE, Potempska A, Kascsak R, Bolton DC (1988) Purification and partial characterisation of the normal cellular homologue of the scrapie agent protein. J Infectious Dis 158: 1198–1208

Bolton DC, McKinley MP, Prusiner SB (1982) Identification of a protein that purifies with the scrapie prion. Science 218: 1309–1311

Borchelt DR, Scott M, Taraboulos A, Stahl N, Prusiner SB (1990) Scrapie and cellular prion proteins differ in their kinetics of synthesis and topology in cultured cells. J Cell Biol 110: 743–752

Braig H, Diringer H (1985) Scrapie: concept of a virus-induced amyloidosis of the brain. EMBO J 4: 2309–2312

Brown HR, Goller NL, Rudelli RD, Merz GS, Wolfe GC, Wisniewski HM, Robakis NK (1990a) The mRNA encoding the scrapie agent protein is present in a variety of non-neuronal cells. Acta Neuropathol 80: 1–6

Brown P, Liberski PP, Wolff A, Gajdusek CD (1990b) Conservation of infectivity in purified fibrillary extracts of scrapie-infected hamster brain after sequential enzymatic digestion or polyacrylamide electrophoresis. Proc Natl Acad Sci USA 87: 7240–7244

Bruce M, McBride PA, Farquhar CF (1989) Precise targeting of the pathology of the sialoglycoprotein, PrP, and vacuolar degeneration in mouse scrapie. Neurosci Lett 102: 1–6

Butler DA, Scott MRD, Bockman JM, Borchelt DR, Taraboulos A, Hsiao K, Kingsbury DT, Prusiner SB (1988) Scrapie-infected murine neuroblastoma cells produce protease-resistant prion proteins. J Virol 62: 1558–1564

Cashman NR, Loertscher R, Nalbantoglu J, Shaw I, Kascsak RJ, Bolton DC, Bendheim PE (1990) Cellular isoform of the scrapie agent protein participates in lymphocyte activation. Cell 61: 185–192

Caughey B, Race R, Cheseboro B (1988) Detection of prion protein mRNA in normal and scrapie-infected tissues and cell lines. J Gen Virol 69: 711–716

Caughey B, Race RE, Ernst D, Buchmeier J, Cheseboro B (1989) Prion protein biosynthesis in scrapie-infected and uninfected neuroblastoma cells. J Virol 63: 175–181

Caughey B, Neary K, Butler R, Ernst D, Perry L, Cheseboro B, Race R (1990) Normal and scrapie-associated forms of prion protein differ in their sensitivities to phospholipase and proteases in intact neuroblastoma cells. J Virol 64: 1093–1101

Cheseboro B, Race R, Wehrly K, Nishio J, Bloom M, Lechner D, Bergstrom S, Robbins K, Mayer L, Keith J, Garon C, Haase A (1985) Identification of scrapie prion protein-specific mRNA in scrapie-infected and uninfected brain. Nature 315: 331–333

Collinge J, Harding AE, Owen F, Poulter M, Lofthouse R, Boughey AM, Shah R, Crow TJ (1989) Diagnosis of Gerstmann–Sträussler syndrome in familial dementia with prion protein gene analysis. Lancet ii: 15–17

Czub M, Braig H, Diringer H (1988) Replication of the scrapie agent in hamsters infected intracerebrally confirms the pathogenesis of an amyloid-inducing virosis. J Gen Virol 69: 1753–1756

DeArmond SJ, Mobley WC, DeMott DL, Barry RA, Beckstead JH, Prusiner SB (1987) Changes in the localisation of brain prion proteins during scrapie infection. Neurology 37: 1271–1281

Dickinson AG (1975) Host-pathogen interactions in scrapie. Genetics 79: 387–395

Diringer H, Gelderblom H, Hilmert H, Ozel M, Edelbluth C, Kimberlin RH (1983) Scrapie infectivity, fibrils and low molecular weight protein. Nature 306: 376–378

Doh-ura K, Tateishi J, Sasaki H (1989) Pro-Leu change at position 102 of prion protein is the most common but not the sole mutation related to Gerstmann–Sträussler syndrome. Biochem Biophys Res Commun 163: 974–979

Dubois-Dalcq M, Rodriquez M, Reese TS (1979) Structural changes in the memrane of cells infected with scrapie and other neurotropic viruses. In: Prusiner SB, Hadlow W (eds) Slow transmissible diseases of the nervous system, vol 2. pp 123–145

Edelman GM (1989) Neural Darwinism: the theory of neuronal group selection. Oxford University Press, Oxford

Eklund CM, Kennedy RC, Hadlow WJ (1967) Pathogenesis of scrapie virus infection in the mouse. J Infect Dis 117: 15–22

Ellis JR, Hemmingsen SM (1989) Molecular chaperones: proteins essential for the biogenesis of some macromolecular structures. Trends Biochem Sci 14: 339–341

Farquhar CF, Somerville RA, Hope J (1991) Effect of *Sinc* genotype and agent isolate on the kinetics of accumulation of protease-resistant PrP in non-CNS tissues during the development of murine scrapie. J Gen Virol (to be published)

Goldfarb LG, Brown P, Goldgaber D, Asher DM, Rubenstein R, Brown WT, Piccardo P, Kascsak RJ, Boellaard JW, Gajdusek DC (1990) Creutzfeldt–Jakob disease and kuru patients lack a mutation consistently found in the Gerstmann–Sträussler syndrome. Exp Neurol 108: 247–250

Goldgaber D, Goldfarb LG, Brown P, Asher DM, Brown WT, Lin S, Teener JW, Feinstone SM, Rubenstein R, Kascsak RJ, Boellaard JW, Gajdusek DC (1989) Mutations in familial Creutzfeldt–Jakob disease and Gerstmann–Sträussler syndrome. Exp Neurol 106: 204–206

Goldmann W, Hunter N, Foster JD, Salbaum JM, Beyreuther K, Hope J (1990a) Two alleles of a neural protein gene linked to scrapie in sheep. Proc Natl Acad Sci USA 87: 2476–2480

Goldmann W, Hunter N, Manson J, Hope J (1990b) The PrP gene of the sheep, a natural host of scrapie. VIIIth International Congress of Virology, Abstract P28–014

Goldmann W, Hunter N, Martin T, Dawson M, Hope J (1991) Different forms of the bovine PrP gene have 5 or 6 copies of a short, G, C-rich element in the protein-coding exon. J Gen Virol 72: 201–204

Hay B, Barry R, Lieberburg I, Prusiner, SB, Lingappa VR (1987) Biogenesis and transmembrane orientation of the cellular isoform of the scrapie prion protein. Mol Cell Biol 7: 914–920

Herskowitz I (1987) Functional inactivation of genes by dominant negative mutations. Nature 329: 219–222

Hope J, Morton LJD, Farquhar CF, Multhaup G, Beyreuther K, Kimberlin RH (1986) The major protein of scrapie-associated fibrils (SAF) has the same size, charge distribution and N-terminal protein sequence as predicted for the normal brain protein (PrP). EMBO J 5: 2591–2597

Hope J, Multhaup G, Reekie LJD, Kimberlin RH, Beyreuther K (1988) Molecular pathology of scrapie-associated fibril protein (PrP) in mouse brain affected by the ME7 strain of scrapie. Eur J Biochem 172: 271–277

Hsiao K, Baker HF, Crow TJ, Poulter M, Owen F, Terwilliger JD, Westaway D, Ott J, Prusiner SB (1989) Linkage of a prion protein missene variant to Gerstmann–Sträussler syndrome. Nature 338: 342–345.

Hsiao KK, Scott M, Foster D, Groth DF, DeArmond SJ, Prusiner SB (1990) Spontaneous neurodegeneration in transgenic mice with mutant prion protein. Science 250: 1587–1590

Hunter GD, Millson GC, Meek G (1964) The intracellular location of the agent of mouse scrapie. J Gen Microbiol 34: 319–325

Hunter GD, Kimberlin RH, Millson GC, Gibbons RA (1971) An experimental examination of the scrapie agent in cell membrane mixtures. I: Stability and physiochemical properties of the scrapie agent. J Gen Microbiol 81: 23–32

Kascsak R, Rubenstein R, Merz PA, Carp RI, Robakis NK, Wisniewski HM, Diringer H (1986) Immunological comparison of scrapie-associated fibrils isolated from animals infected with four different scrapie strains. Virol 59: 676–683

Kimberlin RH, Millson GC, Hunter GD (1971) An experimental examination of the scrapie agent in cell membrane mixtures. II: Studies of the operational size. J Gen Microbiol 81: 383–391

Kitamoto T, Mohri S, Tateishi J (1989) Organ distribution pf proteinase-resistant prion protein in humans and mice with Creutzfeldt–Jakob disease. J Gen Virol 70: 3371–3379

Kretschmar HA, Prusiner SB, Stowring LE, DeArmond SJ (1986) Scrapie prion proteins are synthesized in neurones. Am J Pathol 122: 1–5

Locht C, Cheseboro B, Race R, Keith JM (1986) Molecular cloning and complete sequence of prion protein cDNA from mouse brain infected with the scrapie agent. Proc Natl Acad Sci USA 83: 6372–6376

Lopez CD, Yost CS, Prusiner SB, Myers RM, Lingappa VR (1990) Unusual topogenic sequence direcsts prion protein biogenesis. Science 248: 226–229

Lowe J, McDermott H, Kenward N, Landon M, Mayer RJ, Bruce M, McBride PA, Somerville RA, Hope J (1990) Ubiquitin conjugate immunoreactivity in the brains of scrapie infected mice. J Pathol 162: 61–66

Lowenstein DH, Butler DA, Westaway D, McKinley MP, DeArmond SJ, Prusiner SB (1990) Three hamster species with different scrapie incubation times and neuropathological features encode distinct prion proteins. Mol Cell Biol 10: 1153–1163

McKinley MP, Bolton DC, Prusiner SB (1983) A protease-resistant protein is a structural component of the crapie prion. Cell 35: 57–62

McKinley MP, Hay B, Lingappa VR, Lieberburg I, Prusiner SB (1987) Developmental expression of prion protein gene in brain. Dev Biol 121: 105–110

McKinley MP, DeArmond SJ, Torchia M, Mobley WC, Prusiner SB (1989) Acceleration of scrapie in neonatal Syrian hamsters. Neurology 39: 1319–1324

McKinley MP, Taraboulos A, Kenaga L, Serban D, DeArmond SJ, Stieber A, Prusiner SB (1990) Ultrastructural localication of scrapie prion proteins in secondary lysosomes of infected cultured cells. Proceedings XXXth Annual Meeting of the American Society of Cell Biology Abstract 1760 San Diego, Ca

Manson J, McBride PA, Hope J (1991) Regulation of PrP gene expression in mouse brain. (to be published)

Manuelidis L, Valley S, Manuelidis EE (1985) Specific proteins associated with Creutzfeldt–Jakob disease and scrapie share antigenic and carbohydrate determinants. Proc Natl Acad Sci USA 82: 4263–4267

Manuelidis, L, Sklaviadis T, Manuelidis EE (1987) Evidence suggesting PrP is not the infectious agent in Creutzfeldt–Jakob disease EMBO J 6: 341–347

Merz PA, Somerville RA, Wisniewski HM, Iqbal K (1981) Abnormal fibrils from scrapie-affected brain. Acta Neuropathol 65: 63–74

Merz PA, Rowher RG, Kascsak R, Wisniewski HM, Somerville RA, Gibbs CJ, Gajdusek DC (1984) Infection-specific particle from the unconventional slow virus diseases. Science 225: 437–440

Millson GC, Manning EJ (1979) The effect of selected detergents on scrapie infectivity. In: Prusiner SB Hadlow W (eds) Slow transmissible diseases of the nervous system, vol 2. pp 409–424

Millson GC, Hunter GD, Kimberlin RH (1971) An experimental examination of the scrapie agent in cell membrane mixtures. III: The association of scrapie activity with membrane fractions. J Gen Microbiol 81: 255–265

Mobley WC, Neve RL, Prusiner SB, McKinley MP (1988) Nerve growth factor increases mRNA levels for the prion protein and the B-amyloid protein precursor in developing hamster brains. Proc Natl Acad Sci USA 85: 9811–9815

Oesch B, Westaway D, Walchi M, McKinley MP, Kent SBH, Aebersold R, Barry RA, Teplow DB, Tempst DB, Hood LE, Prusiner SB, Weissmann C (1985) A cellular gene encodes scrapie PrP 27–30 protein. Cell 40: 735–746

Oesch B, Teplow DB, Stahl N, Serban D, Hood LE, Prusiner SB (1990) Identification of cellular proteins binding to the scrapie prion protein. Biochemistry 29: 5848–5855

Olescak EL, Murdoch G, Manuelidis L, Manuelidis EE (1988) Growth factor production by Creutzfeldt–Jakob disease cell lines. J Virol 62: 3103–3108

Outram GW, Dickinson AG, Fraser H (1973) Developmental maturation of susceptibility to scrapie in mice. Nature 241: 536–537

Owen F, Poulter M, Shah T, Collinge J, Lofthouse R, Baker H, Ridley R, McVey J, Crow TJ (1990) An in-frame insertion in the prion protein gene in familial Creutzfeldt–Jakob disease. Mol Brain Res 7: 237–276

Prusiner SB, Scott M, Foster D, Pan KM, Groth D, Mirenda C, Torchia M, Yang SL, Serban D, Carlson GA, Hoppe PC, Westaway D, DeArmond SJ (1990) Transgenetic studies implicate interactions between homologous PrP isoforms in scrapie prion replication. Cell 63: 673–686

Race R, Caughey B, Graham K, Ernst D, Cheseboro B (1988) Analyses of frequency of infection, specific infectivity, and prion protein biosynthesis in scrapie-infected neuroblastoma cell clones. J Virol 62: 2845–2849

Robakis NK, Sawh PR, Wolfe GC, Rubenstein R, Carp RI, Innis MA (1986) Isolation of a cDNA clone encoding the leader peptide of prion protein and expression of the homologous gene in various tissues. Proc Natl Acad Sci USA 83: 6377–6381

Safar J, Wang W, Padgett MP, Ceroni M, Piccardo P, Zopf D, Gajdusek DC, Gibbs CJ (1990) Molecular mass, biochemical composition, and physicochemical behavior of the infectious form of the scrapie precursor protein monomer. Proc Natl Acad Sci USA 87: 6373–6377

Scott M, Foster D, Mirenda C, Serban D, Coufal F, Walchli M, Torchia M, Groth D, Carlson GA, DeArmond SJ, Westaway D, Prusiner SB (1989) Transgenic mice expressing hamster prion protein produce species-specific scrapie infectivity and amyloid plaques. Cell 59: 847–857

Somerville RA, Ritchie LA, (1990) Differential glycosylation of the protein (PrP) forming scrapie-associated fibrils. J Gen Virol 7l: 833–839

Sparkes RF, Simon M, Cohn VH, Fournier REK, Lem J, Klisak I, Heinzmann C, Blatt C, Lucero M, Mohandas T, DeArmond SJ, Westaway D, Prusiner SB, Weiner LP (1986) Assignment of the mouse and human prion protein genes to homologous chromosomes. Proc Natl Acad Sci USA 83: 7358–7362

Stacey A, Bateman J, Choi T, Mascara T, Cole W, Jaenisch R (1988) Perinatal lethal osteogenesis imperfecta in transgenic mice bearing and engineered mutant pro-a1(I) collagen gene. Nature 332: 131–136

Stahl N, Borchelt DR, Hsiao K, Prusiner SB (1987) Scrapie prion protein contains a phosphoinositol glycolipid. Cell 51: 229–240

Stahl N, Borchelt DR, Prusiner SB (1990a) Differential release of cellular and scrapie prion proteins from cellular membranes by phophatidylinositol-specific phospholipase C. Biochemistry 29: 5405–5412

Stahl N, Baldwin MA, Burlingame AL, Prusiner SB (1990b) Identification of glycoinositol phospholipid linked and truncated forms of the scrapie prion protein. Biochemistry 29: 8879–8884

Taraboulos A, Serban D, Prusiner SB (1990a) Scrapie prion proteins accumulate in the cytoplasm of persistently infected cultured cells. J Cell Biol 110: 2117–2132

Taraboulos A, Rogers M, Borhelt DR, McKinley MP, Scott M, Serban D, Prusiner SB (1990b) Acquisition of protease resistance by prion protein in scrapie-infected cells does not require asparagine-linked glycosylation. Proc Natl Acad Sci USA 87: 8262–8266

Turk E, Teplow DB, Hood LE, Prusiner SB (1988) Purification and properties of the cellular and scrapie hamster prion proteins. Eur J Biochem 176: 21–30

Westaway D, Prusiner SB (1986) Conservation of the cellular gene encoding the scrapie prion protein PrP 27–30. Nucl Acids Res 14: 2035–2044

Westaway D, Goodman PA, Mirenda CA, McKinley MP, Carlson GA, Prusiner SB (1987) Distinct prion proteins in short and long scrapie incubation period mice. Cell 51 651–662

Viret J, Dormont D, Molle D, Court L, Leterrier F, Cathala F, Gibbs CJ, Gajdusek DC (1981) Structural modifications of nerve membranes during experimental scrapie evolution in mouse. Biochem Biophys Res Commun 101: 830–836

Yost CS, Lopez CD, Prusiner SB, Myers RM, Lingappa VR (1990) Non-hydrophobic extracytoplasmic determinant of stop transfer in prion protein. Nature 343: 669–672

Ultrastructural Studies of Prions*

M. P. McKinley[1,2] and S. B. Prusiner[1,3]

1 Introduction

The scrapie prion protein denoted PrP 27–30 was discovered by enriching fractions for scrapie infectivity (Prusiner et al. 1982, 1983). This protein migrates during sodium dodecyl sulfate–polyacrylamide gel electrophoresis (SDS-PAGE) as a broad band with an apparent molecular weight of 27–30 kDa (Bolton et al. 1982; Prusiner et al. 1982). It is derived from PrP^{Sc}, a 33- to 35-kDa protein, by limited proteolysis (Oesch et al. 1985). A portion of PrP forms long filamentous structures in scrapie-infected brains (DeArmond et al. 1985), and the purified prions apparently aggregate into rod-shaped structures in brain extracts (Meyer et al. 1986; McKinley et al. 1986). These prion aggregates are ultrastructurally and tinctorially indistinguishable from amyloid (Prusiner et al. 1983). Extensive purification of PrP 27–30 was required before convincing evidence linking the rods and prion infectivity could be obtained. Raising antibodies against PrP

* This work was supported by research grants from the National Institutes of Health (AG02132 and NS14069), the Senator Jacob Javits Center of Excellence in Neuroscience (NS22786), and State of California, Department of Health Services (contract no. 88–94658), as well as gifts from the Sherman Fairchild Foundation and the Fairleigh S. Dickinson, Jr. Foundation, Inc.
Departments of [1]Neurology, [2]Anatomy, and [3]Biochemistry and Biophysics, University of California, San Francisco, CA 94143, USA

Current Topics in Microbiology and Immunology, Vol. 172

27–30 resulted in the demonstration that filaments in tissues and rods in extracts are composed of PrP (DEARMOND et al. 1985; BARRY et al. 1985). By negative staining, the individual rods measure 10–20 nm in diameter and 100–200 nm in length. In highly purified preparations, rods usually are found in large clusters, but individual rods can be dispersed by brief sonication (Fig. 1). Prolonged sonication results in progressive fragmentation of the rods into a variety of small, elongated particles and spheres; however, this profound alteration in ultrastructural morphology is not accompanied by a change in prion infectivity (MCKINLEY et al. 1986) (Table 1). Proteolytic enzyme digestion, alkali treatment, and detergent extraction did not yield an identifiable unit particle associated with infectivity. The lack of correlation between rod morphology and prion titer seems to be consistent with the hypothesis that the

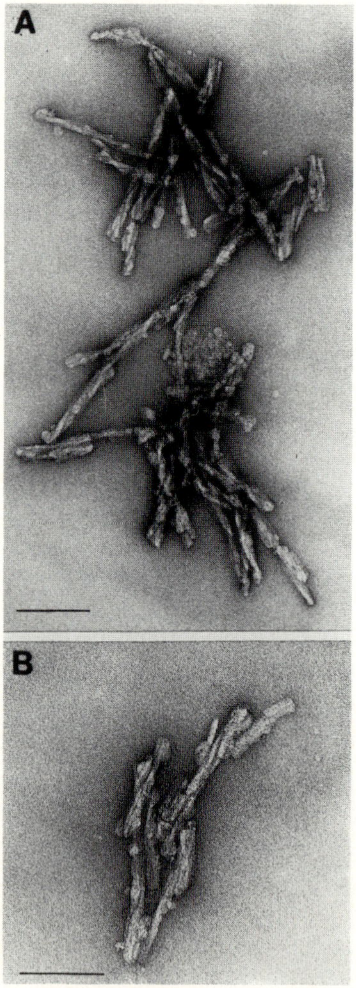

Fig. 1 A, B. Ultrastructure of prion rods isolated from hamster brain extracts on sucrose gradients. Electron micrographs of negatively stained scrapie prion-containing fractions from discontinuous sucrose gradients. Scrapie prion titer was $> 10^{9.5}$ ID_{50} units/ml. *Bars*, 100 nm. (Reproduced from PRUSINER and MCKINLEY 1987, pp 239–275)

Table 1. Multiple molecular forms of prions

Prion form	Prion protein	Method of preparation
1. Membrane	PrP^{Sc} or PrP 27–30	Purification of microsomal membranes
2. Amyloid rod	PrP 27–30	Detergent extraction of membranes combined with limited proteolysis
3. Sphere	PrP 27–30	Sonication of amyloid rods
4. DLPC	PrP^{Sc} or PrP 27–30	Dispersion of prion rods or microsomal membranes with detergents and phospholipids
5. Liposome	PrP^{Sc} or PrP 27–30	Removal of detergents from DLPC by dialysis

DLPC, detergent-lipid-protein complex

rods are aggregates of prions. Clearly, the disruption of prion rods into amorphous globules, smaller rods, and spheres, with no detectable loss of infectivity, establishes that elongated structures the size and shape of prion rods or filaments are not required for infectivity.

1.1 Prion-Infected Tissues

The complete molecular structure of the scrapie agent remains elusive despite many intensive studies during the last three decades. Recent experimental results continue to support the hypothesis that the scrapie agent is different from both viruses and viroids (ALPER et al. 1978; DIENER et al. 1982; PRUSINER 1982). Numerous ultrastructural investigations have attempted to describe a unique particle either in situ or in fractions derived from scrapie-infected brains (Table 2). Most of the early studies focused on the description of structures found in thin section preparations from scrapie-infected rodent and sheep brains. A heterogeneous collection of elongated structures was described in these tissues, including tubules (DAVID-FERREIRA et al. 1968), rods (FIELD et al. 1967; DAVID-FERREIRA et al. 1968; NARANG 1974a), fibrils (DAVID-FERREIRA et al. 1968), filaments (FIELD and RAINE 1966; DEARMOND et al. 1985), and vacuoles/vesicles (PATTISON and SMITH 1963; LAMPERT et al. 1971; BIGNAMI and PARRY 1972). Neurons from scrapie-infected rat brains contained inclusion bodies and minute, rod-like structures with an apparent dense longitudinal core (FIELD et al. 1967; FIELD and NARANG 1972). Additionally, rods were identified by negative staining with ruthenium red or lanthanum nitrate (NARANG 1974a, b) and measured 60 nm in length by 20 nm in diameter. Concurrent investigations identified spherical and virus-like particles in both thin sections and extracts from infected brains. Mouse brain sections revealed enlarged synaptic terminals with spherical particles having a 32- to 36-nm diameter (DAVID-FERREIRA et al. 1968). Spherical particles measuring 35 nm in diameter were found within cytoplasmic projections budding into neuronal vesicles in infected sheep brain

Table 2. Structures in scrapie-infected tissues and extracts

Structures reported	References
Elongated structures—thin sections	
Helical filaments	FIELD and RAINE (1966)
Longitudinal tubules	RAINE and FIELD (1967)
Vesicles and tubules	LAMPERT et al. (1971)
Rod-shaped bodies	FIELD and NARANG (1972)
Rod-shaped particles	NARANG (1974a)
Filaments	DEARMOND et al. (1985)
Spherical particles	
Round particles	BIGNAMI and PARRY (1971)
Spherical particles	SIAKOTOS et al. (1979)
Virus-like particles	CHO and GREIG (1975)
Extracts	
No definable structure	MALONE et al. (1979)
Amorphous material	PRUSINER et al. (1978)

(BIGNAMI and PARRY 1971). Smaller osmiophilic particles (23 nm diameter) were observed in ordered arrays within postsynaptic processes in murine brain only (BARINGER and PRUSINER 1978). In the mid-1970s, many investigators attempted to correlate infectivity with a unique structure. These studies began to focus on samples derived from partially purified extracts of scrapie-infected tissue. The earliest studies on these preparations suggested an association of infectivity with fractions which lacked cellular membranes, although no specific structure could be identified (PRUSINER et al. 1978, 1979; MALONE et al. 1979). Particles measuring 30–60 nm in diameter were observed in negatively stained preparations of murine brain and spleen fractions (SIAKOTOS et al. 1979). Some investigators reported finding virus-like particles measuring 14 nm in diameter in preparations of mouse brains (CHO and GREIG 1975; CHO 1976). Subsequent studies suggested that these particles were probably ferritin molecules (CHO et al. 1977).

1.2 Scrapie-Associated Fibrils

In 1981, long, abnormal fibrils called scrapie-associated fibrils (SAF) were identified in unpurified extracts of scrapie-infected brain. SAF were distinguished from all other filamentous structures by their characteristic and well-defined morphology (MERZ et al. 1981). Published electron micrographs of the SAF showed helically wound structures measuring 300–800 nm in length. Although the chemical composition of these fibrils was unknown (MERZ et al. 1981, 1983a), their ultrastructural morphology was well defined. Two types of SAF were observed: (1) those composed of two helically wound subfilaments measuring

12–16 nm in diameter and having a periodicity of either 40–60 nm or 80–110 nm, and (2) those composed of four helically wound subfilaments measuring 27–34 nm in diameter and having a periodicity of 100–120 nm. At their points of greatest diameter both types of SAF were found to have spaces between the subfilaments of 2–4 nm. Based on their characteristic ultrastructure, the fibrils were stated repeatedly to be different from intermediate filaments and amyloids (MERZ et al. 1981, 1983a, b).

In 1983, SAF were suggested to be either pathologic products of infection or the infectious particle (MERZ et al. 1983a). The fibrils were said to correlate with infectivity because of their presence in both spleens and brains of scrapie-infected rodents. No convincing data have been presented which demonstrate a correlation between scrapie agent titers in brain and concentration of SAF. More important, there is clearly a lack of correlation between titers in spleen and presence of SAF (MERZ et al. 1984).

1.3 Prion Rods and Filaments

Rod-shaped particles measuring 25 nm in diameter and 100–200 nm in length were found in rotary-shadowed fractions which were enriched for scrapie infectivity. These partially purified fractions contained predominantly one protein (PrP 27–30) (Table 2) (PRUSINER et al. 1982). We initially suggested that these rods were either aggregates of infectious prions or pathologic products of infection. The former consideration was shown to be correct when, in 1983, extensively purified hamster brain fractions were produced containing PrP 27–30, high levels of infectivity ($10^{9.5}$ ID^{50} units/ml), and rods (PRUSINER et al. 1983). The morphologic analysis using uranyl formate to stain the rods showed that their diameter varied between 10 and 20 nm, with a mean of 15 nm. No unit structure could be discerned, so no direct correlation between the rods and scrapie prion infectivity could be enumerated. This lack of a unit structure is an important feature which distinguishes prions from viruses at the ultrastructural level. In contrast, the heterogeneous appearance of the rods, as well as their tendency to aggregate and cluster, is reminiscent of many purified amyloids; it provided the clue which led us to ask whether amyloid plaques in prion diseases might be composed of PrP filaments (PRUSINER et al. 1983) (Sect. 2.4). In histochemical studies, the prion rods were shown to be amyloid by their staining with Congo red dye and their green-gold birefringence under polarized light (PRUSINER et al. 1983). Amino acid sequencing studies on both electrophoretically purified PrP 27–30 and sucrose gradient-purified rods revealed that the predominant protein in both samples had identical amino termini (PRUSINER et al. 1984). Subsequently, immunoelectron microscopy studies demonstrated that PrP 27–30 was a component of the rods (BARRY et al. 1985). Recent investigations suggest that these elongated structures are artifacts of the preparations and that maintenance of an elongated structure is not required for transmission of scrapie infectivity (MCKINLEY et al. 1986, 1991; MEYER et al. 1986).

1.4 Structural Features Distinguishing Prion Rods from Scrapie-Associated Fibrils

Shortly after the prion rods were first described, it was assumed by some investigators that these rods and SAF were the same ultrastructural entities (DIRINGER et al. 1983a; MERZ et al. 1984). To date, there is no convincing data to support this assumption. Instead, there are a plethora of reports in which the terms "prion rods" and "SAF" are used interchangeably. It seems clear that based on a series of unsupported assumptions, these investigators claimed that the prion rods are identical to scrapie associated fibrils (SAF) and that SAF are composed of PrP 27–30. They do not address the issue that SAF have been repeatedly distinguished from amyloid based on their ultrastructural and tinctorial properties. Despite claims to the contrary, there is no convincing study which reports the composition of SAF; whether or not authentic SAF contain PrP 27–30 or PrP^{Sc} is unknown (DIENER 1987).

As outlined in Table 3, prion rods and SAF are structurally distinct. Prion rods do not exhibit (1) the substructure, (2) periodicity, or (3) spaces between adjacent filaments that characterize SAF. Equally important are the observations that the prion rods possess no regular or unit structure and that they are indistinguishable from many purified amyloids. In contrast, SAF have a well-defined, regular identifiable structure and repeatedly they have been reported to be different ultrastructurally from amyloids. Attempts to stain SAF with Congo red dye have been unsuccessful (MERZ et al. 1981, 1983b); however, even a positive

Table 3. Comparison of properties of prion rods and SAF

Properties	Prion rods[a]	SAF[b]
Dimensions		
Diameter	10–20 nm	12–16 (type I), 27–34 (type II)
Length	100–200 nm	100–1000 nm
Fine structure	Variable structure with extensive aggregation; indistinguishable from purified amyloids	Helically wound subfilaments with periodicities of 40–60 nm or 80–110 nm (type I); subfilaments with periodicities of 100–120 nm (type II). Subfilaments separated by 2- to 4-nm spaces; readily distinguished from amyloid
Composition	PrP 27–30 in scrapie extracts	Unknown
Tinctorial properties	Congophilic and show green-gold birefringence	Do not stain with Congo red dye
Relationship to infectivity	Composed of PrP 27–30 which has been shown to correlate with titer; no unit structure for correlation with titer; sonically disrupted into spheres or dispersed into liposomes with no loss in infectivity	Found in brain and spleen extracts late in disease; no correlation with spleen titer

[a] Properties of prion rods compiled from the following references: PRUSINER et al. (1982, 1983, 1984); MCKINLEY et al. (1983, 1986); DEARMOND et al. (1985); KITAMOTO et al. (1986); MEYER et al. (1986)
[b] Properties of SAF compiled from the following references: MERZ et al. (1981, 1983a, 1984a, b)

result would have been impossible to interpret because of impurities in the extracts.

To claim that prion rods and SAF are identical is confusing and ignores the published literature. Without any biochemical or ultrastructural basis, some investigators have assumed that any elongated particles in scrapie preparations are SAF (DIRINGER et al. 1983b; ROHWER 1984). For example, DIRINGER et al. (1983b) purified the scrapie agent from rodent brains using a protocol similar to that first developed by us (PRUSINER et al. 1982). Fractions were produced which contained very short structures (some were only 50 nm in length) with no discernible substructure. These structures were called scrapie-associated fibrils in spite of the fact that they did not meet the specific structural criteria used to define SAF and to distinguish SAF from intermediate filaments and amyloids (MERZ et al. 1981). More recently, three studies report the use of SAF fractions in experiments, although neither micrographs showing the presence of SAF nor methods describing electron microscopy procedures to detect SAF are presented in these studies (HOPE et al. 1986; SKLAVIADIS et al. 1986; MANUELIDIS et al. 1987). Although several investigators have stated that prion rods and SAF must be one and the same (ROHWER 1984; SOMERVILLE 1985; MERZ et al. 1987), the inaccuracy of these unsupported assumptions only serves to add confusion to a field already characterized by enigma and unprecedented experimental results.

We suggest that the term SAF is best reserved for structures fulfilling the criteria set forth by MERZ et al. (1981) and that it not be used as a general term for any elongated structure found in extracts of prion-infected tissues. It is noteworthy that during the examination of numerous purified fractions highly enriched for scrapie infectivity, we have consistently failed to see structures fulfilling the ultrastructural criteria used to define SAF. While a few structures might resemble SAF in one respect or another, we do not find any structures which fulfill all the criteria. Furthermore, all purified preparations derived using detergent extractions combined with limited proteolysis display the tinctorial properties of amyloid. Our experience with partially purified fractions derived from microsomal membrane fractions is similar in that we have failed to find SAF but prion rods are readily observed if, as described above, conditions are used which facilitate their formation (MCKINLEY et al. 1991). Lastly, it is important to stress that each prion rod is slightly different; in other words, the ultrastructural morphology of the rods is quite heterogeneous, as is the case for amyloids (PRUSINER et al. 1983) but not for intermediate filaments, filamentous viruses, and SAF. While SAF have been distinguished from amyloids, as noted above, they have been suggested to be the first example of a filamentous animal virus (MERZ et al. 1984).

In a recent communication, SAF were reported to be present in intact, infected rodent brains, but no convincing data were offered to support this contention (SOMERVILLE et al. 1989). Studies performed to demonstrate SAF existence in scrapie-infected brains in vivo did not examine the ultrastructural morphology of whole brains. No elongated particles were found in fractions which contained PrP^{Sc}, but a partially hydrolyzed form of PrP (judged by Western blot analysis) was present in fractions purportedly containing SAF. These fractions were obtained after detergent extraction and centrifugation, but the investigators claim that this is representative of the in vivo state. Clearly, the data presented do not support the claim for SAF existence in vivo.

2 Properties of Prion Rods

2.1 Copurification of Rods, PrP 27–30, and Infectivity

Examination of fractions from sucrose gradients prepared by centrifugation in a vertical rotor revealed several interesting relationships between prion rods, protein levels, and scrapie infectivity (PRUSINER et al. 1982) (Table 1). First, total protein had a bimodal distribution within the gradient while much of the scrapie infectivity was associated with the small peak of protein near the bottom of the gradient. Numerous aggregates of amorphous material and flattened rod-like structures were detected in these highly purified fractions. Second, protein and infectivity had minimum levels in the middle of the gradient. Ultrastructural examination of fractions from the middle of the gradient revealed a few rods but no clusters of rods. Third, there were high titers of scrapie infectivity at the top of the gradient; numerous particles without any distinct shape or size were observed. The significance of the numerous rods found at the bottom of the sucrose gradients containing scrapie infectivity was unclear, since infectivity titers in both top and bottom gradient fractions were similar. These observations raised the possibility that the rods represented either an aggregated form of the infectious agent or a pathologic product of infection.

While our conclusions about the significance of the rods remained tentative initially, studies on the protein components of gradient fractions revealed that a unique protein (PrP 27–30) was present (BOLTON et al. 1982; PRUSINER et al. 1982). The distribution of this protein paralleled that of scrapie infectivity with their bimodal distributions being coincident. Subsequent studies showed an excellent correlation between PrP 27–30 concentration and scrapie infectivity titers (MCKINLEY et al. 1983; PRUSINER et al. 1983). Many lines of evidence now argue that PrP 27–30 is a structural component of the scrapie prion (see GABIZON and PRUSINER 1990).

2.2 Formation of Prion Rods In Vitro

During the development of new purification protocols for scrapie prions from hamster brains, we obtained fractions that were enriched for scrapie infectivity

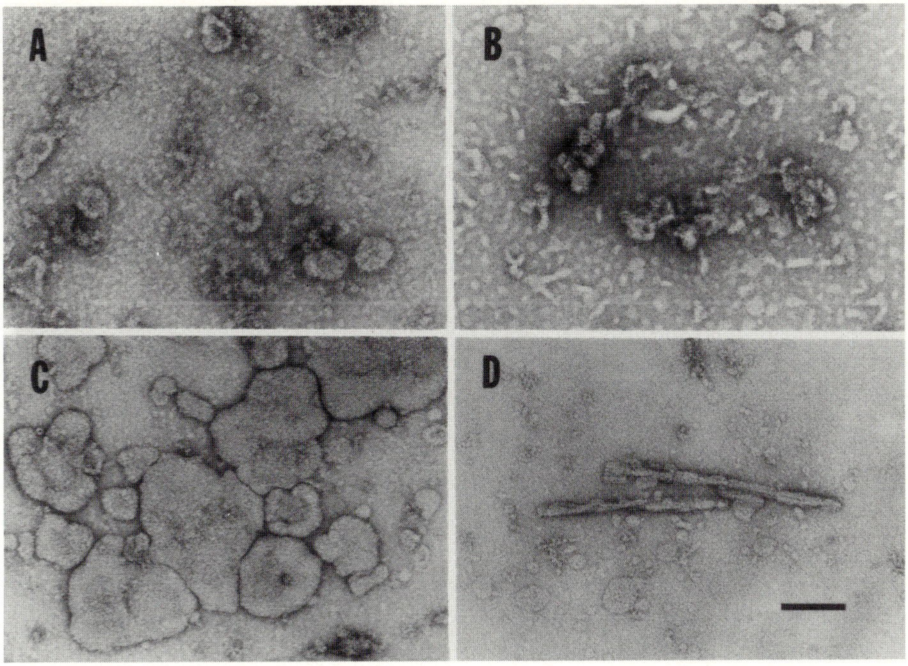

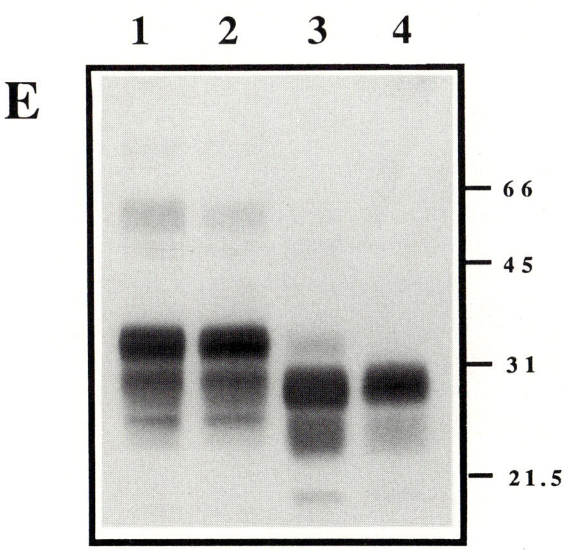

Fig. 2 A–E. Formation of prion rods by detergent extraction and limited proteolysis of membrane-associated PrP^{Sc}. **A** Ultrastructure of total membrane fractions: see also lane 1. **B** Membrane fractions extracted with 2% Sarkosyl for 30 min at 37°C: see also lane 2. **C** Membrane fractions digested with 100 μg/ml proteinase K for 30 min at 37°C: see also lane 3. **D** Rods formed from the membranes extracted with Sarkosyl and digested with proteinase K as described in B and C: see also lane 4. *Bar*, 100 nm. **E** Monoclonal antibody 13A5 was used to detect hamster brain prion proteins. Molecular weight markers are shown on the right

and the protein designated PrP^{Sc} but that lacked prion rods (GABIZON et al. 1988). To prevent proteolysis of PrP^{Sc} to PrP 27–30 during purification, we added a cocktail of proteolysis inhibitors to our homogenization buffers. The lack of prion rods in our purified fractions was puzzling, since we had reported previously that detergent extraction of microsomal membranes resulted in the formation of rods (MEYER et al. 1986). In retrospect, it appears that in addition to the detergent extraction of the membranes, endogenous proteases within the crude microsomal fractions hydrolyzed some PrP^{Sc} to PrP 27–30 (MCKINLEY et al. 1991).

Scrapie prion rods formed only when membrane-associated PrP^{Sc} was subjected to both detergent extraction and limited proteolysis (Table 1). Prior to either extraction or proteolysis, membrane fractions which were highly enriched for scrapie prion infectivity contained only amorphous vesicles (Fig. 2). Detergent extraction dispersed the membrane vesicles into small pleomorphic-shaped fragments with no apparent effect on the size of PrP^{Sc} (Fig. 2). Membrane vesicles hydrolyzed with proteinase K displayed no apparent ultrastructural changes, although PrP^{Sc} was converted to PrP 27–30. Simultaneous detergent extraction and limited proteolysis resulted in the conversion of all PrP^{Sc} to PrP 27–30 with the concomitant elimination of membrane vesicles and formation of prion rods (Fig. 2). There was no loss of scrapie titer following any of these treatments.

To establish the presence or absence of prion rods in hamster brain extracts, ultrastructural comparisons were made between microsomal membrane fractions with and without exogenously added rods obtained from purified prion

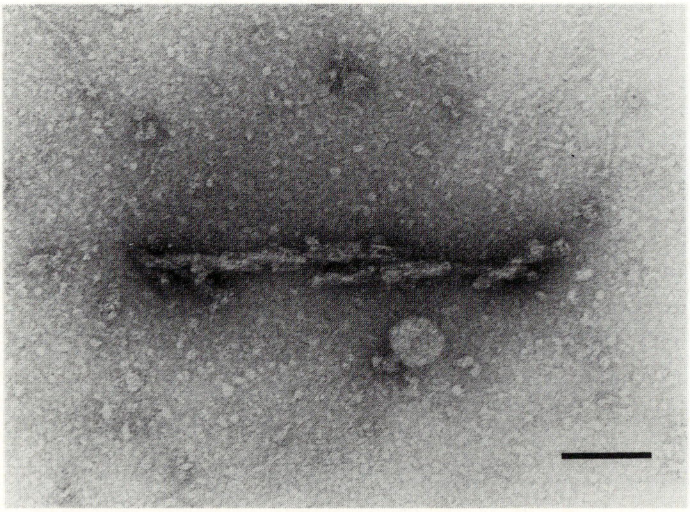

Fig. 3. Detection of prion rods exogenously added to brain membranes. Purified preparations of prion rods ($10^{6.6}$ ID_{50} units/ml) were added at a prion concentration equivalent to ten times less than that in the purified membrane fractions ($10^{7.6}$ ID_{50} units/ml). Rods are detected in the presence of the membranes. *Bar*, 100 nm

preparations. Electron microscopy revealed the presence of prion rods in those fractions to which exogenous rods had been added, demonstrating that the rods could be detected under these conditions (Fig. 3). The detection of exogenous prion rods from purified preparations ($10^{6.6}$ ID_{50} units/ml) at a titer ten times less than that measured for the membrane fraction ($10^{7.6}$ ID_{50} units/ml) suggests that it is unlikely that the membrane vesicles were masking an endogenous population of prion rods. Although prion rods were not observed in microsomal fractions, they were easily detected after detergent extraction and limited proteolysis (Fig. 3).

2.3 Variations in Prion Rod Morphology

The biochemical characteristics and ultrastructural morphology of a single scrapie prion or infectious particle are unknown (Table 1). Because the prion rods are associated with infectivity, we attempted to identify an ultrastructural conformation that is obligatory for infectivity. Our results are consistent with an earlier hypothesis suggesting that the rods represent an aggregated form of scrapie prions (PRUSINER et al. 1982, 1983).

Earlier studies suggested a rough correlation between the concentration of rods and prion titers (PRUSINER et al. 1982, 1983). However, accurate particle counts were not possible because the unit structure of the prion is unknown and conditions to disaggregate rod clusters into uniform particles were undefined. Because of these problems, the relationship of the rods to scrapie prion infectivity was determined indirectly.

Having established the association among prions, PrP 27–30 and rods, we sought to identify the smallest infectious form of the prion by disrupting prion rods using various chemical treatments. Our attempts to disrupt clusters of rods and generate single structures using nondenaturing detergents such as sodium dodecyl sarcosinate (Sarkosyl) and Triton X-100 were unsuccessful. High concentrations of denaturing detergents in the presence of heat destroyed the rods (MCKINLEY et al. 1986). No change in titer or rod structure occurred upon exposure to heat alone. Similarly, at detergent concentrations of 2% SDS and temperature of 65°C or less, no change in prion infectivity or rod morphology could be detected. Fractions exposed to 5% SDS at 37°C contained rods with smooth surfaces. Rods in samples heated to <100°C in 10% SDS were disrupted into filamentous structures 5 nm in diameter (Fig. 4). Exposure of purified preparations of scrapie prions to SDS at 100°C resulted in the complete disruption of rods and reduction of titer by approximately 10^2.

A marked change in pH had contrasting effects on rod morphology and prion titer (MCKINLEY et al. 1986). Reduction of sample pH to 3 by addition of 1 *N* HCl did not alter the ultrastructure of the rods. Earlier studies had demonstrated that prion infectivity was unchanged at pH 3.5 (PRUSINER et al. 1981). However, the ultrastructure of prion rods was markedly reduced at pH 10. While the general rod-like organization persisted, the structure of the individual rods was altered

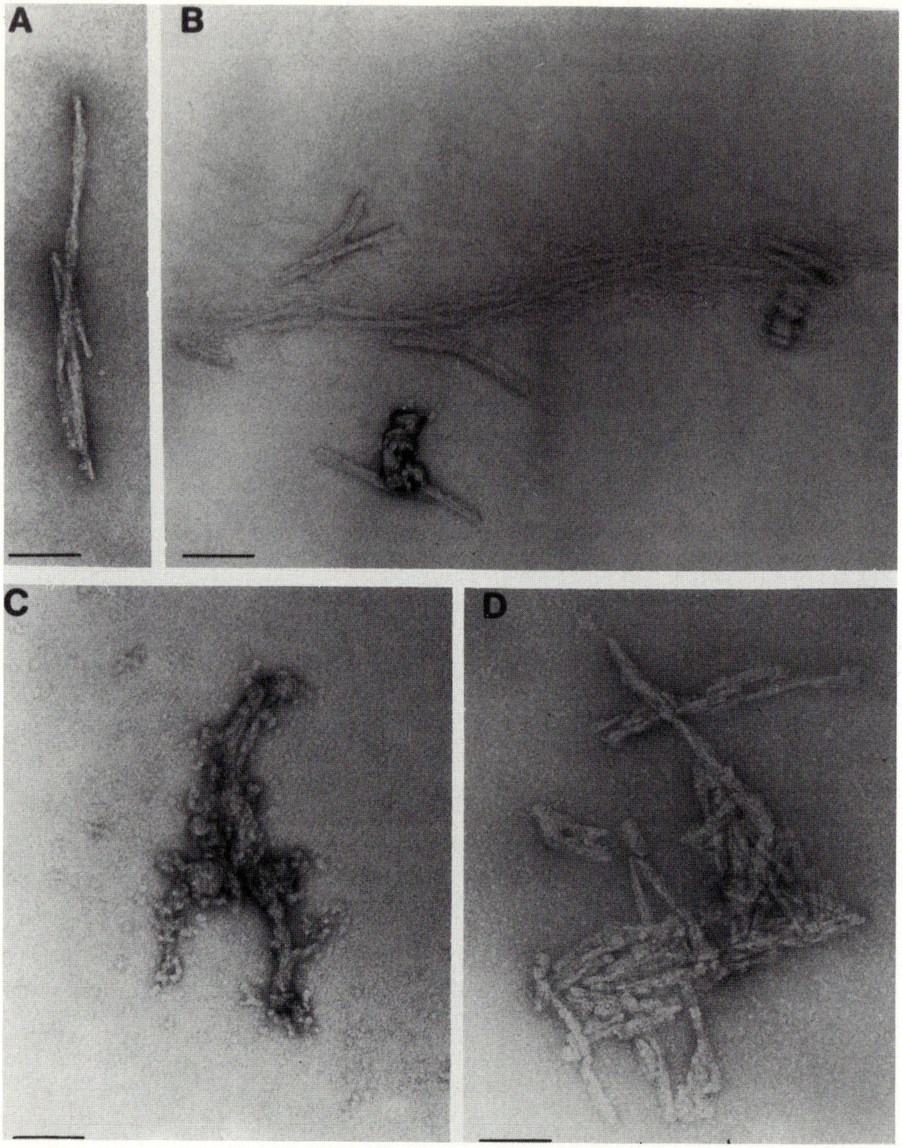

Fig. 4 A–D. Disruption of prion rods by chemical treatments. **A** Effect of 5% SDS at 37°C for 30 min on extensively purified preparations containing scrapie prion rods. **B** Effect of 10% SDS at 85°C for 30 min on extensively purified preparations of prion rods. **C** Alteration of prion rod structure by exposure to pH 10 for 30 min at 15°C. **D** Effect of proteinase K (100 μg/ml for 30 h at 37°C) on structure of prion rods. *Bars*, 100 nm. (Reproduced from McKINLEY et al. 1986)

into a series of globular units. The globular appearance became quite pronounced when the pH was raised to 11. Similar fractions had markedly reduced infectivity titer (PRUSINER et al. 1981).

The relationship between infectivity and the prion rods was also investigated by subjecting purified fractions to protease digestion. Aliquots were held at 37°C in the presence of 100 μg/ml of proteinase K for 0–30 h. Under these conditions, infectivity was decreased by a factor of $> 10^3$ after 30 h of digestion; however, no change in rod morphology or number could be discerned. As observed in earlier studies (McKINLEY et al. 1983), the change in concentration of PrP 27–30 paralleled that of prion titers during proteolytic digestion. Why the rods do not fall apart under these conditions is unclear. Presumably, either intermolecular forces favoring rod formation or other component(s) of the rods maintain rod structure even though the major protein component has been hydrolyzed to smaller polypeptides. Further studies are required to determine the mechanisms by which prion rods are formed and maintained.

Discovering conditions for solubilization of PrP^{Sc} with retention of associated scrapie prion infectivity greatly facilitated studies defining the molecular structure of the infectious particle. To solubilize PrP 27–30, a hydrophobic environment was needed to replace that provided by the rod structure. Phospholipids are the natural environment for a membrane protein (HUANG et al. 1980; BOGONEZ and KOSHLAND 1985) and were used as the needed "detergent" in order to solubilize PrP 27–30 (GABIZON et al. 1987). PrP 27–30 which had polymerized into rods was sedimented completely by low-speed centrifugation even in the presence of detergent. In contrast, most of the PrP 27–30 did not sediment in the presence of phosphatidylcholine (PC) even with ultracertrifugation (170 000 *g* for 30 min). The partitioning of PrP 27–30 between the pellet and the supernatant depends on several factors. Prolonged sonication during formation of detergent-lipid-protein complexes (DLPC) resulted in smaller-sized aggregates of lipids and more protein in the supernatant (Table 1). The ratio of lipid to protein needed to solubilize PrP 27–30 is an important factor. If the molar ratio is above 4000:1 (lipid molecules:protein molecules), most of the PrP 27–30 will be solubilized. Below that ratio, the solubilization will diminish. At a molar ratio of 4000:1 (lipid molecules:protein molecules), no rods were found in the 170 000 *g* pellet by electron microscopy. When we removed most of the detergent by dialysis, we obtained liposomes in which PrP 27–30 was probably in the lipid phase. Bioassays in hamsters demonstrate that scrapie prion infectivity is retained when rods composed of PrP 27–30 undergo dissociation and the protein is incorporated into phospholipid vesicles (GABIZON et al. 1987, 1988). We have frequently observed that prion titers rise at least ten-fold when PrP 27–30 is dispersed into liposomes. Presumably, this increase in prion titer reflects dissociation of prion rod-shaped aggregates.

2.4 Purified Prion Rods Are Amyloid

The ultrastructure of prion rods is indistinguishable from many purified amyloids (PRUSINER et al. 1983). Purified preparations of prions were stained with Congo red dye to investigate further the ultrastructural similarities between prion rods and purified amyloids. By bright-field microscopy, numerous amorphous structures measuring 1–20 μm were observed (PRUSINER et al. 1983). They exhibited a red color, indicating that they had bound the dye. Examination of these structures with crossed polarizers demonstrated a green color and birefringence. A variety of green birefringent structures was observed. It is generally accepted that naturally occurring mammalian protein polymers exhibiting both a rod-like or fibrillary appearance by electron microscopy and green birefringence after Congo red staining should be classified as amyloid (GLENNER 1980). Additionally, sections of scrapie-infected brain stained with Congo red dye have characteristic amyloid plaques which counterstain with antibodies to PrP 27–30 (BENDHEIM et al. 1984; DEARMOND et al. 1985; KITAMOTO et al. 1986; ROBERTS et al. 1988). The green birefringence following Congo red dye binding to amyloid proteins has been attributed to domains within the protein having a high degree of β-structure. The predicted sequence of PrP 27–30 exhibits a few regions which have potential for β-structure (BAZAN et al. 1987). Whether these domains within PrP 27–30 with potential for β-structure are sites for Congo red dye binding remains to be established.

3 Conclusions

Ultrastructural studies over the last decade have played an important role in elucidating the structure of the prion and defining the multiple molecular forms of this particle (Table 1). Fractions highly enriched for scrapie prion infectivity consistently contain PrP^{Sc} (GABIZON and PRUSINER 1990). If these fractions are derived using a protocol involving detergent extraction and limited proteolysis, then linear polymers containing PrP 27–30 are found which display the ultrastructural and tinctorial properties of amyloid. These properties of the prion rods distinguish them from SAF and led to the discovery that the amyloid plaques in the animal and human prion diseases contain linear arrays of prion proteins. Over the past 5 years, there has been a remarkable convergence of information about the role of the scrapie PrP isoform in prion diseases. The multiple molecular forms of the prion particle elucidated through ultrastructural studies offer an important line of evidence in support of the hypothesis that prions are fundamentally different from viruses. Recent results from many other approaches to the study of scrapie also argue that prions are not viruses; whether prions contain a component other than PrP^{Sc} remains to be determined.

Acknowledgments. The authors wish to thank L. Kenaga, M. Braunfeld and F. Rahbar for technical assistance.

References

Alper T, Haig DA, Clarke MC (1978) The scrapie agent: evidence against its dependence for replication on intrinsic nucleic acid. J Gen Virol 41: 503–516

Baringer JR, Prusiner SB (1978) Experimental scrapie in mice: ultrastructural observations. Ann Neurol 4: 205–211

Barry RA, McKinley MP, Bendheim PE, Lewis GK, DeArmond SJ, Prusiner SB (1985) Antibodies to the scrapie protein decorate prion rods. J Immunol 135: 603–613

Bazan JF, Fletterick RJ, McKinley MP, Prusiner SB (1987) Predicted secondary structure and membrane topology of the scrapie prion protein. Protein Eng 1: 125–135

Bendheim PE, Barry RA, DeArmond SJ, Stites DP, Prusiner SB (1984) Antibodies to a scrapie prion protein. Nature 310: 418–421

Bignami A, Parry HB (1971) Aggregations of 35-namometer particles associated with neuronal cytopathic changes in natural scrapie. Science 171: 389–399

Bignami A, Parry HB (1972) Electron microscopic studies of the brain of sheep with natural scrapie. I. The fine structure of neuronal vacuolation. Brain 95: 319–326

Bogonez E, Koshland DE (1985) Solubilization of a vectorial transmembrane receptor in functional form: aspartate receptor of chemotaxis. Proc Natl Acad Sci USA 82: 4891–4895

Bolton DC, McKinley MP, Prusiner SB (1982) Identification of a protein that purifies with the scrapie prion. Science 218: 1309–1311

Cho HJ (1976) Is the scrapie agent a virus? Nature 262: 411–412

Cho HJ, Grieg AS (1975) Isolation of 14-NM virus-like particles from mouse brain infected with scrapie agent. Nature 257: 685–686

Cho HJ, Grieg AS, Crop CR, Kimberlin RH, Chandler RL, Millson GC (1977) Virus-like particles from both control and scrapie-affected mouse brain. Nature 267: 459–460

David–Ferreira JF, David–Ferreira KL, Gibbs CJ, Jr Morris JA (1968) Scrapie in mice: ultrastructural observations in the cerebral cortex. Proc Soc Exp Biol Med 127: 313–320

DeArmond SJ, McKinley MP, Barry RA, Braunfeld MB, McColloch JR, Prusiner SB (1985) Identification of prion amyloid filaments in scrapie-infected brain. Cell 41: 221–235

Diener TO (1987) PrP and the nature of the scrapie agent. Cell 49: 719–721

Diener TO, McKinley MP, Prusiner SB (1982) Viroids and prions. Proc Natl Acad Sci USA 79: 5220–5224

Diringer H, Gelderblom H, Hilmert H, Ozel M, Edelbluth C, Kimberlin RH (1983a) Scrapie infectivity, fibrils and low molecular weight protein. Nature 306: 476–478

Diringer H, Hilmert H, Simon D, Werner E, Ehlers B (1983b) Towards purification of the scrapie agent. Eur J Biochem 134: 555–560

Field EJ, Narang HK (1972) An electron-microscopic study of scrapie in the rat: further observations on "inclusion bodies" and virus-like particles. J Neurol Sci 17: 347–364

Field EJ, Raine CS (1966) Observations on "dense-body" structure in nerve cells with special reference to scrapie. Res Vet Sci 7: 292–295

Field EJ, Raine CS, Joyce G (1967) Scrapie in the rat: an electron-microscope study. II. Glial inclusions. Acta Neuropathol (Berl) 9: 305–315

Gabizon R, McKinley MP Groth DF, Prusiner SB (1988) Immunoaffinity purification and neutralization of scrapie prion infectivity. Proc Natl Acad Sci USA 85: 6617–6621

Gabizon R, McKinley MP, Prusiner SB (1987) Purified prion proteins and scrapie infectivity copartition into liposomes. Proc Natl Acad Sci USA 84: 4017–4021

Gabizon R, Prusiner SB (1990) Prion liposomes. Biochem J 266: 1–14

Glenner GG (1980) Amyloid deposits and amyloidosis. N Engl J Med 302: 1283–1292

Hope J, Morton LJD, Farquhar CF, Multhaup G, Beyreuther K, Kimberlin RH (1986) The major polypeptide of scrapie-associated fibrils (SAF) has the same size, charge distribution and N-terminal protein sequence as predicted for the normal brain protein (PrP). EMBO J 5:2591–2597

Huang K-S, Bayley H, Khorana HG (1980) Delipidation of bacteriorhodopsin and reconstitution with exogenous phospholipid. Proc Natl Acad Sci USA 77: 323–327

Kitamoto T, Tateishi J, Tashima T, Takeshita I, Barry RA, DeArmond SJ, Prusiner SB (1986) Amyloid plaques in Creutzfeldt–Jakob disease stain with prion protein antibodies. Ann Neurol 20: 204–208

Lampert P, Hooks J, Gibbs CJ, Jr Gajdusek DC (1971) Altered plasma membranes in experimental scrapie. Acta Neuropathol (Berl) 19: 81–93

Malone TG, Marsh RF, Hanson RP, Semancik JS (1979) Hamster scrapie agent: properties, partial purification, and a DNA component. In: Prusiner SB, Hadlow WJ (eds) Slow transmissible diseases of the nervous system, vol 2. Academic, New York, pp 479–488

Manuelidis L, Sklaviadis T, Manuelidis EE (1987) Evidence suggesting that PrP is not the infectious agent in Creutzfeldt–Jakob disease. EMBO J 6: 341–347

McKinley MP, Bolton DC, Prusiner SB (1983) A protease-resistant protein is a structural component of the scrapie prion. Cell 35: 57–62

McKinley MP, Braunfeld MB, Bellinger CG, Prusiner SB (1986) Molecular characteristics of prion rods purified from scrapie-infected hamster brains. J Infect Dis 154: 110–120

McKinley MP, Meyer R, Kenaga L, Rahbar F, Serban A, Gabizon R, Prusiner SB (1991) Scrapie prion rod formation in vitro requires both detergent extraction and limited proteolysis. J Virol 65: 1340–1351

Merz PA, Somerville RA, Wisniewski HM, Iqbal K (1981) Abnormal fibrils from scrapie-infected brain. Acta Neuropathol (Berl) 54: 63–74

Merz PA, Somerville RA, Wisniewski HM, Manuelidis L, Manuelidis EE (1983a) Scrapie-associated fibrils in Creutzfeldt–Jakob disease. Nature 306: 474–476

Merz PA, Wisniewski HM, Somerville RA, Masters CL, Iqbal K (1983b) Ultrastructural morphology of amyloid fibrils from neuritic and amyloid plaques. Acta Neuropathol (Berl) 60: 113–124

Merz PA, Kascsak R, Rubenstein R, Carp RI, Wisniewski HM (1984a) Variations in SAF from different scrapie agents. In: Tateishi J (ed) Proceedings of workshop on slow transmissible diseases. Japanese Ministry of Health and Welfare, Tokyo, pp 137–145

Merz PA, Rohwer RG, Kascsak R, Wisniewski HM, Somerville RA, Gibbs CJ Jr, Gajdusek DC (1984b) Infection-specific particle from the unconventional slow virus diseases. Science 225: 437–440

Merz PA, Kascsak RJ, Rubenstein R, Carp RI, Wisniewski HM (1987) Antisera to scrapie-associated fibril protein and prion protein decorate scrapie-associated fibrils. J Virol 61: 42–49

Meyer RK, McKinley MP, Bowman KA, Braunfeld MB, Barry RA, Prusiner SB (1986) Separation and properties of cellular and scrapie prion proteins. Proc Natl Acad Sci USA 83: 2310–2314

Narang HK (1974a) Ruthenium red and lanthanum nitrate a possible tracer and negative stain for scrapie "particles?" Acta Neuropathol (Berl) 29: 37–43

Narang HK (1974b) An electron microscopic study of the scrapie mouse and rat: further observations on virus-like particles with ruthenium red and lanthanum nitrate as a possible trace and negative stain. Neurobiology 4: 349–363

Oesch B, Westaway D, Wälchli M, McKinley MP, Kent SBH, Aebersold R, Barry RA, Tempst P, Teplow DB, Hood LE, Prusiner SB, Weissmann C (1985) A cellular gene encodes scrapie PrP 27–30 protein. Cell 40: 735–746

Pattison IH, Smith K (1963) Histological observations on experimental scrapie in the mouse. Res Vet Sci 4: 269–275

Prusiner SB (1982) Novel proteinaceous infectious particles cause scrapie Science 216: 136–144

Prusiner SB, McKinley MP (eds) (1987) Prions–novel infectious pathogens causing scrapie and Creutzfeldt–Jakob disease. Academic, New York

Prusiner SB, Hadlow WJ, Garfin DE, Cochran SP, Baringer JR, Race RE, Eklund CM (1978) Partial purfication and evidence for multiple molecular forms of the scrapie agent. Biochemistry 17: 4993–4997

Prusiner SB, Garfin DE, Baringer JR, Cochran SP, Hadlow WJ, Race RE, Eklund CM (1979) On the partial purification and apparent hydrophobicity of the scrapie agent. In: Prusiner SB, Hadlow WJ (eds) Slow transmissible diseases of the nervous system, vol 2. Academic, New York, pp 425–464

Prusiner SB, Groth DF, McKinley MP, Cochran SP, Bowman KA, Kasper KC (1981) Thiocyanate and hydroxyl ions inactivate the scrapie agent. Proc Natl Acad Sci USA 78: 4606–4610

Prusiner SB, Bolton DC, Groth DF, Bowman KA, Cochran SP, McKinley MP (1982) Further purification and characterization of scrapie prions. Biochemistry 21: 6942–6950

Prusiner SB, McKinley MP, Bowman KA, Bolton DC, Bendheim PE, Groth DF, Glenner GG (1983) Scrapie prions aggregate to form amyloid-like birefringent rods. Cell 35: 349–358

Prusiner SB, Groth DF, Bolton DC, Kent SB, Hood LE (1984) Purification and structural studies of a major scrapie prion protein. Cell 38: 127–134

Raine CS, Field EJ (1967) Orientated tubules in axoplasm of cerebellar myelinated nerve fibers in the rat. Acta Neuropathol (Berl) 9: 298–304

Roberts GW, Lofthouse R, Allsop D, Landon M, Kidd M, Prusiner SB, Crow TJ (1988) CNS amyloid proteins in neurodegenerative diseases. Neurology 38: 1534–1540

Rohwer RG (1984) Scrapie infectious agent is virus-like in size and susceptibility to inactivation. Nature 308: 658–662

Siakotos AN, Raveed D, Longa G (1979) The discovery of a particle unique to brain and spleen subcellular fractions from scrapie-infected mice. J Gen Virol 43: 417–422

Sklaviadis T, Manuelidis L, Manuelidis EE (1986) Characterization of major peptides in Creutzfeldt–Jakob disease and scrapie. Proc Natl Acad Sci USA 83: 6146–6150

Somerville RA (1985) Ultrastructural links between scrapie and Alzheimer's disease. Lancet 1: 504–506

Somerville RA, Ritchie LA, Gibson PH (1989) Structural and biochemical evidence that scrapie-associated fibrils assemble in vivo. J Gen Virol 70: 25–35

In Vitro Expression and Biosynthesis of Prion Protein

B. CAUGHEY

1 Introduction

Prion protein (PrP) is an endogenous protein of central nervous system and a variety of other mammalian tissues (CHESEBRO et al. 1985; OESCH et al. 1985; HOPE et al. 1986; MEYER et al. 1986; RUBENSTEIN et al. 1986; CHO 1986; HOPE et al. 1988; CAUGHEY et al. 1988a). Although its function is not clear, it has been reported recently that PrP may participate in lymphocyte activation (CASHMAN et al. 1990). In addition, a partially homologous protein of chickens has been reported to be involved in the regulation of the expression of a neurotransmitter receptor at the neuromuscular junction (HARRIS et al. 1989). A common feature of scrapie, Creutzfeldt–Jakob disease, kuru, Gerstmann–Sträussler syndrome and related transmissible neurodegenerative diseases is the accumulation of a protease-resistant form of prion protein (PrP-res[1]) in tissues harboring substantial

Laboratory of Persistent Viral Diseases, National Institute of Allergy and Infectious Diseases, Rocky Mountain Laboratories, Hamilton, Mt 59840, USA

[1] Several names have been used for forms of PrP derived from scrapie-infected tissues [PrP 27–30 (BOLTON et al. 1984), PrP^{Sc} (MEYER et al. 1986), scrapie-associated fibril protein (DIRINGER et al. 1983), Gp34 (SKLAVIADIS et al. 1986), and Sp 33–37 (BOLTON et al. 1987)] and uninfected tissues [PrP^{C} (Meyer et al. 1986), and Cp 33–37 (BOLTON et al. 1987)]. Wishing to avoid terms which assume a relationship of the various forms of this protein to scrapie infectivity or specify a particular size for the protein, we have opted for the operational terms PrP-res and PrP-sen which label two forms of the protein as they are distinguished experimentally by proteinase K digestion and can encompass the multiple molecular masses of the two forms

Current Topics in Microbiology and Immunology, Vol. 172

amounts of the transmissible agent (BOLTON et al. 1982; DIRINGER et al. 1983; McKINLEY et al. 1983; BROWN et al. 1986). Unlike the normal, protease-sensitive PrP (PrP-sen[1]),PrP-res has the tendency to aggregate into amyloid-like fibrils (PRUSINER et al. 1983; DIRINGER et al. 1983; DEARMOND et al. 1985; CHO 1986; WILEY et al. 1987; MERZ et al. 1987) and plaques (BENDHEIM et al. 1984; DEARMOND et al. 1985; ROBERTS et al. 1986; WILEY et al. 1987) and has been identified as a major component of brain fractions enriched for the scrapie agent (BOLTON et al. 1982; DIRINGER et al. 1983). As such, PrP-res, or an aggregate thereof, has been proposed to be the transmissible agent of the scrapie-like neurodegenerative diseases (BOLTON et al. 1982; DIRINGER et al. 1983; McKINLEY et al. 1983). The importance of PrP in scrapie and related diseases has been substantiated by molecular genetic data indicating that variations in the endogenous PrP gene, which encodes both PrP-sen and PrP-res, appear to influence scrapie incubation time (CARLSON et al. 1986, 1988; WESTAWAY et al. 1987; RACE et al. 1990) and susceptibility (HSIAO et al. 1989; SCOTT et al. 1989; DOH-URA et al. 1989; GOLDGABER et al. 1989). However, it is not yet clear whether PrP-res is the transmissible agent itself, a component of the agent, or a by-product of either the infectious process or the disease. Nor are the chemical and biosynthetic bases for the differences between PrP-sen and PrP-res understood. As an approach to defining the structure and biosynthesis of PrP, the nature of its scrapie-associated modification and its role in scrapie, a number of studies of the in vitro expression and biosynthesis of PrP have been undertaken.

2 In Vitro Expression of PrP

PrP genes have been cloned and sequenced from hamster (OESCH et al. 1985; ROBAKIS et al. 1986; BASLER et al. 1986), mouse (CHESEBRO et al. 1985; LOCHT et al. 1986), human (KRETZSCHMAR et al. 1986; LIAO et al. 1986; HSIAO et al. 1989), rat (LIAO et al. 1987) and sheep (GOLDMANN et al. 1990). Studies of the expression of cloned PrP genes in cell-free transcription-translation systems and in tissue culture cells have provided information about the control of PrP gene expression, the biosynthesis of PrP and the issue of whether a scrapie-specific modification of PrP mRNA initiates the synthesis of PrP-res and the scrapie agent.

2.1 Cell-Free Systems

HAY and coworkers have shown that cell-free transcription and translation of a hamster PrP cDNA generates a PrP polypeptide of M_r 28 000 (HAY et al. 1987b) which is close to the predicted molecular mass for the full-length PrP amino acid sequence (ROBAKIS et al. 1986; BASLER et al. 1986). Using a hamster PrP cDNA from a different source (ROBAKIS et al. 1986), we have confirmed that a PrP polypeptide of approximately 28 kDa is generated in a wheat cell-free transcription and translation system (B. Caughey, unpublished observations). A PrP polypeptide of the same size is also translated from native mRNA isolated from both normal and scrapie-infected hamster brains (HAY et al. 1987b). When translated in the presence of dog pancreatic microsomes, glycosylated (M_r 33 000) and unglycosylated (M_r 25 000) PrP forms are generated which apparently lack the signal peptide.

An unusual aspect of PrP biogenesis in cell-free systems supplemented with microsomes is that from a single pool of nascent PrP peptides, two different topological forms of PrP are produced. In a rabbit reticulocyte system the predominant product is completely translocated into the microsomal lumen (HAY et al. 1987a), while in a wheat germ system the predominant PrP species is integral to the microsomal membrane (HAY et al. 1987b). These data suggest that there may be both integral membrane and secreted forms of PrP. However, this interpretation is complicated by the fact that, in brain and neuronally derived cells, PrP contains a phosphatidylinositol (PI) moiety which anchors most of the normal PrP to the cell surface (STAHL et al. 1987; CAUGHEY et al. 1989, 1990; BORCHELT et al. 1990).

A novel nonhydrophobic, luminally disposed stop transfer sequence in the hamster PrP polypeptide is critical for the formation of the integral transmembrane topology in the wheat germ system (YOST et al. 1990). When engineered into heterologous proteins, this sequence can confer transmembrane topology to hydrophobic domains which are normally translocated into the microsomal lumen (YOST et al. 1990; LOPEZ et al. 1990). Since this stop transfer sequence is much more active in wheat germ extracts than reticulocyte lysates, Lopez and coworkers suggest that its activity depends on interaction with one or more cytosolic factors.

2.2 Cell Culture Systems

Since PrP is normally subjected to posttranslational modifications that are not performed in cell-free translation systems, it was important to extend the PrP gene expression studies to eukaryotic cells which might be capable of such modifications. Furthermore, because both PrP-sen and PrP-res appear to be encoded by the same endogenous PrP gene (BASLER et al. 1986), the question arose as to whether a scapie-induced modification in the PrP mRNA could account for the synthesis of PrP-res and perhaps the scrapie agent. To test for this possibility, it was necessary to develop systems for expressing recombinant PrP genes in tissue culture cells.

2.2.1 Expression of Scrapie Mouse Brain PrP cDNA in C127 Cells

Using a fragment of a PrP cDNA cloned from scrapie-infected mouse brain (CHESEBRO et al. 1985; LOCHT et al. 1986), we have constructed a plasmid expression vector which allows the PrP gene to be expressed in mouse C127 epithelial/fibroblast cells (CAUGHEY et al. 1988b). In the plasmid, the PrP gene fragment is located between sequences containing the mouse metallothionine promoter and an SV40 polyadenylation site. The plasmid also contains a portion of the bovine papilloma virus genome which enables it to transform the host cell and replicate stably as an episome. Northern blot analyses of C127 clones containing the plasmid vector have indicated that PrP mRNAs encoded by both the plasmid and the endogenous cellular PrP gene are expressed, with the former being much less abundant.

Despite the relatively low levels of plasmid-encoded PrP mRNA, metabolic labelling and PrP immuoprecipitation experiments have shown that it is much more efficiently translated than the endogenous PrP mRNA. This suggests that, if the scrapie brain and C127 PrP mRNAs are identical initially, removal or replacement of portions of the original noncoding regions with the metallothionine promoter at the 5' end or with SV40 sequences at the 3' end might greatly enhance their translational activity.

Short-term [^{35}S]methionine labeling experiments have identified multiple biosynthetic PrP precursors which appear to be glycosylated to form a major 35- to 41-kDa band within 3 h (Fig. 1). Although the 35- to 41-kDa PrP band is similar in SDS-PAGE mobility to the major product of the endogenous PrP gene in scrapie-competent mouse neuroblastoma cells (Fig. 2; RACE et al. 1988; CAUGHEY et al. 1989), the precursors appear to differ from those encoded by endogenous PrP genes. This casts doubt on whether the recombinant PrP is processed in an authentic manner in the C127 cells. However, like the normal PrP in various brain-derived cells (STAHL et al. 1987; CAUGHEY et al. 1989, 1990; SCOTT et al.

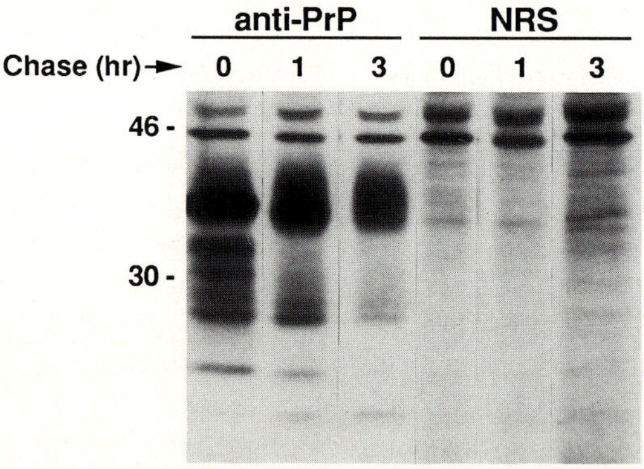

Fig. 1. Expression of scrapie mouse brain PrP cDNA in C127 cells. C127 cells transformed with the PrP-BPV vector were metabolically labeled with [^{35}S]methionine for 1 h, chased with unlabeled methionine medium for the designated period, and analyzed for labeled PrP by immunoprecipitation, SDS-PAGE and fluorography as described (CAUGHEY et al. 1988b). The primary antibodies used in the immunoprecipitation were an *anti-PrP* peptide 1b serum (# 783; WILEY et al. 1987) or normal rabbit serum (NRS). The positions of molecular mass markers are designated in kDa. (Adapted from CAUGHEY et al. 1988b)

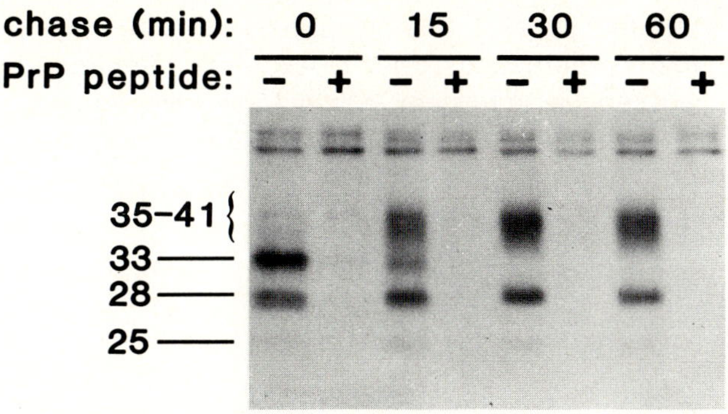

Fig. 2. Pulse-chase metabolic labeling of PrP-sen in mouse neuroblastoma cells. The cells were pulsed for 15 min with [^{35}S]methionine, chased with unlabeled methionine for the designated times, and analyzed for labeled PrP by immunoprecipitation, SDS-PAGE and fluorography as described (CAUGHEY et al. 1989). The primary antibody was the same anti-PrP peptide serum as used in Fig. 1. The PrP bands are those removed from the immunoprecipitate by preabsorption of the antiserum with the synthetic PrP peptide 1b antigen. Before chasing, PrP bands were seen at 25, 28, and 33 kDa. After chases of 15–60 min, additional PrP bands were seen at 30 and 35–41 kDa The 30-kDa PrP band is not labeled in the figure margin but is the band that shifts upward slightly from the 28 kDa band with increasing chase period. (Adapted from CAUGHEY et al. 1989)

1988; BORCHELT et al. 1990), PrP is expressed on the cell surface in the C127 cells, as demonstrated by membrane immunofluorescence experiments (CAUGHEY et al. 1988b). A portion of the 35- to 41-kDa PrP is also released into the medium, indicating that the recombinant PrP is targeted as a secretory protein or a cell surface protein that is released over the course of time. Both cell-associated and secreted forms recombinant PrP have also been observed in *Xenopus* oocytes (HAY et al. 1987a).

Since it has been proposed that PrP is the transmissible agent of scrapie, we have checked the possibility that the infectious scrapie agent is generated by in vitro expression of the PrP gene from scrapie-infected brain (CAUGHEY et al. 1988b). When appropriate cell extracts are inoculated into susceptible mice, none of the mice exhibit symptoms of scrapie. Furthermore, none of the PrP detected in these cells has the proteinase K resistance characteristic of the scrapie-associated form of PrP. Thus, we have concluded that expression in C127 cells of the PrP cDNA cloned from scrapie-infected tissue is not sufficient to generate proteinase K-resistant PrP or the transmissible agent of scrapie.

2.2.2 Expression of Recombinant PrP Gene in Insect and Monkey Cells

Recombinant hamster PrP expression has been accomplished in insect (*Spodoptera*) cells using a baculovirus based vector and in monkey COS-7 cells using an SV40-based vector (SCOTT et al. 1988). As is the case in the mouse cell/bovine papilloma virus vector system (CAUGHEY et al. 1988b), there is SDS-PAGE evidence that the recombinant PrP expressed in both the *Spodoptera* and COS-7 cells is not processed like the endogenous PrP of brain. Nonetheless, the recombinant PrP in the *Spodoptera* cells is full length, and that in COS-7 cells is found on the cell surface. When preparations of the PrP-expressing *Spodoptera* cells are inoculated into susceptible hamsters, nor scrapie symptoms are observed, again indicating that the expression of cloned PrP cDNAs does not generate the scrapie agent.

3 PrP Biosynthesis

The utility of the presently available recombinant PrP expression systems in C127, *Spodoptera* and COS-7 cells is compromised by the facts that these cells have not been found to be capable of relicating the scrapie agent and that the PrP expressed in these cells appears to differ from the endogenously encoded PrP of brain and tissue culture cells. Thus an important step in understanding PrP biosynthesis and scrapie agent replication has been the development of mouse neuroblastoma cells capable of scrapie agent replication (RACE et al. 1987, 1988; BUTLER et al. 1988). Studies of the synthesis of PrP encoded by the endogenous PrP gene of these and other cells have yielded the following information about normal PrP biosynthesis and how it is altered by scrapie agent replication.

3.1 PrP Precursors

Since PrP is a glycoprotein (MANUELIDIS et al. 1985; BOLTON et al. 1985) and has a hydrophobic amino-terminal signal sequence (LOCHT et al. 1986), its biosynthetic processing presumably beings within the endoplasmic reticulum. When both normal (sc^-) and scrapie-infected (sc^+) mouse neuroblastoma cells are [^{35}S] methionine labeled for short periods (2–10 min), PrP precursors of 25, 28, and 33 kDa are detected (Fig. 2; CAUGHEY et al. 1989). Similar precursors have been observed in rat PC12 pheochromocytoma cells (B. Caughey and R. Rubenstein, unpublished data). Treatment of the neuroblastoma cell precursors with endoglycosidase H converts the larger two precursors to species that comigrate with the 25-kDa precursor, indicating that the precursors differ in their content of N-linked high mannose glycan (CAUGHEY et al. 1989). This is confirmed by the fact that the drug tunicamycin, which prevents N-linked glycosylation, allows the synthesis of only the 25-kDa PrP. Since there are two potential N-linked glycosylation sites (Asn-X-Thr) on the mouse PrP polypeptide (LOCHT et al. 1986), it is likely that the 28- and 33-kDa precursors represent the addition of high mannose glycan chains to one and both the glycosylation sites, respectively, of the 25-kDa unglycosylated precursor (Fig. 3).

All of these PrP precursors contain a PI moiety, as indicated by reductions in their SDS-PAGE mobilities upon treatment with PI-specific phospholipase C

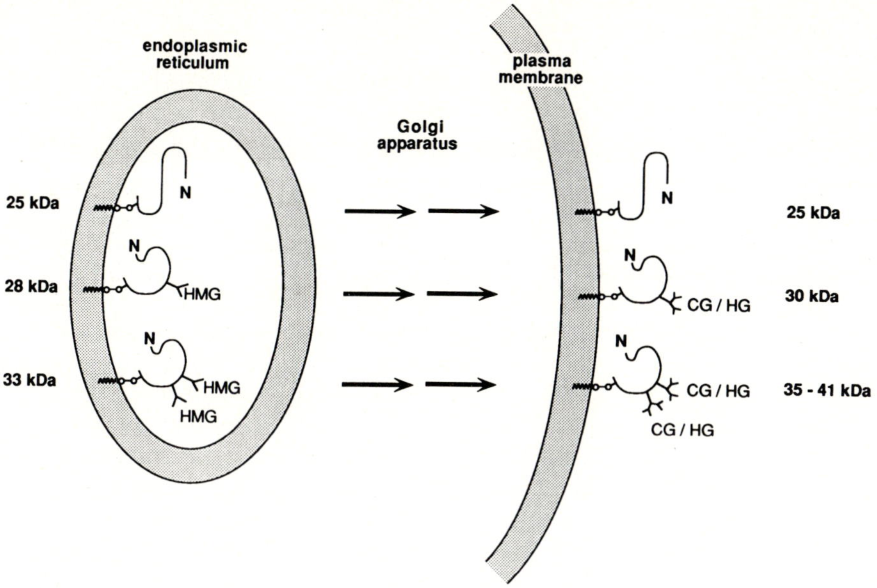

Fig. 3. Model for PrP biosynthesis. PrP precursors in the endoplasmic reticulum and the final PrP products appearing in the plasma membrane after oligosaccharide processing in the Golgi apparatus. The amino termini of the polypeptide chains are designated by *N*. The polypeptides are anchored to the membranes by glycosyl phosphatidylinositol moieties. The symbols for oligosaccharide groups are designated as follows: *HMG*, high mannose glycan; *CG*, complex glycan; *HG*, hybrid glycan. The number of glycan chains shown on the glycosylated PrP forms represents a prediction based on the number of potential N-linked glycosylation sites in the polypeptide

(PIPLC) (CAUGHEY et al. 1989). The simultaneous labeling of the three PrP precursors is consistent with studies of other proteins showing that the addition of glycosyl-PI (LOW and SALTIEL 1988) and high mannose glycans (KORNFELD and KORNFELD 1985) can occur concurrent with or soon after translation and translocation of polypeptides into the lumen of the endoplasmic reticulum.

The fact that the unglycosylated, PIPLC-treated precursor appears smaller in molecular size than the theoretical value of 27.8 kDa calculated for the full length PrP polypeptide suggests that the nascent PrP is rapidly truncated in the endoplasmic reticulum (CAUGHEY et al. 1989). This is likely due to the removal of the N-terminal signal sequence since PrP has a predicted signal peptidase cleavage site (BAZAN et al. 1987) at which PrP is cleaved in brain tissue (HOPE et al. 1986, 1988; BOLTON et al. 1987; TURK et al. 1988).

3.2 Maturation of PrP Precursors

With extended chase periods, the labeled 28- and 33-kDa PrP precursors are posttranslationally processed to 30- and 35- to 41-kDa species within 10–30 min (Fig. 2; RACE et al. 1988; CAUGHEY et al. 1989). These processed PrP species are no longer susceptible to endoglycosidase H, indicating that the high mannose glycans present on the precursors are converted to hybrid or complex glycans, presumably within the Golgi apparatus (Fig. 3). This is consistent with studies indicating that PrP of hamster brain, at least, contains complex carbohydrate moieties (BOLTON et al. 1985; MANUELIDIS et al. 1985; HARAGUCHI et al. 1989; ENDO et al. 1989).

3.3 Subcellular Localization of PrP

Studies with a variety of tissue culture cells have indicated that PrP-sen can be labeled on the cell surface by membrane inmmunofluorescence (STAHL et al. 1987; CAUGHEY et al. 1988b, 1990; BORCHELT et al. 1990), biotinylation (BORCHELT et al. 1990) and radioiodination (B. Caughey, unpublished observations). Cell surface PrP-sen is also susceptible to treatments of intact cells by proteases (Fig. 4; STAHL et al. 1987; CAUGHEY et al. 1989, 1990). Kinetic metabolic labeling experiments have shown that, in mouse neuroblastoma cells, PrP-sen starts to become exposed to extracellular proteases within an hour of the initiation of biosynthesis (CAUGHEY et al. 1989). Within 3 h, approximately 90% of [^{35}S]methionine-labeled 30- and 35- to 41-kDa PrP species can be digested by extracellular trypsin or proteinase K, demonstrating that they have reached the cell surface. Once at the cell surface, the labeled PrP-sen has a half-life of 3–6 h,

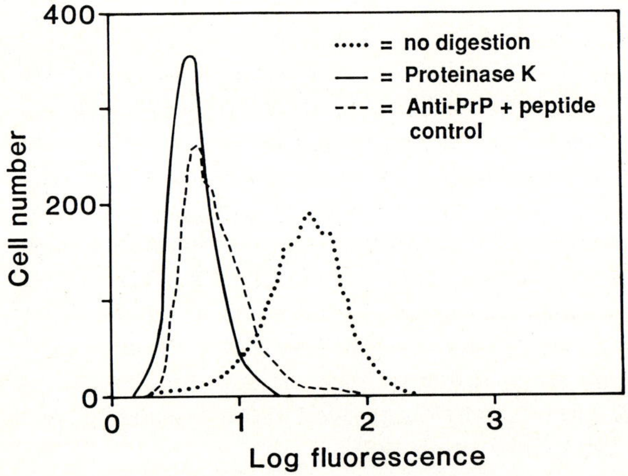

Fig. 4. Flow cytometric analysis of cell surface PrP in scrapie-infected neuroblastoma cells. The cells were treated with medium alone or medium containing 50 μg ml proteinase K, then incubated with affinity purified anti-PrP peptide 1b and fluorescein isothiocyanate-conjugated goat anti-rabbit immunoglobulin and analyzed by flow cytometry as described previously (CAUGHEY et al. 1990). To demonstrate the PrP specificity of the fluorescence of the control cells, the primary antibody was preabsorbed with PrP peptide 1b. Data from several experiments indicated that the specific mean fluorescence intensities of the stained neuroblastoma cell populations were reduced by at least 95% with proteinase K and trypsin, and by 75% with PIPLC. (Adapted from CAUGHEY et al. 1990)

in the absence of exogenous proteases (CAUGHEY et al. 1989; BORCHELT et al. 1990). After chases of 16 h, less than 10% of the labeled PrP is recovered from the medium, suggesting that catabolism, rather than secretion or release from the cell surface, is the major fate of PrP in these cells.

Much of the PrP-sen on the cell surface is anchored by PI (Fig. 3), as shown by treatments of intact cells with PIPLC (STAHL et al. 1987; CAUGHEY et al. 1989, 1990; BORCHELT et al. 1990). Such PIPLC release experiments have indicated that all three mature PrP species (25, 30, and 35–41 kDa) of mouse neuroblastoma cells can be released from the cell surface (CAUGHEY et al. 1989). However, as much as 25% of the cell surface PrP-sen is resistant to PIPLC release (CAUGHEY et al. 1990). This could be due to the presence of a transmembrane topology for PrP-sen, as initially suggested by the cell-free synthesis studies of HAY and colleagues (HAY et al. 1987a) or to variant glycosyl-PI anchors as has been described for another PI-linked protein (ROBERTS et al. 1987).

Although there is much evidence to indicate that PrP-sen is found primarily on the cell surface, recent immunofluorescence experiments performed on fixed and permeabilized cells have shown that PrP can also be stained in discrete perinuclear structures with the morphological appearance of the Golgi body (CAUGHEY et al. 1990). Since PrP undergoes carbohydrate processing in the Golgi body during its biosynthesis, the presence of staining in the Golgi body would not be surprising.

3.4 Effect of Scrapie Infection

An important question has been whether the PrP-res accumulation in scrapie-infected tissues results from in vivo tissue pathology without being directly related to scrapie agent replication. Recent studies of mouse neuroblastoma cells have made it clear that PrP-res can be generated in scrapie-infected cells in vitro in the absence of any signs of cytopathology (BUTLER et al. 1988; CAUGHEY et al. 1990; BORCHELT et al. 1990). This finding narrows the correspondence between PrP-res formation and scrapie infection from the tissue level to the cellular level. However, it does not preclude the possibility that PrP-res formation is secondary to scrapie agent replication.

Comparisons of normal and scrapie-infected mouse neuroblastoma clones have demonstrated that scrapie infection does not affect any of the parameters of PrP-sen biosynthesis that have been described to date (RACE et al. 1988; CAUGHEY et al. 1989, 1990). However, proteinase K-resistant forms of PrP have been identified in scrapie-infected but not uninfected clones by western immunoblot (Fig. 5; BUTLER et al. 1988; CAUGHEY et al. 1990) and by metabolic labeling and immunoprecipitation (BORCHELT et al. 1990; B. Caughey, unpublished data). The PrP-res species observed are similar in apparent molecular mass (19, 23, and ~28 kDa) to the proteinase K-treated, N-terminally truncated PrP-res species from mouse brain, however, they differ in the relative

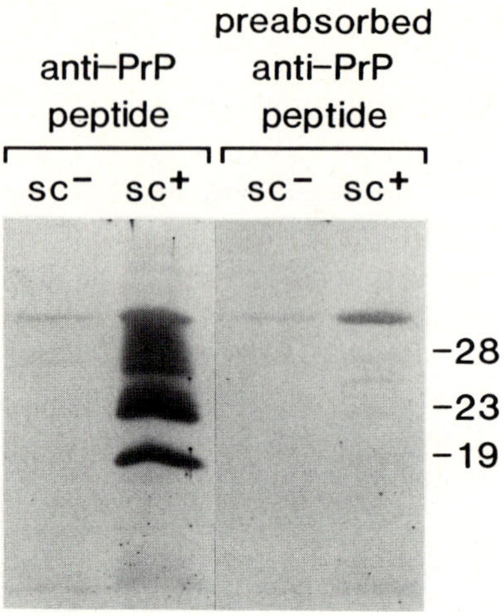

Fig. 5. Detection of PrP-res in scrapie-infected neuroblastoma cells. Normal (sc^-) and scrapie-infected (sc^+) neuroblastoma clones were lysed with detergent, treated with proteinase K, and immunoblotted using anti-PrP peptide 1b, or anti-PrP peptide 1b preabsorbed with synthetic peptide 1b as described (CAUGHEY et al. 1990). (Adapted from CAUGHEY et al. 1990)

intensities of the bands. The relative intensities of the 19, 23, and ~28 kDa neuroblastoma cell PrP-res bands also differ greatly from those of the corresponding full length PrP-sen bands of the same cells (25, 30, and 35–41 kDa) in that the forms with little (23 kDa) or no (19 kDa) asparagine-linked glycans predominate (compare Figs 2 and 5). This indicates that asparagine-linked glycosylation is not essential for PrP-res formation and may make the formation of PrP-res less likely. Analyses using monospecific antipeptide antisera to various parts of the PrP polypeptide have shown that the neuroblastoma cell PrP-res molecules are truncated at the N-terminus within the cells without the addition of exogenous proteases (B. Caughey, unpublished data). Thus the partial proteolysis of PrP-res, which has been observed in PrP-res derived from scrapie-infected mouse brain (HOPE et al. 1988), can be also be observed in PrP-res-producing cells in the absence of scrapie-associated tissue pathology. Velocity gradient centrifugation experiments have indicated that, like the PrP-res of scrapie brain tissue, the neuroblastoma cell PrP-res forms large aggregates in detergent lysates (B Caughey, unpublished data). Although the amount of PrP-res varies between scrapie-infected neuroblastoma clones and passages of individual clones (R. Race, B. Caughey, D.Ernst and B. Chesebro, unpublished observations), it has been estimated that in one clone, at least, PrP-res accumulated to a level that was two to threefold higher than that of PrP-sen (CAUGHEY et al. 1990).

Having identified PrP-res in scrapie-infected clones, the task has become one of determining how its biosynthesis differs from that of the normal PrP-sen. Recent studies have indicated that unlike PrP-sen (Fig. 4), PrP-res is not removed by treatments of intact neuroblastoma cells with PIPLC or proteases (CAUGHEY et al. 1990; BORCHELT et al. 1990). This suggests that PrP-res is primarily intracellular, inherently resistant to these enzymes, or both. Studies with permeabilizied scrapie-infected hamster brain cell cultures have provided

Table 1. Comparison of PrP-sen and PrP-res in mouse neuroblastoma cells[a]

	PrP-sen	PrP-res
Normal cells	+	−
Scrapie-infected cells	+	+
Protease sensitivity (lysates)	+	−
Protease sensitivity (intact cells)	+	−
PIPLC sensitivity (intact cells)	+	−
Apparent molecular masses[b]	25, 30, 35–41 kDa	19, 23, ~28 kDa
Biosynthesis ($t_{1/2}$)	<1 h	15 h
Degradation ($t_{1/2}$)	3–6 h	>24 h
Aggregation	−	+
Cell surface biotinylation	+	−
Cytoplasmic immunofluorescent labeling	+	+

[a] References for the information in this table are presented in the text
[b] Apparent molecular of PrP species isolated without proteinase K treatment

evidence that PIPLC and protease resistance are inherent properties of PrP-res (STAHL et al. 1990). PrP-res also appears to be primarily intracellular because it is not labeled in intact neuroblastoma cells by a membrane impermeant biotinylation reagent (BORCHELT et al. 1990) and can be stained in the cytoplasm of fixed and permeabilized cells (TARABOULOS et al. 1990). The PrP-res staining partially colocalizes with the staining of Golgi markers, suggesting that at least some PrP-res accumulates in the Golgi apparatus. Long-term metabolic labeling studies have shown that PrP-res is synthesized much more slowly ($t_{1/2}$ ~ 15 h) than PrP-sen and thus may be formed by a posttranslational modification of PrP-sen (BORCHELT et al. 1990; B. Caughey and G. Raymond, unpublished data). The characteristics of PrP-sen and PrP-res in scrapie-infected neuroblastoma cells are summarized in Table 1.

4 Summary and Conclusions

In addition to whatever function PrP may have normally, its involvement in scrapie-like neurodegenerative diseases has become clearer in recent years. In vitro studies have made important contributions to the understanding of normal PrP biosynthesis and turnover and how they can be influenced by scrapie infection. Cell-free transcription and translation experiments have indicated that PrP gene translation products are capable of assuming two different topologies, one spanning microsomal membranes and the other completely translocated into the microsomal lumen (HAY et al. 1987a, b). A novel stop transfer signal in the polypeptide is critical to the formation of the transmembrane topology (YOST et al. 1990). Expression of recombinant PrP genes has been accomplished in mouse (CAUGHEY et al. 1988b), monkey (SCOTT et al. 1988), frog (HAY et al. 1987a), and insect (SCOTT et al.1988) tissue culture cells. PrP products encoded by PrP cDNAs cloned from scrapie-infected brain tissues are not infectious and do not have the protease-resistance characteristic of the scrapie-associated form of PrP isolated from diseased tissue (CAUGHEY et al. 1988b; SCOTT et al. 1988). Studies of PrP encoded by the endogenous gene of mouse neuroblastoma cells have identified the precursors (CAUGHEY et al. 1989) and products (RACE et al. 1988; CAUGHEY et al. 1989) of normal PrP biosynthesis and shown that most of the PrP of normal cells is linked to the cell surface by phosphatidylinositol (Fig 3; STAHL et al. 1987; CAUGHEY et al. 1989, 1990; BORCHELT et al. 1990). In scrapie-infected clones, and additional pool of PrP is present which, unlike the normal PrP, aggregates (B. Caughey, unpublished observations) and is partially protease resistant (BUTLER et al. 1988; CAUGHEY et al. 1990; BORCHELT et al. 1990; STAHL et al. 1990). This scrapie-associated pool of PrP differs from the normal PrP in that it is primarily intracellular (CAUGHEY et al. 1990; BORCHELT et al. 1990; TARABOULOS et al. 1990) and resistant to removal from cells by phospholipase or protease (CAUGHEY et al. 1990; BORCHELT et al. 1990; STAHL et al. 1990) treatments. Kinetic studies have shown that while PrP-sen is synthesized and degraded relatively rapidly (CAUGHEY et al. BORCHELT et al. 1990), PrP-res is synthesized slowly and has a very long half-life (BORCHELT et al. 1990).

Further studies with the scrapie-infected mouse neuroblatoma cells should lead toward the elucidation of the molecular details of the scrapie-associated modification of PrP and whether the modification is directly related to scrapie agent replication. In addition, it is hoped that the recently developed technology for recombinant PrP gene expression can be adapted to cells capable of replicating the scrapie agent so that potential effects of PrP gene modification on PrP-sen, PrP-res, and scrapie agent biosynthesis can be investigated.

References

Basler K, Oesch B, Scott M, Westaway D, Walchli M, Groth DF, McKinley MP, Prusiner SB, Weissman C (1986) Scrapie and cellular PrP isoforms are encoded by the same chromosomal gene. Cell 46: 417–428

Bazan JF, Fletterick RJ, McKinley MP, Prusiner SB (1987) Predicted secondary structure and membrane topology of the scrapie prion protein. Protein Eng 1: 125–135

Bendheim PE, Barry RA, DeArmond SJ, Stites DP, Prusiner SB (1984) Antibodies to a scrapie prion protein. Nature 310: 418–421

Bolton DC, McKinley MP, Prusiner SB (1982) Identification of a protein that purifies with the scrapie prion. Science 218: 1309–1311

Bolton DC, McKinley MP, Prusiner SB (1984) Molecular characteristics of the major scrapie prion protein. Biochemistry 23: 5898–5906

Bolton DC, Meyer RK, Prusiner SB (1985) Scrapie PrP 27–30 is a sialoglycoprotein. J Virol 53: 596–606

Bolton DC, Bendheim PE, Marmostein AD, Potempska A (1987) Isolation and structural studies of the intact scrapie agent protein. Arch Biochem Biophys 258: 579–590

Borchelt DR, Scott M, Taraboulos A, Stahl N, Prusiner SB (1990) Scrapie and cellular prion proteins differ in the kinetics of synthesis and topology in cultured cells. J Cell Biol 110: 743–752

Brown P, Coker–Vann M, Pomeroy K, Franko M, Asher DM, Gibbs CJ Jr, Gajdusek DC (1986) Diagnosis of Creutzfeldt–Jakob disease by Western blot identification of marker protein in human brain tissue. N Engl J Med 314: 547–551

Butler DA, Scott MRD, Bockman JM, Borchelt DR, Taraboulos A, Hsiao KK, Kingsbury DT, Prusiner SB (1988) Scrapie-infected murine neuroblastoma cells produce protease-resistant prion proteins. J Virol 62: 1558–1564

Carlson GA, Kingsbury DT, Goodman PA, Coleman S, Marshall ST, DeArmond S, Westaway D, Prusiner SB (1986) Linkage of prion protein and scrapie incubation time genes. Cell 46: 503–511

Carlson GA, Goodman PA, Lovett M, Taylor BA, Marshall ST, Peterson–Torchia M, Westaway D, Prusiner SB (1988) Genetics and polymorphism of the mouse prion gene complex: control of scrapie incubation time. Mol Cell Biol 8: 5528–5540

Cashman NR, Loertscher R, Nalbantoglu J, Shaw I, Kascsak RJ, Bolton DC, Bendheim PE (1990) Cellular isoform of the scrapie agent protein participates in lymphocyte activation. Cell 61: 185–192

Caughey B, Race RE, Chesebro B (1988a) Detection of prion protein mRNA in normal and scrapie-infected tissues and cell lines. J Gen Virol 69: 711–716

Caughey B, Race RE, Vogel M, Buchmeier MJ, Chesebro B (1988b) In vitro expression in eukaryotic cells of the prion protein gene cloned from scrapie-infected mouse brain. Proc Natl Acad Sci USA 85: 4657–4661

Caughey B, Race RE, Ernst D, Buchmeier MJ, Chesebro B (1989) Prion protein (PrP) biosynthesis in scrapie-infected and uninfected neuroblastoma cells. J Virol 63: 175–181

Caughey B, Neary K, Buller R, Ernst D, Perry L, Chesebro B, Race R (1990) Normal and scrapie-associated forms of prion protein differ in their sensitivities to phospholipase and proteases in intact neuroblastoma cells. J. Virol 64: 1093–1101

Chesebro B, Race R, Wehrly K, Nishio J, Bloom M, Lechner D, Bergstrom S. Robbins K, Mayer L, Keith JM, Garon C, Haase A (1985) Identification of scrapie prion protein-specific mRNA in scrapie-infected and unifected brain. Nature 315: 331–333

Cho HJ (1986) Antibody to scrapie-associated fibril protein identifies a cellular antigen. J Gen Virol 67: 243–253

DeArmond SJ, McKinley MP, Barry RA, Braunfeld MB, McColloch JR, Prusiner SB (1985) Identification of prion amyloid filaments in scrapie-infected brain. Cell 41: 221–235

Diringer H, Gelderblom H, Hilmert H, Ozel M, Edelbluth C, Kimberlin RH (1983) Scrapie infectivity, fibrils and low molecular weight protein. Nature 306: 476–478

Doh-ura K, Tateishi J, Sasaki H, Kitamoto T, Sakaki Y (1989) Pro---leu change at position 102 of prion protein is the most common but not the sole mutation related to Gerstmann–Straussler syndrome. Biochem Biophys Res Commun 163: 974–979

Endo T, Groth D, Prusiner SB, Kobata A (1989) Diversity of oligosaccharide structures linked to asparagines of the scrapie prion protein. Biochemistry 28: 8380–8388

Goldgaber D, Goldfarb LG, Brown P, Asher DM, Brown WT, Lin S, Teener JW, Feinstone SM, Rubsenstein R, Kascsak RJ et al. (1989) Mutations in familial Creutzfeldt–Jakob disease and Gerstmann–Straussler–Scheinker's syndrome. Exp Neurol 106: 204–206

Goldmann W, Hunter N, Foster JD, Salbaum JM, Beyreuther K, Hope J (1990) Two alleles of a neural protein gene linked to scrapie in sheep. Proc Natl Acad Sci UA 87: 2476–2480

Haraguchi T, Fisher S, Olofsson S, Endo T, Groth D, Tarentino A, Borchelt DR, Teplow D, Hood L, Burlingame A, Lycke E, Kobata A, Prusiner SB (1989) Asparagine-linked glycosylation of the scrapie and cellular prion proteins. Arch Biochem Biophys 274: 1–13

Harris DA, Falls DL, Walsh W, Fischbach GD (1989) Molecular cloning of an acetylcholine receptor-inducing protein. Soc Neurosci Abstr 15:164

Hay B, Prusiner SB, Lingappa VR (1987a) Evidence of a secretory form of the cellular prion protein. Biochemistry 26: 8110–8115

Hay B, Barry RA, Lieberburg I, Prusiner SB, Lingappa VR (1987b) Biogenesis and transmembrane orientation of the cellular isoform of the scrapie prion protein. Mol Cell Biol 7: 914–920

Hope J, Morton LJD, Farquhar CF, Multhaup G, Beyreuther K, Kimberlin RH (1986) The major polypeptide of scrapie-associated fibrils (SAF) has the same size, charge distribution and N-terminal protein sequence as predicted for the normal brain protein (PrP). EMBO J 5: 2591–2597

Hope J, Multhaup G, Reekie LJD, Kimberlin RH, Beyreuther K (1988) Molecular pathology of scrapie-associated fibril protein (PrP) in mouse brain affected by the ME 7 strain of scrapie. Eur J Biochem 172: 271–277

Hsiao K, Baker HF, Crow TJ, Poulter M, Owen F, Terwilliger JD, Westaway D, Ott J, Prusiner SB (1989) Linkage of a prion protein missense variant to Gerstmann–Straussler syndrome. Nature 338: 342–345

Kornfeld R, Kornfeld S (1985) Assembly of asparagine-linked oligosaccharides. Annu Rev Biochem 54: 631–664

Kretzschmar HA, Stowring LE, Westaway D, Stubblebine WH, Prusiner SB, DeArmond SJ (1986) Molecular cloning of a human prion protein cDNA. DNA 5: 315–324

Liao Y-C, Tokes Z, Lim E, Lackey A, Woo CH, Button JD, Clawson GA (1987) Cloning of rat "prion-related protein" cDNA. Lab Invest 57: 370–374

Liao Y-CJ, Lebo RV, Clawson GA, Smuckler EA (1986) Human prion protein cDNA: molecular cloning, chomosomal mapping, and biological implications. Science 233: 364–367

Locht C, Chesebro B, Race R, Keith JM (1986) Molecular cloning and complete sequence of prion protein cDNA from mouse brain infected with the scrapie agent. Proc Natl Acad Sci USA 83: 6372–6376

Lopez CD, Yost CS, Prusiner SB, Myers RM, Lingappa VR (1990) Unusual topogenic sequence directs prion protein biogensis. Science 248: 226–229

Low MG, Saltiel AR (1988) Structural and functional roles of glycosyl-phosphatidylinositol in membranes. Science 239: 268–275

Manuelidis L, Valley S, Manuelidis EE (1985) Specific proteins associated with Creutzfeldt–Jakob disease and scrapie share antigenic and carbohydrate determinants. Proc Natl Acad Sci USA 82: 4263–4267

McKinley MP, Bolton DC, Prusiner SB (1983) A protease-resistant protein is a structural component of the scrapie prion. Cell 35: 57–62

Merz PA, Kascsak RJ, Rubenstein R, Carp RI, Wisniewski HM (1987) Antisera to scrapie-associated fibril protein and prion protein decorate scrapie-associated fibrils. J Virol 61: 42–49

Meyer RK, Mckinley MP, Bowman KA, Braunfeld MB, Barry RA, Prusiner SB (1986) Separation and properties of cellular and scrapie prion protein. Proc Natl Acad Sci USA 83: 2310–2314

Oesch B, Westaway D, Walchli M, McKinley MP, Kent SBH, Aebersold R, Barry RA, Tempst P, Teplow DB, Hood LE, Prusiner SB, Weissmann C (1985) A cellular gene encodes scrapie PrP 27–30 protein. Cell 40: 735–746

Prusiner SB, McKinley MP, Bowman KA, Bendheim PE, Bolton DC, Groth DF, Glenner GG (1983) Scrapie prions aggregate to form amyloid-like birefringent rods. Cell 35: 349–358

Race RE, Fadness LH, Chesebro B (1987) Characterization of scrapie infection in mouse neuroblastoma cells. J Gen Virol 68: 1391–1399

Race RE, Caughey B, Graham K, Ernst D, Chesebro B (1988) Analyses of frequency of infection, specific infectivity, and prion protein biosynthesis in scrapie-infected neuroblastoma cell clones. J Virol 62: 2845–2849

Race RE, Graham K, Ernst D, Caughey B, Chesebro B (1990) Analysis of linkage between scrapie incubation period and the prion protein gene in mice. J Gen Virol 71: 493–497

Robakis NK, Sawh PR, Wolfe GC, Rubenstein R, Carp RI, Innis MA (1986) Isolation of a cDNA clone encoding the leader peptide of prion protein and expression of the homologous gene in various tissues. Proc Natl Acad Sci USA 83: 6377–6381

Roberts GW, Lofthouse R, Brown R, Crow TJ, Barry RA, Prusiner SB (1986) Prion-protein immunoreactivity in human transmissible dementias. N Engl J Med 315: 1231–1232

Roberts WL, Kim BH, Rosenberry TL (1987) Differences in the glycolipid membrane anchors of bovine and human erythrocyte acetylcholinesterases. Proc Natl Acad Sci USA 84: 7817–7821

Rubenstein R, Kascsak RJ, Merz PA, Papini MC, Carp RI, Robakis NK, Wisniewski HM (1986) Detection of scrapie-associated fibril (SAF) proteins using anti-SAF antibody in non-purified tissue preparations. J Gen Virol 67: 671–681

Scott M, Foster D, Mirenda C, Serban D, Coufal F, Walchli M, Torchia M, Groth D, Carlson G, DeArmond SJ, Westaway D, Prusiner SB (1989) Transgenic mice expressing hamster prion protein produce species-specific scrapie infectivity and amyloid plaques. Cell 59: 847–857

Scott MRD, Butler DA, Bredesen DE, Walchli M, Hsiao KK, Prusiner SB (1988) Prion protein gene expression in cultured cells. Protein Eng 2: 69–76

Sklaviadis T, Manuelidis L, Manuelidis EE (1986) Characterization of major peptides in Creutzfeldt–Jakob disease and scrapie. Proc Natl Acad Sci USA 83: 6146–6150

Stahl N, Borchelt DR, Hsiao K, Prusiner SB (1987) Scrapie prion protein contains a phosphatidylinositol glycolipid. Cell 51: 229–240

Stahl N, Borchelt DR, Prusiner SB (1990) Differential release of cellular and scrapie prion proteins from cellular membranes by phosphatidylinositol-specific phospholipase C. Biochemistry 29: 5405–5412

Taraboulos A, Serban D, Prusiner SB (1990) Scrapie prion proteins accumulate in the cytoplasm of persistently infected cultured cells. J Cell Biol 110: 2117–2132

Turk E, Teplow DB, Hood LE, Prusiner SB (1988) Purification and properties of the cellular and scrapie hamster prion proteins. Eur J Biochem 176: 21–30

Westaway D, Goodman PA, Mirenda CA, McKinley MP, Carlson GA, Prusiner SB (1987) Distinct prion proteins in short and long scrapie incubation period mice. Cell 51: 651–662
Wiley CA, Burrola PG, Buchmeier MJ, Wooddell MK, Barry RA, Prusiner SB, Lampert PW (1987) Immuno-gold localization of prion filaments in scrapie-infected hamster brains. Lab Invest 57: 646–655
Yost CS, Lopez CD, Prusiner SB, Myers RM, Lingappa VR (1990) Non-hydrophobic extracytoplasmic determinant of stop transfer in the prion protein. Nature 343: 669–672

Prion Protein Genes: Evolutionary and Functional Aspects*

B. OESCH[1], D. WESTAWAY[1], and S. B. PRUSINER[1,2]

1 Introduction

Over the past decade, a wealth of new information about the biological and physical properties as well as genetics of the particle causing scrapie has accumulated. A host-encoded protein denominated prion protein (PrP) is transformed upon infection into a scrapie-specific isoform (PrP^{Sc}). Considerable evidence argues that PrP^{Sc} is a major and necessary component of the infectious scrapie particle termed prion (PRUSINER 1982; OESCH et al. 1985; GABIZON et al. 1988; GABIZON and PRUSINER 1990). Genetically, the PrP gene has been linked to a gene controlling the incubation time and species specificity (CARLSON et al. 1986, 1988; HUNTER et al. 1987; MOHRI and TATEISHI 1989; SCOTT et al. 1989; RACE et al. 1990). Unfortunately, we still lack molecular probes directed towards the scrapie-specific portion of prions. The only convenient markers known to date are the protease resistance and detergent insolubility of PrP^{Sc}. The identification of different scrapie "strains" or isolates which appear to breed true in the same inbred host suggests that inheritable information independent of the host must be encoded within the infectious particle (BRUCE and DICKINSON 1987; KIMBERLIN et al. 1987). This hypothetical informational molecule has been equated with a nucleic acid genome which, however, is still elusive (ALPER et al. 1967; ROHWER

* This work was supported by research grants from the National Institutes of Health (AG02132 and NS14069) and by a Senator Jacob Javits Center of Excellence in Neuroscience award (NS22786) as well as by a gift from the Sherman Fairchild Foundation. B. Oesch was supported by a postdoctoral fellowship from the European Molecular Biology Organization (EMBO)
Departments of [1]Neurology and [2]Biochemistry and Biophysics, University of California, San Francisco, CA 94143, USA

1984; DUGUID et al. 1988; OESCH et al. 1988; AIKEN et al. 1989). As an unusual possibility, inheritable information may be encoded in molecules other than nucleic acid. It has been proposed that posttranslational modification/ translational mistakes of the proteinaceous component of the prion or other molecules associated with the infectious particle, like lipids or carbohydrates, might specify different properties of prions which may then be interpreted as genetic information (GIBBONS and HUNTER 1967; WILLS 1989). A candidate is the entity which causes the altered physical properties of PrP^{Sc} as compared to the cellular PrP isoform (PrP^{C}). Experimental studies have suggested that PrP^{Sc} is generated by a posttranslational process possibly involving chemical modifications or changes in conformation (BASLER et al. 1986; BORCHELT et al. 1990). The most prominent alterations are the protease resistance of PrP^{Sc} and its propensity to aggregate in detergents (PRUSINER et al. 1983; OESCH et al. 1985; MEYER et al. 1986). The identification of PrP^{Sc} and the finding of its association with the prion represent an important step which has allowed the analysis of its role in prion diseases and may ultimately lead us to an understanding of the pathological changes occurring upon infection.

2 Cloning of the Hamster Prion Protein Gene

Purification of scrapie prion infectivity from Syrian hamster brain led to the identification of PrP 27-30, the protease-resistant core of PrP^{Sc} (BOLTON et al. 1982; PRUSINER et al. 1982; MCKINLEY et al. 1983). Its N-terminus was sequenced, which allowed the design of the corresponding oligonucleotides to be used as probes in the search for the gene encoding PrP 27–30 (PRUSINER et al. 1984). In a cDNA library from scrapie-infected hamster brain a clone encoding the N-terminus of PrP 27–30 was identified (OESCH et al. 1985). The cDNA was also found to encode internal tryptic peptides. Using the same approach a Swiss mouse PrP cDNA was cloned (CHESEBRO et al. 1985). Subsequently, full-length cDNAs and genomic fragments of PrP genes from a number of species were cloned: Armenian and Chinese hamster (AHaPrP, CHaPrP; LOWENSTEIN et al. 1990), Syrian hamster (SHaPrP; BASLER et al. 1986; ROBAKIS et al. 1986), Swiss, NZW and I/LnJ mice (MoPrP; LOCHT et al. 1986; WESTAWAY et al. 1987), rats (RaPrP; LIAO et al. 1987), human (HuPrP; KRETZSCHMAR et al. 1986b; LIAO et al. 1986) and sheep (ShePrP; GOLDMAN et al. 1990). The alignment of the nucleotide sequence of the open reading frames (ORFs) is shown in Fig. 1a. The deduced amino acid sequences are shown in Fig. 1b. Comparison of the PrP sequences to those of other known genes did not reveal any homologies. A candidate cDNA encoding an acetylcholine receptor inducing activity (ARIA) from chicken brain showed high homology in the hydrophobic domain and the amphipathic helix (HARRIS et al. 1989; Fig. 3) but was less conserved otherwise. It remains to be shown whether mammalian PrP has ARIA activity. The ORF predicted from the human PrP cDNA described by LIAO et al. (1986) differs from the amino terminal consensus sequence of 12 published cDNA and genomic clones indicating that this particular clone may be due to a polymorphism or technical difficulties in cloning. It is noteworthy that features of the PrP mRNA may interfere with the fidelity of cDNA synthesis (BASLER et al. 1986) resulting in aberrant 5′ sequences.

The expression of the Syrian hamster PrP gene has been analyzed by Northern and dot blots as well as in situ hybridization. A 2.1-kb PrP mRNA was found to be present in normal and scrapie-infected hamster brain at the same level, indicating that the expression of the PrP gene is not changed upon infection (OESCH et al. 1985). PrP mRNA with the same electrophoretic mobility was also detected in lung, heart and spleen, though at a 10- to 50-fold lower level than in brain (OESCH et al. 1985; ROBAKIS et al. 1986; SHINAGAWA et al. 1986). PrP-related sequences were also detected in pancreas, testes and kidney by dot blot analysis; no reproducible hybridization signal was detected in liver RNA. In situ hybridization of brain sections revealed that PrP is expressed predominantly in neurons at about 10–50 mRNA copies per cell; glial cells contain less than 3 copies per cell (KRETZSCHMAR et al. 1986a). By immunocytochemistry, PrP^C was expressed by most neurons in the hippocampus, the septal nuclei and the thalamus, while fewer neurons were positive in the caudate nucleus or the neocortex (DEARMOND et al. 1987). In the cerebellum, most of the Purkinje cells stained while neurons of the granular cell layer were mostly negative. In peripheral tissues PrP^C has not been localized by immunocytochemistry, except for peripheral blood cells, where PrP^C was found on the surface of lymphocytes but not granulocytes (CASHMAN et al. 1990); in situ hybridization of spleen tissue did not reveal any signal above background (KRETZSCHMAR et al. 1986a).

The genomic organization of the Syrian hamster PrP gene has been analyzed in detail (BASLER et al. 1986). A small non-coding 5′ exon is separated from the 3′ exon by a 10-kb intron. The 3′ exon contains the ORF encoding PrP. The promoter of the hamster PrP gene lacks a TATA box, which usually is important to specify a single transcription initiation site. It was found that PrP mRNA from hamster brain starts at multiple sites spanning about 20 nucleotides. This was confirmed in an in vitro transcription system, indicating that it represents an intrinsic feature of the promoter. The region approximately −50 to +20 around the first transcription initiation site contains three repeats of the sequence GCCCCGCCC, which resembles the site utilized by the transcription factor Sp1. The organization of the PrP promoter is reminiscent of a "housekeeping" gene, but in light of the predominantly neuronal expression of the PrP gene there may be other unrecognized transcriptional control elements in the PrP promoter or in the intron that directs cell-type-specific expression. Expression may also be regulated by differential stability of the mRNA in different tissues. It is interesting to note that most established cell lines express PrP independent of their origin (D. Borchelt, unpublished observation). Tissue- or cell-specific expression of the PrP gene may therefore be controlled by factors which are not active in established cell lines.

```
          1       10        20        30        40        50        60
          .        .         .         .         .         .         .
 AHaPrP : ATGGCGAA......CCTTAGCTACTGGCTGCTGGCACTCTTTGTGGCTACTTGGACTGAT
 CHaPrP : -----A--......------------------------------------G---------
 SHaPrP : --------......-----------------------------------TG---------
  HuPrP : --------......----G---G----A------TT-----------C--A----G---C
MoPrP a : --------......----G----------------C---------A---TG---------
MoPrP b : --------......----G----------------C---------A---TG---------
  RaPrP : ............................................................
 ShePrP : ----T---AAGCCA-A-AG--AGT---A-C----TT-----------C-TG----G---C

                  70        80        90       100       110       120
                   .         .         .         .         .         .
 AHaPrP : GTTGGCCTCTGCAAGAAGCGGCCAAAGCCT...GGAGGGTGGAACACTGGTGGAAGCCGA
 CHaPrP : ------------------------------...---------------------------
 SHaPrP : ------------------------------...-----------------C---------
  HuPrP : C-G-----------------C--G------...-----A-----------G--C------
MoPrP a : --C-----------A---------------...--------------C-----------G
MoPrP b : --C-----------A---------------...--------------C-----------G
  RaPrP : .................................--------------------------G
 ShePrP : --G-----------------A-----A---GGC-----A-----------G--G------

                 130       140       150       160       170       180
                   .         .         .         .         .         .
 AHaPrP : TACCCTGGGCAGGGCAGCCCTGGAGGCAACCGTTACCCACCTCAGGGTGGTGGCACCTGG
 CHaPrP : ------------------------------------------------------------
 SHaPrP : --------------------------------------------------C-----A---
  HuPrP : -----G--------------------------C--------------C-----TGG----
MoPrP a : --T--C--------A------------------------------...------------
MoPrP b : --T--C--------A------------------------------...------------
  RaPrP : --------------A------------------------------A-------T------
 ShePrP : -----G--A--------T--------------C--T-----------A--G--TGG----

                 190       200       210       220       230       240
                   .         .         .         .         .         .
 AHaPrP : GGGCAACCCCATGGTGGTGGCTGGGGACAGCCCCATGGTGGTGGCTGGGGACAACCTCAT
 CHaPrP : --------------------A---------------------------------------
 SHaPrP : -----------------------------------------------------G--C---
  HuPrP : -----G--T-----------------G-----T-----------------G--G--C---
MoPrP a : -----G-----C-----------------A--------G--CA-----------------
MoPrP b : -----G-----C-----------------A--------G--CA-----------------
  RaPrP : -----G-----------------------A--T---------------------------
 ShePrP : --T--G--------A-----------C--A--T-----A-----------T--G--C---

                 250       260       270       280       290       300
                   .         .         .         .         .         .
 AHaPrP : GGTGGTGGTTGGGGTCAGCCCCATGGTGGTGGCTGGGGTCAAGGAGGTGGCACCCACAAT
 CHaPrP : ------------------------------------------------------------
 SHaPrP : --------C---------------------------------------------------
  HuPrP : --------C-----A-----T-------------------------------------G-
MoPrP a : ------A-------------------C-----A-----C--------G--T-----T---
MoPrP b : ------A-------------------C-----A-----C--------G--T-----T---
  RaPrP : --------C-----------------C--G------A----------G--T-----T---
 ShePrP : --------C-----A-----A-----------AG-CT-GGGTCA------T-G-----G-
```

Fig. 1 (Continued)

```
                 310       320       330       340       350       360
                   .         .         .         .         .         .
 AHaPrP : CAGTGGAACAAGCCCAATAAGCCAAAAACCAGCATGAAGCATATGGCAGGTGCTGCTGCG
 CHaPrP : ----------------G--------------A----------G----------G------
 SHaPrP : ----------------G--------------A---------C-----C--C---------
  HuPrP : --------------G-G--------------A---------C-----T--------A--A
MoPrP a : ----------------GC--A----------A-C-C------G-------G-----G--A
MoPrP b : ----------------GC--A----------A-T-C------G-------G-----G--A
  RaPrP : ----------------GC-------------A-C-C------G-------G-----C--A
 ShePrP : ----------------G--------------A----------G-------A--------A

                 370       380       390       400       410       420
                   .         .         .         .         .         .
 AHaPrP : GCTGGGGCAGTGGTGGGGGGCCTTGGGGGCTACATGCTGGGGAGTGCCATGAGCAGACCC
 CHaPrP : --------G-----------------------------------C-----------G---
 SHaPrP : --A-----C-----------------T-----------------------------G---
  HuPrP : --------------------------C--------------A--------------G---
MoPrP a : -----------A--------------T-----------------C-----------G---
MoPrP b : -----------A--------------T-----------------C-----------G---
  RaPrP : -----------A--------------T---------T-------------------G---
 ShePrP : -----A--------A-----------T--------------A--------------G--T

                 430       440       450       460       470       480
                   .         .         .         .         .         .
 AHaPrP : ATGCTCCATTTTGGCAATGACTGGGAGGACCGCTACTACCGTGAAAACATGAACCGCTAC
 CHaPrP : ------------------------------------------------------------
 SHaPrP : ---A-G------------------------------------------------------
  HuPrP : --CA-A-----C----G-----AT--------T-----T------------C----T---
MoPrP a : ---A-------------C---------------------------------T--------
MoPrP b : ---A-------------C---------------------------------T--------
  RaPrP : -----------------C-----------------------A---------T----T---
 ShePrP : C-TA-A----------------AT--------T-----T------------T----T---

                 490       500       510       520       530       540
                   .         .         .         .         .         .
 AHaPrP : CCTAACCAAGTGTACTACCGGCCAGTGGACCAGTACAACAACCAGAACAACTTCGTGCAC
 CHaPrP : -----------------------------------------------------T------
 SHaPrP : --------------T--------------------------------------T------
  HuPrP : --C---------------A----CA----TG------G---------------T------
MoPrP a : ------------------A----------T-------G----------------------
MoPrP b : ------------------A----------T-------G----------------------
  RaPrP : ------------------A----G-----T-------G----------------------
 ShePrP : --C---------------A-A--------T-G---T-GT--------------T-----T

                 550       560       570       580       590       600
                   .         .         .         .         .         .
 AHaPrP : GACTGTGTCAATATCACGATCAAACAGCATACAGTCACCACCACCACCAAGGGGGAGAAC
 CHaPrP : -----------------------G------------------------------------
 SHaPrP : --T--------C-----C-----G-----C------------------------------
  HuPrP : -----C-----------A-----G-----C--G--------A------------------
MoPrP a : -----C-----------C-----G-----C--G---------------------------
MoPrP b : -----C-----------C-----G-----C--G---GT----------------------
  RaPrP : -----------------C-----G------------------------------------
 ShePrP : -----------C-----AG----G--A--C------------------------------
```

Fig. 1 (Continued)

```
                    610       620       630       640       650       660
                     .         .         .         .         .         .
   AHaPrP : TTCACGGAGACCGACGTCAAGATGATGGAGCGCGTGGTGGAGCAGATGTGTGTCACCCAG
   CHaPrP : ------------------------------------------------------------
   SHaPrP : ---------------A-------A---------------------------AC-------
    HuPrP : -----C-----------T--------------------T------------A--------
  MoPrP a : -----C--------T--G--------------------------------C---------
  MoPrP b : -----C--------T--G--------------------------------C---------
    RaPrP : -----------------G--------------T-----------------C---------
   ShePrP : -----C--A--T---A-------A--------A-----------A-----CA--------

                    670       680       690       700       710       720
                     .         .         .         .         .         .
   AHaPrP : TATCAGAAGGAGTCCCAGGCCTACTATGACGGAAGAAGATCCAGC...GCGGTGCTCTTC
   CHaPrP : --------------------------C------------------...------------
   SHaPrP : --------------------------C--T--------G------...--------G---
    HuPrP : --CG---G---A--T--------T--CC-...G---G----G---...AT---C------
  MoPrP a : --C--------------------T--C-----G------------AGCA-C-T---T---
  MoPrP b : --C--------------------T--C-----G------------AGCA-C-T---T---
    RaPrP : -----------------------T--C-----G--------T---...--C-----T---
   ShePrP : --C----GA--A--------T--T--CC-...---GG-GG-A--T...-T-A-C-----T

                    730       740       750       760       770
                     .         .         .         .         .
   AHaPrP : TCCTCCCCTCCTGTGATCCTCCTCATTTCCTTCCTCATCTTCCTGATAGTGGGA
   CHaPrP : -----T------------------------------------------------
   SHaPrP : --------------------------------T--------------G------
    HuPrP : -----T--A--------------G--C--T------------------------
  MoPrP a : --------------C-----------C--------------------C------
  MoPrP b : --------------C-----------C--------------------C------
    RaPrP : --------------------------C--------------------C------
   ShePrP : --T-----------------------C--T--------T--T--C-----A---
```

Fig. 1 (Continued)

The genomic organization of the PrP gene(s) in species other than hamster is not well studied. The mouse gene for the *PrP*^{b} allele (from I/LnJ mice) has been cloned in cosmids; however, the promoter has not yet been identified. As in the hamster, the mouse PrP ORF is contained within a single exon (Westaway et al. 1987, 1990). The same is true for the PrP genes of man as well as Armenian and Chinese hamsters (Hsiao et al. 1989; Lowenstein et al. 1990). Hybridization of genomic DNA on Southern blots has also indicated that PrP-related sequences are present in goats and rabbits (Westaway and Prusiner 1986; Robakis et al. 1986). Under low stringency conditions, PrP-related sequences were also detected in nematodes, fruitflies and possibly yeast (Westaway and Prusiner 1986). The only signal arising from hamster and mouse DNA analyzed in parallel corresponded to the authentic PrP gene sequences. Two overlapping *Drosophila melanogaster* genomic clones were retrieved under the same conditions and sequenced. Homology was restricted to nucleotides 355–380 (Fig. 1a) of the hamster ORF, a tract with high G + C content (Basler et al. 1986), but these clones do not appear to encode a PrP-related peptide. Two other clones isolated using a technique that reduces the effect of G + C base

```
             1        10        20        30        40        50        60
             .         .         .         .         .         .         .
 AHaPrP  : MA..NLSYWLLALFVATWTDVGLCKKRPKP.GGWNTGGSRYPGQGSPGGNRYPPQGGGTW
 CHaPrP  : --..--------------------------.-----------------------------
 SHaPrP  : --..------------M-------------.-----------------------------
   HuPrP : --..--GC-M-V------S-L---------.---------------------------G-
 MoPrP a : --..--G--------TM-------------.--------------------------.--
 MoPrP b : --..--G--------TM-------------.--------------------------.--
   RaPrP : ...............................------------------------S----
 ShePrP  : -VKSHIGS-I-V----M-S-----------G---------------------------G-

                      70        80        90       100       110       120
                       .         .         .         .         .         .
 AHaPrP  : GQPHGGGWGQPHGGGWGQPHGGGWGQPHGGGWGQGGGTHNQWNKPNKPKTSMKHMAGAAA
 CHaPrP  : ---------------------------------------------S----N---V-----
 SHaPrP  : ---------------------------------------------S----N---------
   HuPrP : ---------------------------------------S-----S----N---------
 MoPrP a : --------------S-------S----------------------S----NL--V-----
 MoPrP b : --------------S-------S----------------------S----NF--V-----
   RaPrP : --------------------------------S------------S----NL--V-----
 ShePrP  : -------------------------------GWGQ--S-S-----S----N---V-----

                     130       140       150       160       170       180
                       .         .         .         .         .         .
 AHaPrP  : AGAVVGGLGGYMLGSAMSRPMLHFGNDWEDRYYRENMNRYPNQVYYRPVDQYNNQNNFVH
 CHaPrP  : ------------------------------------------------------------
 SHaPrP  : ---------------------M--------------------------------------
   HuPrP : --------------------II---S-Y---------H----------M-E-S-------
 MoPrP a : --------------------I----------------Y--------------S-------
 MoPrP b : --------------------I----------------Y--------------S-------
   RaPrP : -------------------------------------Y--------------S-------
 ShePrP  : --------------------LI-----Y---------Y------------R-S-------

                     190       200       210       220       230       240
                       .         .         .         .         .         .
 AHaPrP  : DCVNITIKQHTVTTTTKGENFTETDVKMMERVVEQMCVTQYQKESQAYYDGRRSS.AVLF
 CHaPrP  : -------------------------------------------------------.----
 SHaPrP  : -------------------------I-I---------T-----------------.----
   HuPrP : -------------------------------------I---ER------Q.-G--.M---
 MoPrP a : -------------------------------------------------------ST---
 MoPrP b : ------------V------------------------------------------ST---
   RaPrP : -------------------------------------------------------.----
 ShePrP  : ------V------------------I-I---------I----R------Q.-GA-.VI--

                     250
                       .
 AHaPrP  : SSPPVILLISFLIFLIVG
 CHaPrP  : ------------------
 SHaPrP  : ---------------M--
   HuPrP : ------------------
 MoPrP a : ------------------
 MoPrP b : ------------------
   RaPrP : ------------------
 ShePrP  : ------------------
```

Fig. 1 a, b. Alignment of **a** nucleotide and **b** predicted protein sequences of PrP genes from hamster (Armenian, *AHaPrP*; Chinese, *CHaPrP*; Syrian, *SHaPrP*), human (*HuPrP*), mouse (*NZW* allele a, *MoPrP* a; I/LnJ allele b, *MoPrP b*), rat (*RaPrP*) and sheep (Suffolk, *ShePrP*). The nucleotide sequence of the open reading frame is shown starting with the first nucleotide of the initiation codon

content upon hybrid stability (WOOD et al. 1985) show homology to a different area of the PrP ORF centered around a run of four threonines (nucleotides 577–590, Fig 1a). Using the ORF analogous to that used in mammalian PrP genes, the clones are predicted to encode threonine-rich proteins unrelated to PrP (D. Westaway and C. Mirenda, unpublished result). Casual inspection of the literature reveals many threonine (and serine)-rich proteins (TOMITA and MARCHESI 1975; HUNZIKER et al. 1986; JOHNSON et al. 1986), and it is likely that a formal database search would yield many more examples. Failure to retrieve the invertebrate homologues of mammalian genes by reduced stringency hybridization has been noted by other workers, and alternative strategies such as antibody screening of expression libraries or genetic complementation (LEE and NURSE 1987) may need to be invoked for these experiments. Another strategy may be the synthesis of the regions homologous to chicken ARIA (HARRIS et al. 1989) as degenerate oligonucleotide probes allowing the detection of genes differing by silent substitutions in more distant species. Alternatively, bona fide PrP genes may be restricted to vertebrates.

3 Conservation of PrP Genes Among Species

The lengths of the ORF in different species vary from 759 to 768 nucleotides. The alignment of the protein sequences indicates deletions and insertions of single codons (see below) yielding prion proteins with 254 aa (hamster, mouse), 253 aa (human) and 256 aa (sheep). The N-terminus of rat PrP has not yet been cloned; the known portion of the rat sequence corresponds most closely to that of mouse.

The relatedness of the cloned PrP genes was further analyzed by aligning the nucleotide sequences and counting the number of mismatches. Mutations leading to a replacement in the amino acid sequence were counted independently from silent mutations. Silent mutations should evolve independent of the protein sequence; in fact, silent mutations can be viewed as a molecular clock and used to calculate the time of divergence. The ratio of replacement to silent mutations allows us to evaluate the constraint imposed by the protein product. Table 1 shows the pairwise comparison of the PrP genes of Armenian, Chinese and Syrian hamsters, humans, mice (the NZW and I/LnJ alleles gave the same values), rats and sheep; the corrected divergence values (d) were calculated as described by PERLER et al. (1980) using a computer program written by J.D. Weissmann. The Armenian and Chinese hamsters are very closely related, as indicated by the low divergence at silent sites ($d_s = 0.09$; Table 1); the Syrian hamster is more distant from the other hamsters ($d_s = 0.25$). Comparison of rats and mice shows a similar divergence at silent sites ($d_s = 0.25$); however, the mouse and rat PrP genes differ much less at replacement sites then the hamsters ($d_r = 0.01$ and 0.05 respectively). This may point to a more stringent control of the

Table 1. Divergence of PrP genes[a]

	AHaPrP		CHaPrP		SHaPrP		HuPrP		MoPrP		RaPrP	
	d_s	d_r	d_s	d_r	d_s	d_r	d_s	d_r	d_s	d_r	d_s	d_r
AHaPrP	- -	- -										
CHaPrP	0.09	0.01	- -	- -								
SHaPrP	0.25	0.05	0.21	0.05	- -	- -						
HuPrP	0.51	0.05	0.50	0.05	0.51	0.09	- -	- -				
MoPrP	0.47	0.03	0.43	0.02	0.61	0.10	0.64	0.05	- -	- -		
RaPrP	0.35	0.02	0.36	0.01	0.51	0.08	0.58	0.05	0.24	0.01	- -	- -
ShePrP	0.93	0.09	0.88	0.08	0.83	0.09	0.59	0.06	0.98	0.09	0.82	0.00

[a] Divergence values for silent sites (d_s) or replacement sites (d_r) were calculated for pairs of genes according to Perler et al. (1980). For abbreviations of PrP genes from different species see legend to Fig. 1

PrP amino acid sequence in mice and rats than in hamsters. The human PrP gene is about equally related to the rodent and the sheep PrP genes ($d_s = 0.5$), while the sheep PrP gene is clearly the most distant of all PrP genes [$d_s = 0.9$ (sheep-rodents) and 0.59 (sheep-human)]. The constraint imposed upon the protein sequence can be measured by the ratio of silent to replacement mutations. Averaged over all the species this ratio is 12.25 ± 7.4. This value is higher than those found for the insulin C chain (5 ± 1), mammalian alpha- and beta-globin genes (7 ± 4 and 6 ± 2, respectively) and interferon-alpha genes (2.2 ± 0.3) but lower than those for the insulin A and B chains (16 ± 2); values are taken from WEISSMANN and WEBER (1986) and PERLER et al. (1980). Closer inspection of d_s/d_r ratios reveals that high values (15 and higher) occur for the comparisons among rodents (with the Syrian hamster as an exception) while the d_s/d_r ratios for the combinations of the other species range from 4.2 to 12. These numbers suggest that PrP may be more constrained in mice, rats and Armenian and Chinese hamsters, assuming that the mutation fixation rate at silent sites is comparable for all PrP genes. It has been proposed previously that the rate of silent substitutions is similar for different genes and may therefore be used as an evolutionary clock (MIYATA and HAYASHIDA 1982). It was estimated that the rate of substitution is about 5.1×10^{-9} per site per year. Using the equation described by MIYATA and HAYASHIDA (1982), we calculated that the time since the divergence of man and rodent is 9.8×10^7 years, which is in good agreement with the divergence time estimated from other genes (7.5×10^7 years; DAYHOFF 1978).

The phylogenetic tree built by progressive alignment according to FENG and DOOLITTLE (1987) is shown in Fig. 2. The numbers show the percentage difference from next the node. The root represents an average of all aligned sequences. The tree found by this method is similar to what has been described earlier (GOODMAN 1976) In order to maximize matches of the amino acid sequences, gaps had to be introduced representing insertions or deletions of blocks of three nucleotides into a putative ancestral PrP gene. The algorigthm employed for alignment

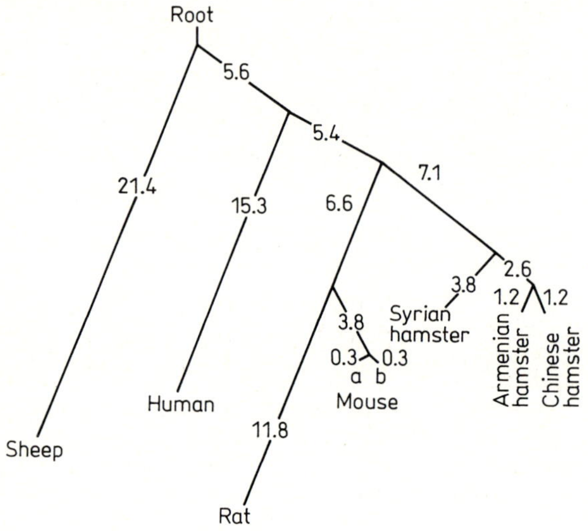

Fig. 2. Phylogenetic tree derived from the nucleotide sequence of the PrP open reading frames. The numbers show the percentage difference from the next branch. The root represents an average of all PrP sequences

emphasizes the rule "once a gap, always a gap" assuming that insertions or deletions at a specific site would be propagated within a given branch of the tree (FENG and DOOLITTLE 1987). In the mouse PrP alleles, one of the glycines at amino acids 56–58 was deleted compared to the other sequences. For Fig. 1 we arbitrarily chose gly 56. Since the nucleotide sequence is repeated, a deletion of any triplet between nucleotides 166 and 173 would result in the observed sequence. It is therefore not possible to determine exactly which nucleotides have been deleted. Serine 236 is only present in mice and was probably inserted after the speciation of rodents. As for the deletion described above, the exact point of insertion cannot be determined due to repeated sequences. The sheep PrP sequence has been altered by insertions of two amino acids in the signal sequence (lysine 2, serine 3) and a single glycine at position 31. We further postulate that glycine 92 in the sheep sequence was inserted and glycine 95 of the ancestral PrP was deleted, resulting in the shift of the sequence Trp-Gly-Gln by one amino acid (Fig. 1b). Alternatively, it is possible that the same sequence may have been produced by mutagenesis, even though it would take more independent events to arrive at the same result. Glycine 231 is absent from the human and sheep PrP genes. This may either come about by a deletion of the codon in a common ancestor of sheep and humans or, more likely, glycine 231 may have been inserted into the rodent lineage prior to their speciation. All of the insertion or deletions seem to occur in repeated regions, which prevents us from determining the precise events at the nucleotide level. A distinct possibility is that these changes may have occurred through unequal crossing over between sister chromatids. This may also account for a mutation linked to the occurrence of familial CREUTZFELDT-JAKOB disease (CJD), a prion disease of humans, where the octa repeat region (codons 54–95) is partially duplicated (COLLINGE et al. 1989; OWEN et al. 1989; Fig. 3).

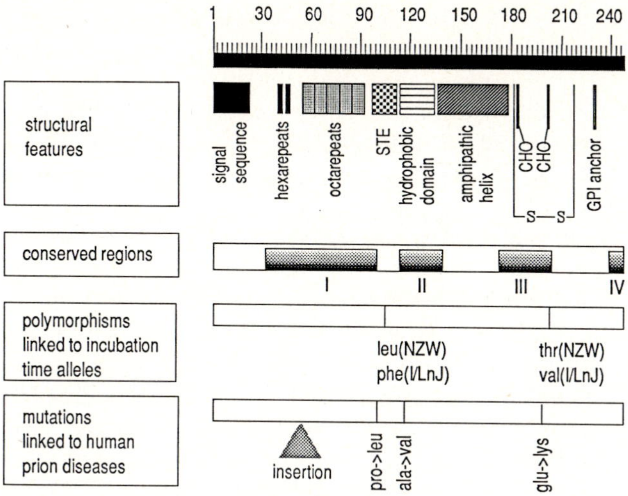

Fig. 3. Spatial relationships of structural and functional elements of the PrP polypeptide. For detailed description see text. The ruler indicates the codon positions. *STE*, stop transfer effector; *CHO*, *N*-linked carbohydrate attachment site; *GPI anchor*, glycan phosphoinositol attachment site; *-S-S-*, proposed disulfide bond

4 Functional and Structural Aspects of the PrP Polypeptide

Comparison of the amino acid sequence of all species shows that 78% of the positions are invariant. The changes in amino acid sequence tend to be clustered. The highest degree of divergence is found in the signal sequence (33% conserved, amino acids 1–24). This is not surprising for signal sequences, where the secondary rather than primary structure is important for function (VON HEIJNE 1985). Four major regions are highly conserved and are designated I (codons 32–111), II (116–140), III (174–205) and IV (239–255) (Fig. 3).

Region I is characterized by the occurrence of two repeats with the hexapeptide sequence GG(S/N)RYP and five repeats with the octapeptide sequence P(H/Q)GGG(-/T)WGQ. The octapeptide sequence has been noted to bear some similarity with a 50-kDa keratin protein which polymerizes into intermediate filaments (MARCHUK et al. 1984; OESCH et al. 1985), yet this region is absent from rods composed of aggregated PrP 27–30 (PRUSINER 1984). Removal of region I by limited proteolysis in fact appears to favor rod formation (MCKINLEY et al. 1991) yet it does not affect infectivity. This would indicate that this region does not need to be present in the infectious particle; however, an insertion into this region has been found to be associated with the occurrence of familial CJD (COLLINGE et al. 1989; OWEN et al. 1989). Since the mechanism of prion formation is unknown it is difficult to assess the role of this region. The formation of infectious particles may be facilitated by mutations in region I, which is then dispensable. It will be interesting to see whether the disease-specific isoform of PrP from patients with this mutation is predominantly of the mutant or normal allelic product.

Region II was predicted to be a hydrophobic domain which could potentially span the membrane (codons 115–134; BAZAN et al. 1987). In vitro translation of synthetic PrP mRNA confirmed this prediction; however, a second, fully translocated form was also found (HAY et al. 1987a, b). In vivo, PrP^{C} was found at the cell surface with a glycolipid anchor (STAHL et al. 1987), which would correspond to the fully translocated form of PrP. Interestingly, PrP^{C} but not PrP^{Sc} could be released from cells by phosphatidylinositol-specific phospholipase (BORCHELT et al. 1990; STAHL et al. 1990). One possible interpretation is that the scrapie and the cellular forms of PrP differ in their topological orientation, resulting in the release of glycoinositol phospholipid (GPI)-linked PrP^{C} but not of transmembrane PrP^{Sc}. The sequence between codons 97 and 114 (designated stop tansfer effector, STE) was subsequently shown to govern the alternative fates of PrP (LOPEZ et al. 1990; YOST et al. 1990). Mutations in the STE region might therefore influence the topology of the PrP polypeptide. A polymorphism in STE [leucine 112 in NZW mice to phenylalanine in I/LnJ mice in Fig. 1b; codon 108 in the contiguous mouse PrP sequence (WESTAWAY et al. 1987)] has been linked to a host gene controlling the incubation time (WESTAWAY et al. 1987). I/LnJ mice have a longer incubation time than NZW mice. However, there is a second polymorphism in the PrP gene (threonine 193 is a valine in I/LnJ mice; codon 189 in the contiguous mouse PrP sequence; WESTAWAY et al. 1987). To date, it is not resolved whether either of these amino acid changes alone or both together are responsible for prolongation of the incubation time. It is not clear whether the changes in the amino acid sequence of PrP are responsible for altered incubation times. As an alternative, the developmental expression of the two PrP alleles may differ. It is certainly possible that a gene near the PrP gene controls incubation times.

Two mutations of the human PrP gene near or in region II ($Pro_{105} \rightarrow Leu$ and $Ala_{120} \rightarrow Val$) are linked to the occurrence of familial prion disease (DOH-URA et al. 1989; HSIAO et al. 1989, 1991). The Pro to Leu change occurs within the STE domain, while the Ala to Val mutation is within the hydrophobic domain. These different alleles may alter the topology of PrP and in turn influence the efficiency with which the infectious prion particles are formed. The spontaneous occurrence of Gerstmann-Sträussler syndrome (GSS) or CJD in humans with mutations in the PrP gene poses an interesting conundrum: Do these mutations in the PrP gene cause the disease, or are they just dominant "incubation time alleles" allowing the replication of a universally present infectious particle? In the first case, a mutant protein would lead to the generation of an infectious agent, which is without precedent. Alternatively, subclinical prion diseases might be common in the general population, giving rise to sporadic CJD and causing familial CJD and GSS in individuals rendered susceptible by a PrP mutation.

Region II is also highly homologous to putative ARIA (USDIN and FISCHBACH 1986; HARRIS et al. 1989). The sequence of the candidate ARIA is identical with the I/LnJ allele of PrP between amino acids 108 and 131. The overall homology of ~40% between PrP and ARIA, as well as similar structural predictions for the polypeptide, suggests that PrP may be a mammalian homologue of candidate chicken ARIA. ARIA induces increased transcription of the gene encoding the α subunit of the acetylcholine receptor (AChR) (HARRIS et al. 1988), which is limiting for the expression of receptor. ARIA therefore increases the number of AChR at the cell surface of muscle cells. Future experiments will show whether PrP has ARIA activity and whether PrP has similar activity in brain.

Region III is characterized by two sites for *N*-linked glycosylation which conform to the consensus sequence *NXT*. Both sites are utilized in vivo, yet there are lower molecular weight forms of PrP which are only glycosylated at one site or not at all (TARABUOLOS et al. 1991). There are also multiple potential *O*-glycosylation sites, especially a run of threonines (193–196) one of which is changed to a valine in the I/LnJ allele of PrP. However, all of the carbohydrates of PrP seem to be either *N*-linked or in the GPI anchor at the *C*-terminus (HARAGUCHI et al. 1989).

Region IV is at the *C*-terminus of the predicted polypeptide. It is cleaved off upon addition of the glycolipid anchor at serine 234 (codon 231 of contiguous HaPrP; STAHL et al. 1987, 1990). It has been predicted to be hydrophobic (BAZAN et al. 1987). Detailed analysis of the signal for the addition of a GPI anchor to alkaline phosphatase or decay accelerating factor has shown that a serine at the attachment site and a *C*-terminal hydrophobic domain of 17 amino acids are important (BERGER et al. 1988; CARAS et al. 1989). Region IV may therefore be involved in the specification of the precise site of GPI anchor addition.

5 Conclusions

While the cloning and analysis of the PrP gene has not resolved all questions concerning the nature of scrapie, it has unveiled some of the elusive properties of the infectious particle. The structure of the PrP gene makes it unlikely that the scrapie isoform is encoded by a different gene other than cellular PrP (for a more complete discussion, see BASLER et al. 1986) leaving the possibility that the altered properties of PrP are due to some posttranslational modification, change of conformation or association with other molecule(s). Evidence for this postulate has recently been obtained by metabolic radiolabeling experiments in scrapie-infected neuroblastoma cells (BORCHELT et al. 1990). It is tempting to speculate that the same entity which encodes genetic information for different scrapie strains is also responsible for the physical properties of the prion. Manipulation of the amino acid sequence of the prion protein in conjunction with the construction of transgenic animals (SCOTT et al. 1989) may reveal the role that PrP plays in disease. Using similar strategies, we may also learn about the normal function of PrP in brain as well as other tissues.

Acknowledgments. We would like to thank Dr. L. Taylor for assistance in the alignment of the PrP sequences and Dr. J.D. Weissmann for providing the program for divergence calculation. We also thank Prof. C. Weissmann for critically reading the manuscript.

References

Aiken JM, Williamson JL, Marsh RF (1989) Evidence of mitochondrial involvement in scrapie infection. J Virol 63: 1686–1694

Alper T, Cramp WA, Haig DA, Clarke MC (1967) Does the agent of scrapie replicate without nucleic acid? Nature 214: 764–766

Basler K, Oesch B, Scott M, Westaway D, Wälchli M, Groth DF, McKinley MP, Prusiner SB, Weissmann C (1986) Scrapie and cellular PrP isoforms are encoded by the same chromosomal gene. Cell 46: 417–428

Bazan JF, Fletterick RJ, McKinley MP, Prusiner SB (1987) Predicted secondary structure and membrane topology of the scrapie prion protein. Protein Eng 1: 125–135

Berger J, Howard AD, Brink L, Gerber L, Hauber J, Cullen BR, Udenfriend S (1988) COOH–terminal requirements for the correct processing of a phosphatidylinositol–glycan anchored membrane protein. J Biol Chem 263: 10016–10021

Bolton DC, McKinley MP, Prusiner SB (1982) Identification of a protein that purifies with the scrapie prion. Science 218: 1309–1311

Borchelt DR, Scott M, Taraboulos A, Stahl N, Prusiner SB (1990) Scrapie and cellular prion proteins differ in their kinetics of synthesis and topology in cultured cells. J Cell Biol 110: 743–752

Bruce ME, Dickinson AG (1987) Biological evidence that the scrapie agent has an independent genome. J Gen Virol 68: 79–89

Caras IW, Weddell GN, Williams SR (1989) Analysis of the signal for attachment of a glycophospholipid membrane anchor. J Cell Biol 108: 1387–1396

Carlson GA, Goodman P, Kingsbury DT, Prusiner SB (1986) Scrapie incubation time and prion protein genes are linked. Clin Res 34: 675A

Carlson GA, Goodman PA, Lovett M, Taylor BA, Marshall ST, Peterson–Torchia M, Westaway D, Prusiner SB (1988) Genetics and polymorphism of the mouse prion gene complex: the control of scrapie incubation time. Mol Cell Biol 8: 5528–5540

Cashman NR, Loertscher R, Nalbantoglu J, Shaw I, Kascsak RJ, Bolton DC, Bendheim PE (1990) Cellular isoform of the scrapie agent protein participates in lymphocyte activation. Cell 61: 185–192

Chesebro B, Race R, Wehrly K, Nishio J, Bloom M, Lechner D, Bergstrom S, Robbins K, Mayer L, Keith JM, Garon C, Haase A (1985) Identification of scrapie prion protein-specific mRNA in scrapie-infected and uninfected brain. Nature 315: 331–333

Collinge J, Harding AE, Owen F, Poulter M, Lofthouse R, Boughey AM, Shah T, Crow TJ (1989) Diagnosis of Gerstmann–Sträussler syndrome in familial dementia with prion protein gene analysis. Lancet 2: 15–17

Dayhoff MO (1978) In: Atlas of protein sequence and structure, Vol 5, suppl 3. National Biochemical Research Foundation, Washington, DC

DeArmond SJ, Mobley WC, Demott DL, Barry RA, Beckstead JH, Prusiner SB (1987) Changes in the localization of brain prion proteins during scrapie infection. Neurology 37: 1271–1280

Doh-ura K, Tateishi J, Sasaki H, Kitamoto T, Sakaki Y (1989) Pro → Leu change at position 102 of prion protein is the most common but not the sole mutation related to Gerstmann-Sträussler syndrome. Biochem Biophys Res. Commun 163: 974–979

Duguid JR, Rohwer RG, Seed B (1988) Isolation of cDNAs of scrapie-modulated RNAs by subtractive hybridization of a cDNA library. Proc Natl Acad Sci USA 85:5738–5742

Feng D-F, Doolittle RF (1987) Progressive sequence alignment as a prerequisite to correct phylogenetic trees. J Mol Evol 25: 351–360

Gabizon R, Prusiner SB (1990) Prion liposomes. Biochem. J 266: 1–14

Gabizon R, McKinley MP, Groth DF, Prusiner SB (1988) Immunoaffinity purification and neutralization of scrapie prion infectivity. Proc Natl Acad Sci USA 85: 6617–6621

Gibbons RA, Hunter GD (1967) Nature of the scrapie agent. Nature 215: 1041–1043

Goldman W, Hunter N, Foster JD, Salbaum JM, Beyreuther K, Hope J (1990) Two alleles of a neural protein gene linked to scrapie in sheep. Proc Natl Acad Sci USA 87: 2476–2480

Goodman M (1976) Protein sequences in phylogeny. In: Molecular evolution. Sinauer Associates, Massachusetts, Ayala FJ (ed), pp 141–153

Haraguchi T, Fisher S, Olofsson S, Endo T, Groth D, Tarantino A, Borchelt DR, Teplow D, Hood L, Burlingame A, Lycke E, Kobata A, Prusiner SB (1989) Asparagine-linked glycosylation of the scrapie and cellular prion proteins. Arch Biochem Biophys 274: 1–13

Harris DA, Falls DL, Dill-Devor RM, Fischbach GD (1988) Acetylcholine receptor-inducing factor from chicken brain increases the level of mRNA encoding the receptor alpha subunit. Proc Natl Acad Sci USA 85: 1983–1987

Harris DA, Falls DL, Walsh W, Fischbach GD (1989) Molecular cloning of an acetylcholine receptor-inducing protein. Soc Neurosci Abstr 15: 70.7

Hay B, Barry RA, Lieberburg I, Prusiner SB, Lingappa VR (1987a) Biogenesis and transmembrane orientation of the cellular isoform of the scrapie prion protein. Mol Cell Biol 7: 914–920

Hay B, Prusiner SB, Lingappa VR (1987b) Evidence for a secretory form of the cellular prion protein. Biochemistry 26: 8110–8115

Hsiao K, Baker HF, Crow TJ, Poulter M, Owen F, Terwilliger JD, Westaway D, Ott J, Prusiner SB (1989) Linkage of a prion protein missense variant to Gerstmann–Sträussler syndrome. Nature 338: 342–345

Hsiao KK, Cass C, Schellenberg GD, Devine-Gage E, Wisniewski H, Prusiner SB (1991) A prion protein variant in a family with the telencephalic form of Gerstmann–Sträussler–Scheinker syndrome. Neurology 41: 681–684

Hunter N, Hope J, McConnell I, Dickinson AG (1987) Linkage of the scrapie-associated fibril protein (PrP) gene and Sinc using congenic mice and restriction fragment length polymorphism analysis. J Gen Virol 68: 2711–2716

Hunziker W, Spiess M, Semenza G, Lodish HF (1986) The sucrase-isomaltase complex: primary structure, membrane-orientation, and evolution of a stalked, intrinsic brush border protein. Cell 46: 227–234

Johnson D, Lanahan A, Buck CR, Sehgal A, Morgan C, Mercer E, Bothwell M, Chao M (1986) Expression and structure of the human NGF receptor. Cell 47: 545–554

Kimberlin RH, Cole S, Walker CA (1987) Temporary and permanent modifications to a single strain of mouse scrapie on transmission to rats and hamsters. J Gen Virol 68: 1875–1881

Kretzschmar HA, Prusiner SB, Stowring LE, DeArmond SJ (1986a) Scrapie prion proteins are synthesized in neurons. Am J Pathol 122: 1–5

Kretzschmar HA, Stowring LE, Westaway D, Stubblebine WH, Prusiner SB, DeArmond SJ (1986b) Molecular cloning of a human prion protein cDNA. DNA 5: 315–324

Lee MG, Nurse P (1987) Complementation used to clone a human homologue of the fission yeast cell cycle control gene cdc2. Nature 327: 31–35

Liao Y-C, Lebo RV, Clawson GA, Smuckler EA (1986) Human prion protein cDNA: molecular cloning, chromosomal mapping, and biological implication. Science 233: 364–367

Liao Y-C, Tokes Z, Lim E, Lackey A, Woo CH, Button JD, Clawson GA (1987) Cloning of rat "prion-related protein" cDNA. Lab Invest 57: 370–374

Locht C, Chesebro B, Race R, Keith JM (1986) Molecular cloning and complete sequence of prion protein cDNA from mouse brain infected with the scrapie agent. Proc Natl Acad Sci USA 83: 6372–6376

Lopez CD, Yost CS, Prusiner SB, Myers RM, Lingappa VR (1990) Unusual topogenic sequence directs prion protein biogenesis. Science 248: 226–229

Lowenstein DH, Butler DA, Westaway D, McKinley MP, DeArmond SJ, Prusiner SB (1990) Three hamster species with different scrapie incubation times and neuropathological features encode distinct prion proteins. Mol Cell Biol 10: 1153–1163

Marchuk D, McCrohon S, Fuchs E (1984) Remarkable conservation of structure among intermediate filament genes. Cell 39: 491–498

McKinley MP, Bolton DC, Prusiner SB (1983) A protease-resistant protein is a structural component of the scrapie prion. Cell 35: 57–62

McKinley MP, Meyer RK, Kenaga L, Rahbar F, Cotter R, Serban A, Prusiner SB (1991) Scrapie prion rod formation in vitro requires both detergent extraction and limited proteolysis. J Virol 65: 1340–1351

Meyer RK, McKinley MP, Bowman KA, Braunfeld MB, Barry RA, Prusiner SB (1986) Separation and properties of cellular and scrapie prion proteins. Proc Natl Acad Sci USA 83: 2310–2314

Miyata T, Hayashida H (1982) Recent divergence from a common ancestor of human IFN-α genes. Nature 295: 165–168

Mohri S, Tateishi J (1989) Host genetic control of incubation periods of Creutzfeldt–Jakob disease in mice. J Gen Virol 70: 1391–1400

Oesch B, Westaway D, Wälchli M, McKinley MP, Kent SBH, Aebersold R, Barry RA, Tempst P, Teplow DB, Hood LE, Prusiner SB, Weissmann C (1985) A cellular gene encodes scrapie PrP 27–30 protein. Cell 40: 735–746

Oesch B, Groth DF, Prusiner SB, Weissmann C (1988) Search for a scrapie-specific nucleic acid: a progress report. Ciba Found Symp 135

Owen F, Poulter M, Lofthouse R, Collinge J, Crow TJ, Risby D, Baker HF, Ridley RM, Hsiao K, Prusiner SB (1989) Insertion in prion protein gene in familial Creutzfeldt–Jakob disease. Lancet 1: 51–52

Perler F, Efstratiadis A, Lomedico P, Gilbert W, Kolodner R, Dodgson J (1980) The evolution of genes: the chicken preproinsulin gene. Cell 20: 555–566

Prusiner SB (1982) Novel proteinaceous infectious particles cause scrapie. Science 216: 136–144

Prusiner SB (1984) Prions. Sci Am 251: 50–59

Prusiner SB, Bolton DC, Groth DF, Bowman KA, Cochran SP, Mckinley MP (1982) Further purification and characterization of scrapie prions. Biochemistry 21: 6942–6950

Prusiner SB, McKinley MP, Bowman KA, Bolton DC, Bendheim PE, Groth DF, Glenner GG (1983) Scrapie prions aggregate to form amyloid-like birefringent rods. Cell 35: 349–358

Prusiner SB, Groth DF, Bolton DC, Kent SB, Hood LE (1984) Purification and structural studies of a major scrapie prion protein. Cell 38: 127–134

Race RE, Graham K, Ernst D,Caughey B, Chesebro (1990) Analysis of linkage between scrapie incubation period and the prion protein gene in mice. J Gen Virol 71: 493–497

Robakis NK, Sawh PR, Wolfe GC, Rubenstein R, Carp RI, Innis MA (1986) Isolation of a,cDNA clone encoding the leader peptide of prion protein and expression of the homologous gene in various tissues: Proc Natl Acad Sci USA 83: 6377–6381

Rohwer RG (1984) Scrapie infectious agent is virus-like in size and susceptibility to inactivation. Nature 308: 658–662

Scott M. Foster D, Mirenda C, Serban D, Coufal F, Wälchli M, Torchia M, Groth D, Carlson G, DeArmond SJ, Westaway D, Prusiner SB (1988) Transgenic mice expressing hamster prion protein produce species-specific scrapie infectivity and amyloid plaques. Cell 59: 847–857

Shinagawa M, Munekata E, Doi S, Takahashi K, Goto H, Sato G (1986) Immunoreactivity of a synthetic pentadecapeptide corresponding to the N-terminal region of the scrapie prion protein. J Gen Virol 67: 1745–1750

Stahl N, Borchelt DR, Hsiao K, Prusiner SB (1987) Scrapie prion protein contains a phosphatidylinositol glycolipid. Cell 51: 229–240

Stahl N, Borchelt DR, Prusiner SB (1990) Differential release of cellular and scrapie prion proteins from cellular membranes by phosphatidylinositol-specific phospholipase C. Biochemistry 29: 5405–5412

Taraboulos A, Rogers M, Borchelt DR, McKinley MP, Scott M, Serban D, Prusiner SB (1990) Acquisition of protease resistance by prion proteins in scrapie-infected cells does not require asparagine-linked glycosylation. Proc Natl Acad Sci USA 87: 8262–8266

Tomita M, Marchesi VT (1975) Amino-acid sequence and oligosaccharide attachment sites of human erythrocyte glycophorin. Proc Natl Acad Sci USA 72: 2964–2968

Usdin TB, Fischbach GD (1986) Purification and characterization of a polypeptide from chick brain that promotes the accumulation of acetylcholine receptors in chick myotubes. J Cell Biol 103: 493–507

von Heijne G (1985) Signal sequences, the limits of variation. J Mol Biol 184: 99–105

Weissmann C, Weber H (1986) The interferon genes. Prog Nucleic Acid Res Mol Bio 33: 251–300

Westaway D, Prusiner SB (1986) Conservation of the cellular gene encoding the scrapie prion protein. Nucleic Acids Res 14: 2035–2044

Westaway D, Goodman PA, Mirenda CA, McKinley MP, Carlson GA, Prusiner SB (1987) Distinct prion proteins in short and long scrapie incubation period mice. Cell 51: 651–662

Westaway D, Mirenda C, Foster D, Carlson G, Hoppe P, Scott M, Wälchli M, Prusiner S (1990) Molecular genetics of the mouse prion gene complex. Neurobiol Aging 11(1): 86–87

Wills PR (1989) Induced frameshifting mechanism of replication for an information-carrying scrapie prion. Microb Pathogen 6: 235–249

Wood WI, Gitschier J, Lasky LA, Lawn RM (1985) Base composition-independent hybridization in tetramethylammonium chloride: a method for oligonucleotide screening of highly complex gene libraries. Proc Natl Acad Sci USA 82: 1585–1588

Yost CS, Lopez CD, Prusiner SB, Meyers RM, Lingappa VR (1990) A non-hydrophobic extra-cytoplasmic determinant of stop transfer in the prion protein. Nature 343: 669–672

Scrapie Strain Variation and Its Implications

M. E. BRUCE and H. FRASER

1 Introduction

Strain variation in scrapie was first recognised about 30 years ago, when experimental sheep scrapie was transmitted to and serially passaged in goats. Two separate passage lines were maintained which produced strikingly different clinical signs in unselected goats from the same herd, either a "drowsy" or a "scratching" syndrome (PATTISON and MILLSON 1961). This difference between the two isolates remained constant on repeated goat-to-goat passage. Since then, overwhelming evidence for scrapie strain variation has accumulated, mainly from the long-term studies conducted in Edinburgh by Dickinson and coworkers (including ourselves), who have serially passaged scrapie from a wide range of natural and experimental sources in inbred mice (DICKINSON 1976). About 20 strains of scrapie have been identified so far in mice, each with its own distinct and repeatable set of disease characteristics.

In the absence of serological or biochemical markers for different scrapie strains, it has been necessary to develop novel strain discrimination methods. The most obvious way in which strains differ is in their incubation periods in inbred mice of defined genotypes (OUTRAM 1976; DICKINSON and FRASER 1977). There are also striking differences between scrapie strains in the type, severity

Institute for Animal Health, AFRC and MRC Neuropathogenesis Unit, Ogston Building, West Mains Road, Edinburgh EH9 3JF, UK

Current Topics in Microbiology and Immunology, Vol. 172

and distribution of pathological changes they produce in the brain (FRASER 1976). These two aspects of scrapie strain variation have been exploited to develop quantitative methods for strain discrimination. Scrapie strains also differ in many other respects, for example in their clinical characteristics (CARP et al. 1984), their ease of transmission between species (KIMBERLIN and WALKER 1978) and their resistance to thermal inactivation (DICKINSON and TAYLOR 1978; KIMBERLIN et al. 1983). In addition, there is ample evidence of strain variation in scrapie in other experimental species, such as sheep (FOSTER and DICKINSON 1988) and hamsters (KIMBERLIN and WALKER 1978; KIMBERLIN et al. 1989) and in experimentally transmitted Creutzteldt–Jakob disease (MORI et al. 1989; KITAMOTO et al. 1990) and transmissible mink encephalopathy (KIMBERLIN et al. 1986). The following chapter reviews the evidence for scrapie strain variation based on incubation periods and pathology and explores the important implications this has for molecular models of the scrapie agent.

2 Incubation Periods and the *Sinc* Gene

In general, when a single strain of scrapie is injected intracerebrally (i.c.) at high dose into a group of mice of a single inbred strain, the incubation period between initial infection and the terminal stage of the disease is remarkably uniform between individuals in the group and is also highly repeatable. For many years we have been using a standardised system in which incubation period is measured up to a clinically defined end-point when mice are showing severe neurological signs (DICKINSON et al. 1968); this method has yielded precise and reproducible results for a wide range of mouse scrapie models, generally giving standard errors of less than 2% of the mean incubation period.

It is perfectly possible to identify some differences between scrapie strains simply from their incubation periods in a single inbred mouse strain (Fig. 1). However, the potential for strain discrimination is considerably expanded by the use of mice carrying different alleles of the *Sinc* gene, the major host gene controlling incubation period (*Sinc* from *s*crapie *inc*ubation) (DICKINSON et al. 1968; DICKINSON and MEIKLE 1971; DICKINSON and FRASER 1977). Two alleles of the *Sinc* gene have been identified, s7 and p7, and most strain typing has been carried out in the C57BL (*Sinc*^{s7}) and VM (*Sinc*^{p7}) inbred mouse strains and in the F_1 cross between them. Figure 1 shows the incubation periods of 11 of the best-characterised scrapie strains in these three *Sinc* genotypes following i.c. infection with a high standard dose. Recently some of these strains have been tested in *Sinc* congenic mouse lines, giving similar results to those presented here (BRUCE et al. 1991).

Each scrapie strain has a characteristic and highly repeatable pattern of incubation periods in the three genotypes. Scrapie strains differ, firstly, in their absolute incubation periods within a single *Sinc* genotype. Secondly, they differ in the relative incubation periods in the two homozygotes; the incubation period

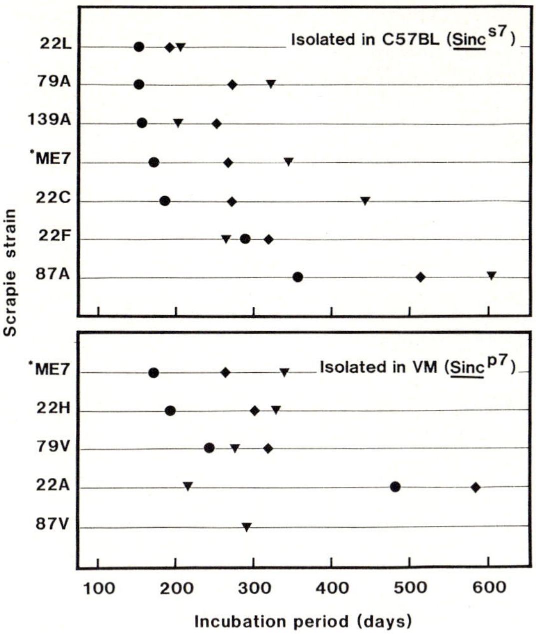

Fig. 1. Incubation periods of 11 scrapie strains in C57BL (●), VM (▼) and C57BLxVM F_1 (◆) mice, injected i.c. with 1% brain inocula. For 87V, incubation periods in C57BL and C57BLxVM mice are longer than life-span. *Note that ME7 has been isolated by passage in both C57BL and VM mice

is shorter in *Sinc*^{s7} than in *Sinc*^{p7} mice for some scrapie strains (e.g. ME7), but this ranking is reversed for other strains (e.g. 22A) (DICKINSON and MEIKLE 1971). Thirdly, the dominance characteristics of two alleles appear to differ according to the scrapie strain, with incubation periods in the F_1 heterozygote lying either between (e.g. ME7) or beyond (e.g. 22A) those in the parental homozygotes.

3 Brain Pathology

Scrapie strains also differ strikingly in the pathological changes they produce in the brain (FRASER 1976). The most obvious change in mouse scrapie is a vacuolar degeneration of the neuropil, which is precisely targeted to different areas of brain, depending primarily on the strain of scrapie and to a lesser extent on *Sinc* and other mouse genes (FRASER and DICKINSON 1968; FRASER 1976). Thus, in any single mouse strain each scrapie strain produces a characteristic distribution of vacuolation which differs predictably from that produced by other scrapie strains. A dramatic example is the selective targeting of vacuolation to the cerebellar cortex seen only with strain 22L (Fig. 2).

To detect more subtle differences we use a quantitative representation of the severity and distribution of vacuolar changes in the brain, the "lesion profile" (FRASER and DICKINSON 1968). Vacuolation is scored from coded sections in nine

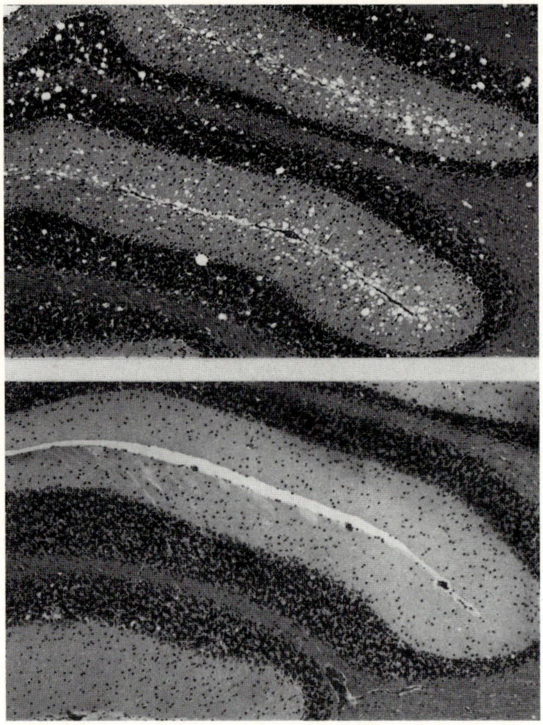

Fig. 2. a Vacuolar degeneration in cerebellar cortex of RIII mouse with 22L scrapie, contrasted with **b** absence of vacuolation in same brain area in RIII mouse with 87A scrapie. Both scrapie strains were injected i.c. as 1% brain homogenates and mice were killed at the terminal stage of the disease H&E, ×60

grey matter and three white matter areas of brain (Fig. 3a). There are striking and highly repeatable differences between scrapie strains, each having a characteristic height and shape of lesion profile for a given mouse genotype and route of infection (FRASER and DICKINSON 1973; FRASER 1976) (Fig. 3b). This discrimination is in complete agreement with that based on incubation periods. However, lesion profiles have the great advantage that they are not dose-dependent and can therefore be used to identify scrapie strains in samples containing low levels of infectivity. They can even be used to identify which strain kills the mouse when a mixture of strains is injected.

Recently we have investigated the topographical distribution of pathology involving PrP, a host glycoprotein which accumulates in modified forms during scrapie pathogenesis (MCKINLEY et al. 1983; HOPE et al. 1986; DEARMOND et al. 1987). Immunostaining with antisera to PrP demonstrates extensive deposition of this protein in the brains of mice with scrapie, diffusely in areas of vacuolar degeneration and focally in amyloid plaques (MCBRIDE et al. 1988; BRUCE et al. 1989). Again, there are dramatic and reproducible differences between scrapie strains in the distribution of these changes. For example, with the 87V strain, diffuse PrP accumulation is precisely targeted to particular anatomically defined areas of brain, mostly in the brainstem; in contrast, ME7 in the same mouse strain produces widespread diffuse changes throughout the brain, although some

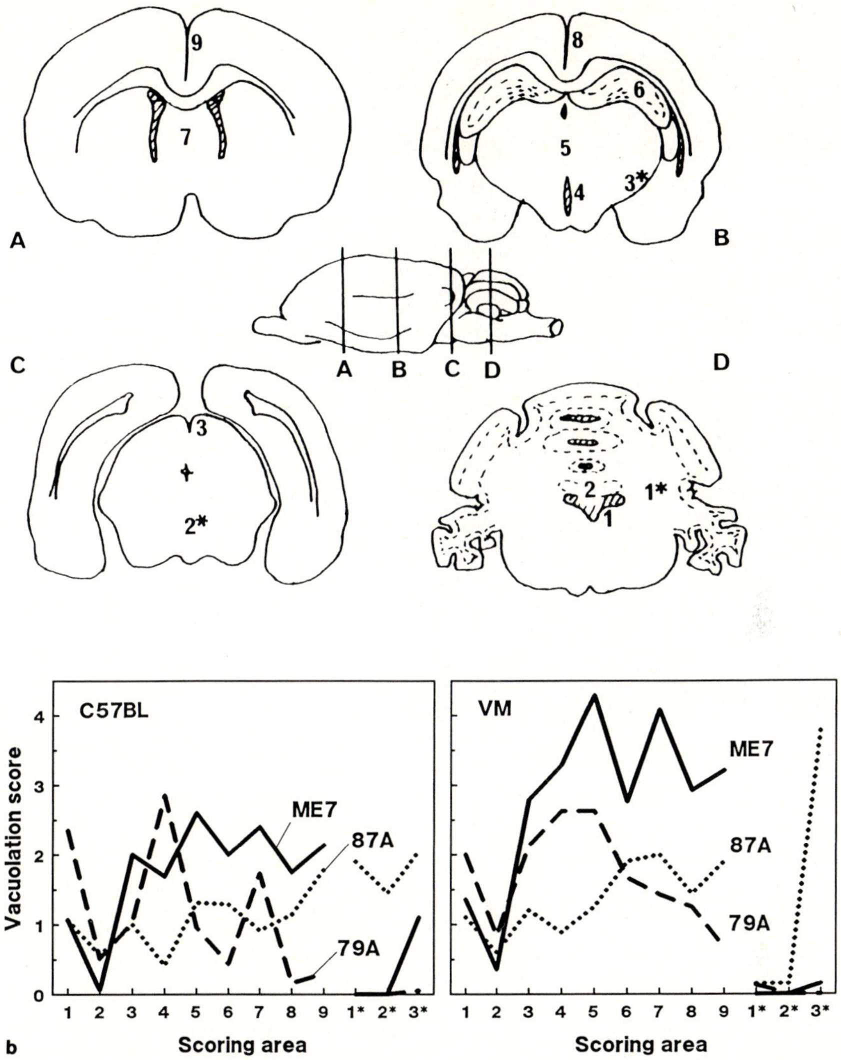

Fig. 3. **a** Brain areas in which vacuolation is scored to construct lesion profiles; 1 to 9 are grey matter areas and 1* to 3* are white matter areas. **b** Lesion profiles for three scrapie strains in C57BL and VM mice, injected i.c. with 1% inocula

areas are predictably more severely affected than others (Fig. 4). Other pathological features which depend on the strain of scrapie are the occurrence and distribution of amyloid plaques (BRUCE et al. 1976), the degeneration of hippocampal neurones (SCOTT and FRASER 1984) and the loss of retinal photoreceptor cells (FOSTER et al. 1986).

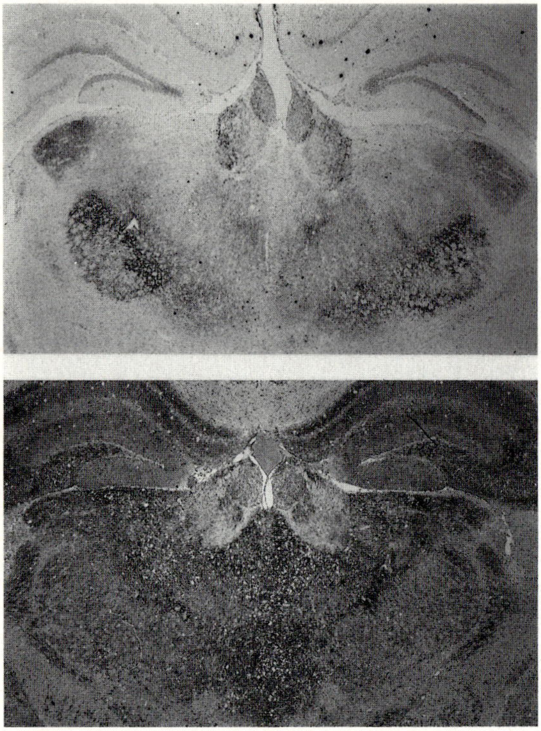

Fig. 4 a, b. Different patterns of immunostaining with antiserum to PrP in VM mice infected with **a)** 87V and **b)** ME7 scrapie strains. Both scrapie strains were injected i.c. as 1% brain homogenates and mice were killed at the terminal stage of the disease. PAP, × 15

4 Isolation of Scrapie Strains

When scrapie from a sheep or goat source is initially transmitted to mice the incubation period is usually very long and there may be survivors. On further serial passage in a single mouse genotype the incubation period shortens and stabilises, after about two or three passages, to give a recognisable strain with its own characteristic set of properties (DICKINSON et al. 1984). These properties remain stable indefinitely as long as the conditions of passage, in particular the mouse genotype, are kept constant.

The range of strains isolated can be expanded by passaging in mice of different *Sinc* genotypes (DICKINSON 1976; DICKINSON et al. 1984). It has been possible to isolate two distinct strains from most primary sources by setting up separate i.c. brain passage lines in C57BL ($Sinc^{s7}$) and VM ($Sinc^{p7}$) mice; for example, 79A and 79V were derived from "drowsy" goat scrapie and 22C and 22H from "scratching" goat scrapie (Fig. 5). It is clear that scrapie strain differences are not simply imposed by passage through different mouse genotypes, by some type of host-induced modification, as numerous different strains have been isolated in each single *Sinc* genotype, but the same strain (e.g. ME7) has sometimes been isolated independently in both genotypes (Fig. 1). Rather, extensive passaging from many primary sources has shown that this resolution is entirely consistent with the selection of scrapie strains with shorter incubation periods in the particular mouse genotype used for passage. Thus 79A has a shorter incubation period than 79V in C57BL mice and 79V has a shorter incubation period than 79A in VM mice (Fig. 5). The isolation of 22C and 22H follows the same principle. An interesting point here is that, although 79V and 22H were isolated in VM mice, their incubation periods are actually shorter in C57BL mice. Further to this, certain primary sources have given rise to more than two strains, as described below.

In the case of transmissions from natural scrapie, it is not known to what extent the strains isolated in mice are representative of field strains. The fact that more than one strain is obtained from most primary sources may suggest that mixed infections are common in the field (DICKINSON 1976). However, there is a great deal of evidence from studies of scrapie in laboratory animals that strains readily mutate and that one of the ways of selecting variants is to change the species used for passage (see below). When scrapie is transmitted to mice from another species, the new selection pressures may favour minor mutant strains,

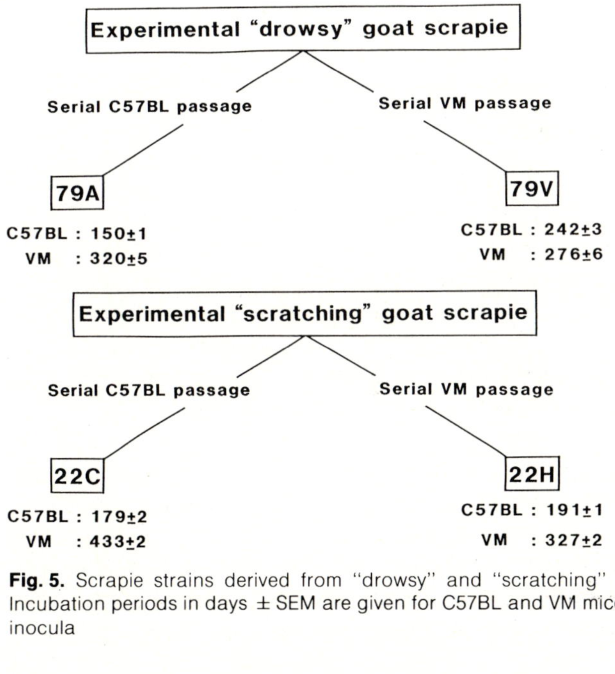

Fig. 5. Scrapie strains derived from "drowsy" and "scratching" experimental goat scrapie. Incubation periods in days ± SEM are given for C57BL and VM mice injected i.c. with 1% brain inocula

rather than the major strain present in the source brain, and different mutants may be selected according to the *Sinc* genotype used. The suggestion that the two strains isolated in the different *Sinc* genotypes may be derived by mutation from a common ancestral field strain is supported by the fact that pairs of strains are repeatedly isolated together. For example, on all of the six separate occasions when 87A has been isolated from unrelated natural sheep sources by passaging in C57BL mice, the strain obtained by VM passage has been 87V (DICKINSON et al. 1984).

5 Biological Cloning of Scrapie Strains

There is no guarantee that strains obtained by passaging at high dose are "pure", as two or more strains can be copassaged in mice as a stable mixture, particularly if they have similar incubation periods. To remove any minor strains originally present in the isolate it is necessary to clone the scrapie strain by performing several sequential passages using inoculum at the limiting dilution for infectivity (DICKNSON and OUTRAM 1983). This involves injecting serial dilutions of homogenate into groups of mice and passaging from relatively long incubation period cases within the highest dilution group to produce scrapie. There is a good chance that these mice will have been infected with a single operationally infective particle of scrapie. The confidence that this has been achieved is increased by repeating the procedure several times. However, even after cloning new mutants may accumulate in the isolate as described below.

The criticism of this cloning method, of course, is that we do not know the state of aggregation of infective particles in an homogenate. Scrapie infective units might clump together, either due to aggregation after homogenisation or because they are already closely associated in the living tissue. This would become a problem for cloning particularly if the isolate contained high concentrations of two scrapie strains with very similar incubation periods under the particular passaging conditions used. Nevertheless, it has been demonstrated that cloning works, even under these difficult circumstances. In one particular passage line, ME7 and another strain, now identified as 22C, were serially passaged together as a stable mixture about 10 times, 22C representing an estimated 5%–10% of the total infectivity in the isolate (DICKINSON and OUTRAM 1983; Outram, personal communication). The incubation periods and lesion profiles of the mixed isolate when injected i.c. were identical to those of ME7. It was only following intraperitoneal (i.p.) infection that 22C could be detected, as it gave shorter incubation periods than those expected for ME7 and different lesion profiles. We know from other studies that 22C has a shorter incubation period than ME7 when injected peripherally and is able to "overtake" ME7, even though it has been injected at lower dose. Cloning of the isolate demonstrably removed 22C from the mixture and subsequently the properties were indistinguishable from ME7 from other sources, even using the i.p. route.

6 Stability of Scrapie Strains: Mixtures and Mutations

The incubation period characteristics, lesion profiles and other properties of any particular scrapie strain are stable on serial passage as long as the details of passage, such as host genotype and route and dose of infection, are kept constant. However, a subsequent change in these passage conditions may lead to a change in properties. Whenever such changes have occurred they have been consistent with the selection of a new strain with a shorter incubation period than the original strain under the new passage conditions. With an uncloned isolate, a possible explanation is simply the separation of strains which have been propagated as a mixture since the primary isolation in mice. However, there are now several known examples of such changes taking place in scrapie strains which have been cloned, a procedure which would remove any minor strains orginally present in the isolate. Under these circumstances the most likely explanation is mutation in the replicating informational molecule of the scrapie strain involved, followed by the selection of a mutant with a shorter incubation period under the particular passage conditions used. Because of the time involved in such studies the stability characteristics of only a few scrapie strains have been investigated in any detail. Even so, several different types of stability have been recognised.

The most stable strains in mice, ME7 and 22C, retain their properties irrespective of the *Sinc* genotype in which they are passaged. For example, ME7 has remained unchanged during 11 serial C57BL or 10 serial VM passages (Bruce and Dickinson 1979). Either the informational molecules of these scrapie strains are remarkably stable during repeated replication cycles or any mutants which do arise are at a selective disadvantage in the mice currently used for strain discrimination. In the case of 22C the second alternative appears to be true, as a new strain has been selected by passaging the cloned strain several times through hamsters and then back to mice (Kimberlin et al. 1989). On the other hand the properties of ME7 in mice have been completely unchanged by several passages through hamsters.

At the other extreme, 87A is unstable, even when passaged in the mouse genotype in which it was originally isolated. The properties of 87A remain stable indefinitely if passaged i.c. at low dose in C57BL mice, but often suddenly change when passaged at higher dose in the same mouse genotype (Bruce and Dickinson 1987). This invariably results in the isolation of the same new strain, which has much shorter incubation periods than 87A in all three *Sinc* genotypes and very different lesion profiles. The new strain is stable on passage, even using high doses, and is identical in all ways to ME7. 87A has been independently isolated from six natural scrapie sheep of different breeds and from different flocks; in each case it behaves in exactly the same way, giving rise to ME7. This change is seen even after 87A has been cloned, strongly suggesting that ME7 is a mutant strain derived from 87A.

A different type of instability is shown by 22A, which is perfectly stable when passaged in VM mice, the *Sinc* genotype in which it was originally isolated (BRUCE and DICKINSON 1979; DICKINSON et al. 1984), or in hamsters (KIMBERLIN et al. 1989). However, if 22A which has been cloned in VM mice is then serially passaged in C57BL mice, the incubation period characteristics and lesion profiles gradually change, eventually stabilising after four or five passages to give a new strain, 22F (BRUCE and DICKINSON 1979; DICKINSON et al. 1984). 22F is quite distinct from 22C, the strain obtained by C57BL passage from the same primary source as 22A. Many different C57BL-passaged lines of 22A have been studied and, although the number of passages required to regain stability has varied considerably, the resulting strain has always been 22F. These results have been interpreted in terms of the gradual selection of shorter incubation mutant strains.

Another striking change in properties has been observed with the "Chandler" isolate, derived from the "drowsy" goat source and used widely in scrapie research (DICKINSON and OUTRAM 1983; DICKINSON et al. 1984). 139A, the major strain from the uncloned main "Chandler" line, has had stable properties on passage at high dose in both C57BL and VM mice and also during many tissue culture passages. However, under certain circumstances C57BL-passaged 139A has suddenly changed, giving another strain, 79A. This has occurred once during a cloning passage, even though 139A has survived cloning on other occasions, and in two out of four experiments when uncloned 139A inoculum was boiled. These results suggest that the "Chandler" isolate contains both 139A and 79A, but is not clear whether 79A was already present during the earliest mouse passages or whether it was derived later by mutation from 139A. Whatever the explanation, the practical implication is that the "Chandler" isolate used in different laboratories is not uniform, and this could lead to discrepancies when comparing results. The "Chandler"-derived isolate used in at least one of the laboratories in the United States is in fact closer in its properties to 79A than to 139A (CARLSON et al. 1986). The other problem is that an unrecognised mixture of strains can lead to confusing incubation period results. Therefore studies using the uncloned "Chandler" isolate should be interpreted with extreme caution.

7 Interspecies Transmission: The "Species Barrier"

Because of the possible epidemiological implications for scrapie and related diseases, it is important to undertand the events which take place on the transmission of scrapie from one species to another. This has become particularly urgent in Britain, where the inclusion of scrapie-contaminated rendered offal in high-protein feed supplements has led to a catastrophic outbreak of a scrapie-like disease in cattle (WILESMITH et al. 1988). As already noted, the incubation period at the first passage of scrapie from its natural host into an experimental species is usually very long compared with later passages, and there are often survivors. Similarly, comparatively long incubation periods are often seen on interspecies transmission of experimentally passaged scrapie, a phenonmenon referred to as the "species barrier" effect (DICKINSON 1976; KIMBERLIN 1979). It is important to realise that the species barrier is highly variable among different scrapie isolates, even for the same change of passage species. At opposite ends of the spectrum, the 431K strain has been passaged from hamsters to *Sinc*^{s7} mice with only a minimal species barrier effect, whereas the hamster-passaged 263K strain is difficult to transmit to mice, producing extremely long incubation periods (KIMBERLIN and WALKER 1978).

The basis of the species barrier when it occurs is by no means fully understood, but studies on the passage of cloned strains among mice, hamsters and rats (KIMBERLIN et al. 1987, 1989) have shown that it is likely to depend on two major factors. The first of these is the selection of strains in the new host species, other than the major ones present in the original host. This leads to a permanent change in properties when the isolate is repassaged in the original species. The second is the effect that some foreign host tissue components have on the efficiency of infection or on the pathogenesis of the disease when they are associated with infective agent. This results in a temporary change in properties which is only seen at the first passage in a new species and is unrelated to any permanent changes in the scrapie strain.

8 Implications of Scrapie Strain Variation

We have shown that there are many distinct strains of scrapie which have stable, reproducible properties over many serial mouse passages, that scrapie strains retain their identity when copassaged as mixtures with other strains and that certain strains give rise to mutants which can be selected by changing the conditions of passage. As numerous different scrapie strains have been isolated by passaging in the same single mouse genotype, the specificity for strain variation must reside in the agent, not the host—in other words, scrapie strains carry their own information which is independent of the host. The nature of the molecule which specifies this information is still a matter for speculation.

In recent years scrapie research has centred on a host-coded protein, PrP (BOLTON et al. 1982; MCKINLEY et al. 1983). Relatively protease-resistant forms of PrP accumulate in the brain during scrapie pathogenesis and aggregate into scrapie-associated fibrils (SAF) and amyloid (MERZ et al. 1981; DIRINGER et al. 1983; HOPE et al. 1986; DEARMOND et al. 1987; MCBRIDE et al. 1988; BRUCE et al.

1989). Infectivity in tissue extracts copurifies with SAF, at least to some extent (McKinley et al. 1983; Diringer et al. 1983; Somerville et al. 1986), but no scrapie-specific nucleic acids have yet been identified in these fractions (Oesch et al. 1988). The failure so far to detect scrapie-specific nucleic acids and the resistance of scrapie to treatments which would usually inactivate nucleic acids have prompted the "prion hypothesis"—that the infectious agent is a replicating protein, PrP, devoid of nucleic acid (Prusiner 1982). The crucial challenge for the prion hypothesis is to account for the existence of many distinct strains of scrapie. With ingenuity it is possible to construct theoretical models in which proteins replicate and code for some form of strain specificity (Bolton and Bendheim 1988; Wills 1989). However, there is absolutely no direct experimental evidence for replicating proteins, nor have any specific differences been detected in PrP from the same host species and genotype infected with different strains of scrapie (Somerville and Ritchie 1986). We prefer the simpler and more conventional explanation, that the informational molecule is an as yet undetected scrapie-specific nucleic acid which is protected by its close association with host tissue components, the "virino hypothesis" of Dickinson and Outram 1988; PrP (or modified PrP) could well be the host component in such a model.

There is mounting evidence that PrP is involved in the host's control of pathogenesis, as well as accumulating pathologically in the course of the disease. Several studies have indicated a close genetic linkage in mice between the gene coding for PrP and the *Sinc* gene, which controls incubation period; there are also consistent amino acid differences between PrPs from $Sinc^{s7}$ and $Sinc^{p7}$ mice (Carlson et al. 1986; Westaway et al. 1987; Hunter et al. 1987). This has raised the possibility that PrP is, in fact, the *Sinc* gene product. Further evidence that PrP (or its gene) controls incubation period has come from a recent study in which transgenic mice carrying multiple copies of the hamster PrP gene had hamster-like incubation periods when infected with a hamster-passaged scrapie isolate (Scott et al. 1989).

It is clear that scrapie strain and host genotype do not act independently in determining the timing of events in pathogenesis—rather, the incubation period depends on specific and precise interactions between the two (Fig. 1). It is reasonable to suggest that, in the course of this interaction, the *Sinc* gene product (probably PrP) actually binds the scrapie informational molecule, perhaps acting as some type of receptor. We can speculate further that this interaction interferes with the normal turnover of the protein, leading to its accumulation. We have shown that scrapie strains also show characteristic tropisms within the brain, targeting changes precisely to particular groups of neurones. The recognition and selection of different target cells by different scrapie strains is one of the most intriguing and least understood aspects of scrapie pathogenesis.

References

Bolton DC, Bendheim PE (1988) A modified host protein model of scrapie. Ciba Found Symp 135: 164–181

Bolton DC, McKinley MP, Prusiner SB (1982) Identification of a protein that purifies with the scrapie agent. Science 218: 1309–1311

Bruce ME, Dickinson AG (1979) Biological stability of different classes of scrapie agent. In: Prusiner SJ, Hadlow WJ (eds) Slow transmissible diseases of the nervous system, vol 2. Academic, New York, pp 71–86

Bruce ME, Dickinson AG (1987) Biological evidence that scrapie agent has an independent genome. J Gen Virol 68: 79–89

Bruce ME, Dickinson AG, Fraser H (1976) Cerebral amyloidosis in scrapie in the mouse: effect of agent strain and mouse genotype. Neuropathol Appl Neurobiol 2: 471–478

Bruce ME, McBride PA, Farquhar CF (1989) Precise targeting of the pathology of the sialoglyoprotein, PrP, and vacuolar degeneration in mouse scrapie. Neurosci Lett 102: 1–6

Bruce ME, McConnell I, Fraser H, Dickinson AG (1991) The disease characteristics of different strains of scrapie in *Sinc* congenic mouse lines: implications for the nature of the agent and host control of pathogenesis. J Gen Virol 72: 595–603

Carlson GA, Kingsbury DT, Goodman PA, Coleman S, Marshall ST, DeArmond S, Westaway D, Prusiner SB (1986) Linkage of prion protein and scrapie incubation time genes. Cell 46: 503–511

Carp RI, Callahan SM, Sersen EA, Moretz RC (1984) Preclinical changes in weight of scrapie-infected mice as a function of scrapie agent-mouse strain combination. Intervirology 21: 61–69

DeArmond SJ, Mobley WC, DeMott DL, Barry RA, Beckstead JH, Prusiner SB (1987) Changes in the localization of brain prion proteins during scrapie infection. Neurology 37: 1271–1281

Dickinson AG (1976) Scrapie in sheep and goats. In: Kimberlin RH (ed) Slow virus diseases of animals and man. North-Holland, Amsterdam, pp 209–241

Dickinson AG, Fraser H (1977) Scrapie pathogenesis in inbred mice: an assessment of host control and response involving many strains of agent. In: Ter Meulen V, Katz M (eds) Slow virus infections of the central nervous system. Springer, Berlin Heidelberg New York, pp 3–14

Dickinson AG, Meikle VMH (1971) Host-genotype and agent effects in scrapie incubation: change in allelic interaction with different strains of agent. Mol Gen Genet 112: 73–79

Dickinson AG, Outram GW (1983) Operational limitations in the characterisation of the infective units of scrapie. In: Court LA, Cathala F (eds) Virus nonconventionnels et affections du systeme nerveux central, Masson, Paris, pp 3–16

Dickinson AG, Outram GW (1988) Genetic aspects of unconventional virus infections: the basis of the virino hypothesis. Ciba Found Symp 135: 63–83

Dickinson AG, Taylor DM (1978) Resistance of scrapie to decontamination. New Eng J Med 299: 1413–1414

Dickinson AG, Meikle VMH, Fraser H (1968) Identification of a gene which controls the incubation period of some strains of scrapie in mice. J Comp Pathol 78: 293–299

Dickinson AG, Bruce ME, Outram GW, Kimberlin RH (1984) Scrapie strain differences: the implications of stability and mutation. In: Tateishi J (ed) Proceedings of workshop on slow transmissible diseases. Japanese Ministry of Health and Welfare, Tokyo, pp 105–118

Diringer H, Gelderblom H, Hilmert H, Ozel M, Elderbluth C, Kimberlin RH (1983) Scrapie infectivity, fibrils and low molecular weight protein. Nature 306: 476–478

Foster JD, Dickinson AG (1988) The unusal properties of CH1641, a sheep-passaged isolate of scrapie. Vet Rec 123: 5–8

Foster JD, Fraser H, Bruce ME (1986) Retinopathy in mice with experimental scrapie. Neuropathol Appl Neurobiol 12: 185–196

Fraser H (1976) The pathology of natural and experimental scrapie. In: Kimberlin RH (ed) Slow virus diseases of animals and man. North-Holland, Amsterdam, pp 267–305

Fraser H, Dickinson AG (1968) The sequential development of the brain lesions of scrapie in three strains of mice. J Comp Pathol 78: 301–311

Fraser H, Dickinson AG (1973) Scrapie in mice: agent-strain differences in the distribution and intensity of grey matter vacuolation. J Comp Pathol 83: 29–40

Hope J, Morton LJD, Farquhar CF, Multhaup G, Beyreuther K, Kimberlin RH (1986) The major polypeptide of scrapie-associated fibrils (SAF) has the same size, charge distribution and N-terminal protein sequence as predicted for the normal brain protein. EMBO J 5: 2591–2597

Hunter N, Hope J, McConnell I, Dickinson AG (1987) Linkage of the scrapie-associated fibril protein (PrP) gene and *Sinc* using congenic mice and restriction fragment length polymorphism analysis. J Gen Virol 68: 2711–2716
Kimberlin RH (1979) Early events in the pathogenesis of scrapie in mice: biological and biochemical studies. In: Prusiner SB, Hadlow WJ (eds) Slow transmissible diseases of the nervous system, vol 2. Academic, New York, pp 33–54
Kimberlin RH, Walker CA (1978) Evidence that the transmission of one source of scrapie agent to hamsters involves separation of agent strains from a mixture. J Gen Virol 39: 487–496
Kimberlin RH, Walker CA, Millson GC, Taylor DM, Robertson PA, Tomlinson AH, Dickinson AG (1983) Disinfection studies with two strains of mouse-passaged scrapie agent, J Neurol Sci 59: 355–369
Kimberlin RH, Cole S, Walker CA (1986) Transmissible mink encephalopathy (TME) in Chinese hamsters: identification of two strains of TME and comparisons with scrapie. Neuropathol Appl Neurobiol 12: 197–206
Kimberlin RH, Cole S, Walker CA (1987) Temporary and permanent modifications to a single strain of mouse scrapie on transmission to rats and hamsters. J Gen Virol 68: 1875–1881
Kimberlin RH, Walker CA, Fraser H (1989) The genomic identity of different strains of mouse scrapie is expressed in hamsters and preserved on reisolation in mice. J Gen Virol 70: 2017–2025
Kitamoto T, Yi R, Mohri S, Tateishi J (1990) Cerebral amyloid in mice with Creutzfeldt–Jakob disease is influenced by the strain of the infectious agent. Brain Res 508: 165–167
McBride PA, Bruce ME, Fraser H (1988) Immunostaining of scrapie cerebral amyloid plaques with antisera raised to scrapie-associated fibrils (SAF). Neuropathol Appl Neurobiol 14: 325–336
McKinley MP, Bolton DC, Prusiner SB (1983) A protease-resistant protein is a structural component of the scrapie prion. Cell 35: 57–62
Merz PA, Somerville RA, Wisniewski HM, Iqbal K (1981) Abnormal fibrils from scrapie–infected brain. Acta Neuropathol 54:63–74
Mori S, Hamada C, Kumanishi T, Fukuhara N, Ichihashi Y, Ikuta F, Miyatake T, Tsubaki T (1989) A Creutzfeldt–Jakob disease agent (Echigo-1 strain) recovered from brain tissue showing the "panencephalophathic type" disease. Neurology 39: 1337–1342
Oesch B, Groth DF, Prusiner SB, Weissmann C (1988) Search for a scrapie-specific nucleic acid: a progress report. Ciba Found Symp 135: 209–233
Outram GW (1976) The pathogenesis of scrapie in mice. In: Kimberlin RH (ed) Slow virus diseases of animals and man. North-Holland, Amsterdam, pp 325–357
Pattison IH, Millson GC (1961) Scrapie produced experimentally in goats with special reference to the clinical syndrome. J Comp Pathol 71: 101–108
Prusiner SB (1982) Novel protemaceous infectious particles cause scrapie. Science 216: 136–144
Scott JR, Fraser H (1984) Degenerative hippocampal pathology in mice with scrapie. Acta Neuropathol 65: 62–68
Scott M, Foster D, Mirenda C, Serban D, Coufal F, Walchii M, Torchia M, Groth D, Carlson G, DeArmond SJ, Westaway D, Prusiner SB (1989) Transgenic mice expressing hamster prion protein produce species-specific scrapie infectivity and amyloid plaques. Cell 59: 847–857
Somerville RA, Ritchie LA (1986) Are scrapie associated fibrils a pathological product of infection? In: Court LA, Dormont D, Brown P, Kingsbury DT (eds) Unconventional virus diseases of the central nervous system. Commissariat à l'Énergie Atomique, Fontenay-aux-Roses, pp 521–535
Somerville RA, Merz PA, Carp RI (1986) Partial copurification of scrapie-associated fibrils and scrapie infectivity. Intervirology 25: 48–55
Westaway D, Goodman PA, Mirenda CA, McKinley MP, Carlson GA, Prusiner SB (1987) Distinct prion proteins in short and long scrapie incubation period mice. Cell 51: 651–662
Wilesmith JW, Wells GAH, Cranwell MP, Ryan JBM (1988) Bovine spongiform encephalopathy: epidemiological studies. Vet Rec 123: 638–644
Wills PR (1989) Induced frameshifting mechanism of replication for an information-carrying scrapie prion. Microb Pathogen 6: 235–249

Evidence for Biological and Structural Diversity Among Scrapie Strains*

R. J. KASCSAK, R. RUBENSTEIN, and R. I. CARP

1 Introduction

The uniqueness of individuals within any species is perpetuated through genetically programmed characteristics. This controlling influence on phenotypic traits is clearly important in multicellular and single celled organisms down to the simplest infectious agents, viruses and viroids. This phenomenon extends even to the infectious agents of scrapie and the other unconventional slow infections. Indeed, as documented by Bruce and Fraser in this volume, a variety of biological parameters are influenced by the inherent properties of scrapie agents independent of the host species. Additional information on the genetic interaction between host and scrapie agent as well as agents of the other unconventional slow infections has been detailed in several recent reviews (CARP et al. 1989b, c). With regard to those parameters under genetic control, our published work has concentrated on analyzing five of the mouse-adapted scrapie strains; ME7, 22A, 22L, 139A and 87V (CARP et al. 1984, 1985a, 1987; CARP and CALLAHAN 1986; KASCSAK et al. 1985, 1986, 1987; RUBENSTEIN et al. 1986; KIM et al. 1987a, b) and three hamster adapted strains, 263K, 139-H and 22C-H (KASCSAK et al. 1985, 1986; CARP et al. 1990).

It is worth emphasizing that when different laboratories have examined the same parameters in specific scrapie strain-host combinations, the findings have

* This work was sponsored in part by the New York State Office of Mental Retardation and Development Disabilities and by Public Health Service Grants NS21349 and NS25308 from the National Institutes of Health.
New York State Institute for Basic Research in Developmental Disabilities, Department of Virology, 1050 Forest Hill road, Staten Island, NY 10314, USA

been remarkably consistent. Our results with the five mouse-adapted strains noted above closely parallel and extend the findings of the Edinburgh group and of other workers. Biological parameters examined include incubation period (DICKINSON and MEIKLE 1969; BRUCE et al. 1976; DICKINSON and FRASER 1979; KINGSBURY et al. 1983), vacuolation patterns (FRASER 1979), concentration of amyloid deposits (FRASER and BRUCE 1973; BRUCE et al. 1976; BRUCE 1985) and induction of obesity (OUTRAM 1972; CARP et al. 1984, 1989a; KIM et al. 1987a, 1988).

The cause of biological differences among scrapie strains has not been resolved. There are several possible explanations of strain diversity which are based upon theories related to the nature of the agent. In order to set the stage for further discussion, the three theories, termed virus, prion and virino, will be discussed for relevance to scrapie strain diversity. These theories have been the subject of a number of recent reviews (CARP et al. 1985b, 1989c; PRUSINER 1987; PRUSINER et al. 1987; MANUELIDIS et al. 1988) and are discussed in depth by ROHWER and PRUSINER in this volume. The theories can be summarized as follows:

1. The virus theory states that scrapie agent is a virus with unconventional characteristics that account for the remarkable physical–chemical properties (ROHWER 1984; CZUB et al. 1986, 1988; MANUELIDIS et al. 1988).
2. The prion (PRUSINER 1982, 1987) or modified host protein (BOLTON and BENDHEIM 1988) theory postulates that the only component of the agent is the host-coded glycoprotein termed protease resistant protein (PrP; also known as prion protein).
3. The virino theory (DICKINSON and OUTRAM 1979; KIMBERLIN 1982) states that the agent embodies two components: (a) a host-derived protein coat, with PrP being one of the candidates for this protective protein, and (b) a small non-coding regulatory nucleic acid.

One explanation for the existence of scrapie strains, as with all known living organisms, is that differences in the sequence of nucleic acid of the infectious agent lead to distinctive traits. This explanation is consistent with the virus and virino theories of the nature of the agent. However, despite numerous attempts (WIETGREFE et al. 1985; BORRAS and GIBBS 1986; BORRAS et al. 1986; BELLINGER-KAWAHARA et al. 1987a, b; DUGUID et al. 1988) this putative nucleic acid has not been identified, and the results of several studies have suggested that such a molecule may not exist (PRUSINER 1982, 1987). This formed the basis of the prion theory, in which there are two additional mechanisms which have been postulated as the cause of scrapie strain diversity. One of these involves differences in post-translational modifications to PrP in which the distinctive modifications are self-perpetuating. This possibility will be discussed in Sect. 3 below. The other postulated mechanism is that differences in the coding sequence for the PrP molecule account for all of the known scrapie strains. If the coding sequence for PrP were the sole determinant of strain characteristics, differences in the nucleic acid sequences for PrP could yield strains but repeated passages in a single host or hosts with the same PrP coding sequence should eliminate differences and yield a single scrapie strain that would be specific for that host, i.e., for that particular PrP coding sequence. Examination of the characteristics of "strains" after different passage histories should provide information on the validity of this proposed mechanism.

2 Biological Parameters

In the present analysis of our work on biological parameters of scrapie strains, comparisons were made among "strains" that had been passaged at least four times in the same strain of mouse or hamster. Results from each of these three experiments show that strain differences are maintained after repeated passages in the same host.

In the first experiment, 10% brain homogenates were prepared from three Suffolk sheep sick with scrapie. On first passage, 10% homogenate was injected intracerebrally (i.c.) into C57BL/6J mice. The isolates were termed C600, C602, and C608 and their incubation periods on first passage were remarkably similar: 408, 396, and 416 days respectively. Subsequent passages were done in C57BL/6J mice with either 5% or 10% homogenates, except the passage just prior to that described in Table 1, in which 0.01% homogenates were injected. As shown in Table 1, incubation periods in C57BL/6J and IM/DK mice differed considerably for the three isolates. For example, the values for the C602 isolate were more than twice those for the C608 isolate, with values for the C600 isolate in an intermediate range. The C600 and C602 isolates caused an increase in weight in C57BL/6J mice (compared to mice injected with normal mouse brain homogenate) during the preclinical and early clinical phases of disease, whereas the C608 isolate did not.

Differences were seen (Table 1) with early passage isolates from a natural host for scrapie. In a second set of experiments, three strains that have long passage histories since their initial isolation (ME7, 22L, and 139A) were cloned by

Table 1. Incubation periods and weight changes in C57BL/6J mice injected with three mouse passaged isolates of sheep scrapie

Isolate	Passage no.	Incubation period in			Weight increase in C57BL/6J
		C57BL/6J 10%[a]	C57BL/6J 0.01%[a]	IM/Dk 1%[a]	
C600	4	169 ± 2	377 ± 20	258 ± 6	Yes
C602	5	244 ± 14	384 ± 8	> 450	Yes
C608	4	113 ± 0	170 ± 9	219 ± 3	No

[a]Percent homogenate used as inoculum

three limiting dilution passages in C57BL/6J mice (for ME7 and 22L) or in another s7s7 mouse strain, Compton White (for 139A) (KIMBERLIN et al. 1987, 1989). These cloned scrapie strains were then passaged at least three times in C57BL/6J mice and a large pool of brain homogenate produced. These homogenates were then used in experiments employing either routine i.c. injection or stereotaxic injection into various brain regions. The data summarized in Table 2 were derived from a series of studies in which four parameters were examined: (1) incubation periods (DICKINSON and MEIKLE 1969; DICKINSON et al. 1984; CARP and CALLAHAN 1986; KIM et al. 1987b), (2) histopathological changes (FRASER and DICKINSON 1973; BRUCE et al. 1976; FRASER 1979), (3) effects on body weight (OUTRAM 1972; CARP et al. 1984; KIM et al. 1987a, 1988) and (4) glucose tolerance (CARP et al. 1989a). As shown, there are marked differences among the three scrapie strains that are manifest even after a number of passages in the same mouse strain. Assertions have been made that the reported strain differences are based upon subjective

Table 2. Comparison of scrapie strains ME7, 22L and 139A

Characteristic	ME7	139A	22L
Incubation period in s7s7	122	106	110
Incubation period in p7p7 mice	300	160	200
Brain areas with shortest incubation period	Thalamus cerebellum	Caudate Nucleus, cerebellum substantia nigra and thalamus	Cerebellum
Brain vacuolation intensity after i.c. injection	Highest in anterior	Similar in various regions	Highest in posterior
Brain vacuolation after i.e. injection	Throughout brain	Throughout brain	Posterior areas only
White matter vacuolation in C57BL/6J mice	No	Yes	No
White matter vacuolation in IM/Dk mice	No	Yes	Yes
Plaque formation in p7p7 mice	Yes	No	No
Weight increase in SJL mice	Yes	No	No
Weight increase in CBA mice	Yes	No	No
Weight increase after hypothalamic injection in C57BL/6J mice	Yes	No	No
Altered glucose tolerance in SJL mice	Yes	No	Yes

assessments of criteria that are open to interpretation, e.g., the assessment of incubation period differences. These arguments cannot apply in those instances in which objective criteria have been employed (e.g., measurements of weight and glucose tolerance). Furthermore, it is unlikely that error due to subjective bias can play a role when differences are dramatic. For example, the incubation period for ME7 in IM/Dk mice extends well beyond the time when IM/Dk mice injected with 22L or 139A would have died. Other examples are those in which there are positive or negative findings with regard to a specific pathological change, such as the difference in vacuolation in forebrain regions between mice injected stereotaxically in the cerebellum (see Table 2). Even in those instances in which subjective criteria are used, such as in lesion profile analysis, many studies have been blinded to eliminate observer bias (FRASER and DICKINSON 1973; CARP et al. 1987; KIM et al. 1987b). In summary, the differences between the standard scrapie strains are real, repeatable and unambiguous.

The third protocol involved comparisons of three scrapie strains that had been passaged at least eight times in LVG/LAK hamsters. The 263K strain is used extensively in molecular biology studies of scrapie because it reaches extremely high titers and because it yields the shortest incubation periods of any known scrapie strain-host combination. In a series of studies, KIMBERLIN and colleagues took several well-characterized, mouse-passaged scrapie strains and transferred them to hamsters (KIMBERLIN et al. 1987, 1989). As per the phenomenon known as "species barrier," the incubation period for the initial passage in the new species was extremely long (KIMBERLIN et al. 1987). On subsequent passages the length of the incubation period decreased, and by the second or third passage, the incubation period stabilized and remained the same on all subsequent passages. We obtained several of these hamster-adapted strains from Dr. Richard Kimberlin and passaged them two more times in LVG/LAK hamsters. Two of these strains, 139-H and 22C-H, were then characterized and the results compared to those obtained with the 263K strain. As stated above, the 263K, 139-H, and 22C-H strains had been passaged a total of at least eight times in hamsters. Table 3 shows that there were marked differences between 263K and the other two strains. The incubation periods for 139-H and 22C-H were approximately twice as long as the incubation period for 263K. In addition, 139-H

Table 3. Comparison of the biological properties associated with scrapie strains 263K, 139-H and 22C-H in hamsters

Characteristic	263K	139-H	22C-H
Incubation period (in days)	65	130	125
Weight increase	No	Yes	Yes
Hypoglycemia-hyperinsulinemia	No	Yes	Yes
Enlargement of liver, kidney, endocrine organs	No	Yes	Yes
Pancreatic lesions	No	Yes	Yes

and 22C-H caused an increase in total body weight in the preclinical phase of disease, produced hyperinsulinemia-hypoglycemia and caused histopathological changes in a variety of visceral organs, particularly the pancreas (CARP et al. 1990). In contrast, 263K did not cause a weight increase, nor did it produce hyperinsulinemia-hypoglycemia, nor did it cause changes in visceral organs (CARP et al., unpublished work).

In summary, in both mice and hamsters repeated passages of scrapie strains in the same host did not eliminate differences between the strains. Distinctive differences were present in a variety of parameters, including incubation period, induction of increases in weight in the preclinical period, histopathological changes in the central nervous system (CNS), changes in insulin levels and changes in glucose tolerance. The fact that differences can be seen in scrapie strains that have been passaged in a single host strain strongly supports the concept that a component independent of the coding sequence for PrP is involved in specifying the characteristics of scrapie strains.

3 Structural Parameters

As discussed in the previous section, many scrapie strains have been identified which can be distinguished through specific biological parameters such as incubation period and histopathological lesion profile (FRASER 1979; BRUCE and DICKINSON 1979); also see Table 2). The phenotypic expression of disease is influenced both by the host genotype and by the specific scrapie strain. These distinct phenotypic properties among scrapie strains suggest the existence of other differences at the biochemical and molecular levels. Scrapie-associated fibrils (SAF) are specific markers of unconventional slow infectious diseases (MERZ et al. 1984). As such, this unique class of structures was examined for characteristics which might reflect the distinct biological properties of the scrapie strains with which they are associated. Indeed, it has been shown that SAF from various scrapie agent-host strain combinations can exhibit subtle differences in morphology (KASCSAK et al. 1985; MERZ et al. 1987), differences in sensitivity to proteinase K digestion (KASCSAK et al. 1985) and differences in the molecular weight profile of their PrP components (RUBENSTEIN et al. 1987). The major component of SAF is a host-coded glycoprotein termed Sp33–35 (BOLTON et al. 1987; BOLTON and BENDHEIM 1988) or PrP^{Sc} (BOLTON et al. 1982; MCKINLEY et al. 1983), which represents a modified form of the constitutive host precursor protein referred to as Cp33–35 (BENDHEIM et al. 1988) or PrP^{C} (OESCH et al. 1985). Sp33–35 differs from Cp33–35 in several characteristics: Sp33–35 aggregates to form fibrils (SAF) and is readily sedimentable under conditions where Cp33–35 remains soluble (MEYER et al. 1986); Cp33–35 is completely degraded by protease treatment, while Sp33–35 breaks down to lower molecular weight components known as PrPs (DIRINGER et al. 1983; OESCH et al. 1985). As

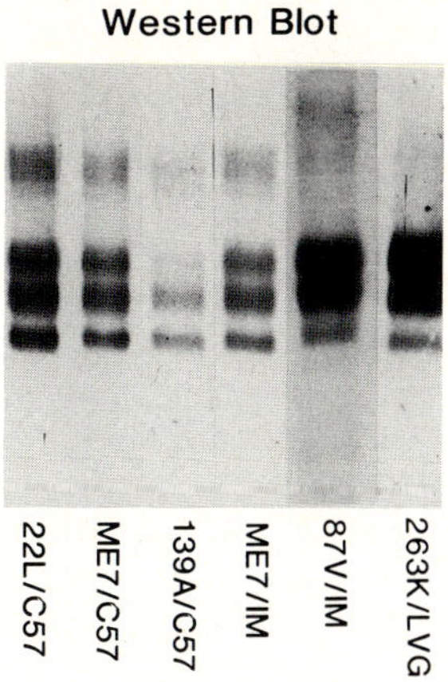

Fig. 1. Western blot analysis of PrPs associated with different scrapie strains. PrPs were isolated as previously described (Kascsak et al. 1986) from C57BL/6J mice infected with scrapie strains 22L, ME7 or 139A, IM/Dk mice infected with scrapie strains ME7 or 87V or Syrian LVG/LAK hamsters infected with scrapie agent 263K. Approximately 0.5 μg purified PrPs was electrophoresed on a 12% polyacrylamide gel and western blotted using rabbit antibody to mouse PrPs (Kascsak et al. 1986)

discussed previously, Sp33–35 is closely associated with infectivity and may indeed represent a major component of the infectious agent (McKinley et al. 1983; Merz et al. 1987). Subtle but unique differences have been identified in the western blot (WB) profiles of PrPs associated with SAF isolated from the brains of animals infected with different scrapie strains (Kascsak et al. 1986) (Fig. 1). PrPs associated with ME7 and 22L scrapie strains exhibited a profile consisting of three bands having equal staining intensity with molecular weights of 26–28 kDa, 23–24 kDa and 21–22 kDa. PrPs associated with scrapie strain 139A displayed a slightly different profile, with the 26–28 kDa band being less immunoreactive than the two lower molecular weight bands. PrPs associated with 87V from IM/Dk mice and with 263K from LVG/LAK hamsters presented a third type of WB profile, with the 26–28 kDa band intensely stained, the 23–24 kDa band moderately stained and a lightly stained, less diffuse band at 19 kDa. Preliminary results suggested that these profiles were independent, at least in part, of the host species or genotype. The profile for ME7 PrPs was identical in both C57BL/6J (s7s7) and IM/Dk (p7p7) mice. The profile for 87V PrPs from IM/Dk mice differed from that of ME7 PrPs in the same mouse strain but was very similar to that of 263K PrPs from Syrian LVG/LAK hamsters. These differences in WB profiles appear not to have been dependent upon the antisera employed, in that a variety of polyclonal and monoclonal antibodies in addition to affinity purified antibodies have yielded similar results.

Additional studies were conducted to further analyze the role of both host and scrapie strain in determining the characteristics of SAF and their component protein. The generation of various antibodies to PrP indicated the presence of species-specific epitopes which could be used to identify the origin of SAF components. Mouse antihamster PrP antisera recognizes an epitope present on hamster and human PrPs (Kascsak et al. 1987). Similarly, hamster anti-mouse PrP antisera recognizes an epitope present on mouse and rat PrPs that is absent from hamster and human PrPs. WB profiles generated using these antibodies

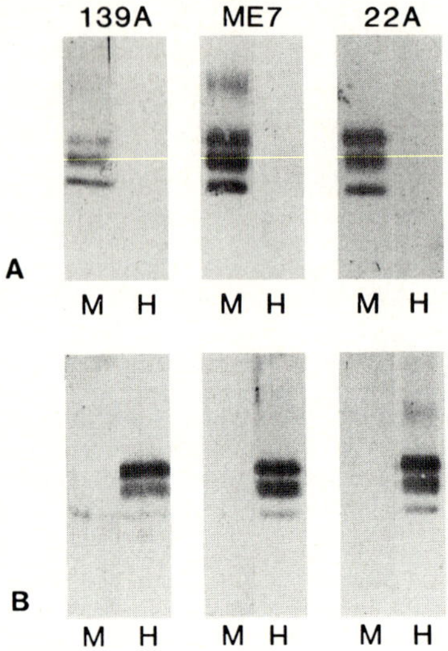

Fig. 2a, b. Comparison of PrPs associated with different scrapie strains passaged in mice or hamsters. PrPs were isolated from 139A- or ME7-infected C57BL/6J mice, from 22A-infected IM/Dk mice or from 139A-, ME7- or 22A-infected LVG/LAK hamsters as previously described (KASCSAK et al. 1986). PrPs were western blotted as described in Fig. 1 using either **a** hamster anti-mouse PrPs or **b** mouse anti- hamster PrPs as antisera. *M*, mice; *H*, hamsters

served both to identify species-directed epitopes present on PrPs and the pattern of PrPs associated with a specific host and agent strain combination.

Host influences on the properties of PrPs were investigated using individual scrapie strains passaged in a variety of host species. Such scrapie strains were provided by Dr. Richard Kimberlin and are described in publications by his group (KIMBERLIN et al. 1987, 1989). Scrapie strains 139A, ME7 and 22A were passaged in mice and hamsters under conditions described in the previous section. SAF were isolated from each agent-host strain combination and the PrPs associated with each combination were examined by WB analysis (Fig. 2). These studies indicate host influences on both the WB profile and the species-specific epitopes present on the PrPs. The WB profile associated with all three scrapie strains is clearly different in mice and hamsters. In addition, the species-directed epitopes present on the PrPs change depending upon the source of the protein, i.e., PrPs derived from mice react only with hamster anti-mouse antisera whereas PrPs derived from hamsters react only with the mouse anti-hamster antisera (Fig. 2). Since PrPs are derived from a host-coded precursor protein, the strong influence of the host on their antigenic and molecular weight characteristics was not surprising. This requisite change in the host component of SAF when agent is passaged from one host to another may well be responsible for what is commonly referred to as the "species barrier" phenomenon (see discussion above). This phenomenon could be explained by the PrP component of the infectious agent being less efficient in initiating infection in a host species different from the one from which it was derived. Recent studies employing transgenic mice (SCOTT et al. 1989) harboring the hamster PrP gene support the conclusion that the species barrier between mice and hamsters is maintained by differences in primary structure of the PrP molecules of the donor and recipient hosts.

While the host strain was shown to play a major role in determining PrP characteristics, additional specific influences of scrapie strains were investigated in another set of experiments. In these studes, the host genotype was held constant and infection was carried out using several different scrapie strains. The profiles of PrPs associated with infection by scrapie strains 263K, 139-H, ME7-H, and 22A-H were examined in Syrian LVG/LAK hamsters. These agents were chosen because of the variation in their incubation periods in LVG/LAK hamsters: approximately 65 days for 263K, 130 days for 139-H, 150 days for 22A-H and 260 days for ME7-H (KIMBERLIN and WALKER 1977; KIMBERLIN et al. 1989). A purification protocol for the isolation of PrPs from infected brain was employed which did not require the use of proteolytic enzymes (HOPE et al. 1988); this protocol permitted examination of the protease senstivity of PrPs associated with each scrapie strain. WB profiles and the protease senstivities of PrPs associated with infections with each of these strains were indistinguishable (results not shown). The WB profiles were identical to that shown for 263K PrPs in Fig. 1. Thus, despite variation in both incubation periods and the pathological changes

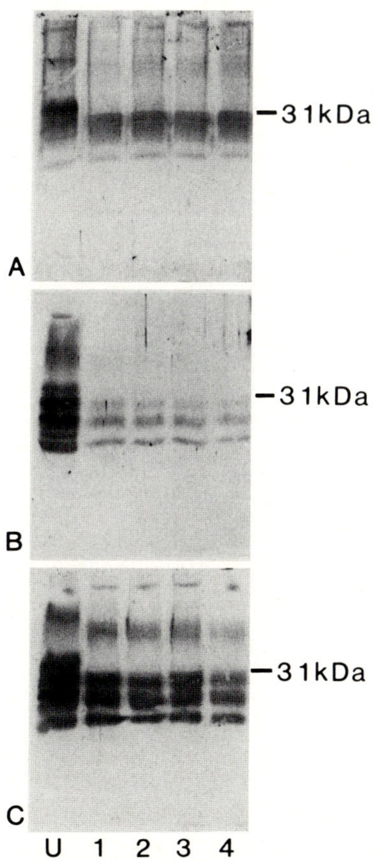

Fig. 3 a–c. Comparison of western blot profile and protease sensitivity of PrPs associated with three different scrapie strains in IM/Dk mice. PrPs were isolated by a modification of the procedure of HOPE et al. (1988) from IM/Dk mice infected with scrapie strains **a** 87V, **b** 139A or **c** ME7. Western blotting was carried out as described in Fig. 1. Samples were either untreated (*U*) or treated with proteinase K (50 μg/ml) for 1, 2, 3 or 4 h

(KIMBERLIN and WALKER 1977; KIMBERLIN et al. 1989; CARP et al. 1990) seen in these hamster models, the characteristics of PrPs from the four strains were identical.

A similar study was carried out in IM/Dk mice infected with scrapie strains 139A, 87V and ME7. In this instance, variations were evident in both WB profile and protease sensitivity of PrPs (Fig. 3). Following protease treatment, PrPs associated with each scrapie strain exhibited a distinct WB profile. While three PrP bands were associated with each scrapie strain, variation in band size, staining instensity and molecular weight resulted in distinctive pattern differences. PrPs associated with 139A were clearly more sensitive to protease degradation than PrPs associated with 87V or ME7. The use of a single inbred mouse strain suggests that these differences are a direct result of the individual scrapie strain. It should be reiterated that differences in WB profiles of PrPs associated with differet scrapie strains were also observed in C57BL/6J mice (Fig. 1), which are s7s7, whereas IM/Dk mice are p7p7. In fact, WB profiles of PrPs for a particular scrapie strain were identical for that scrapie strain in mice with different *Sinc* genotypes. Two amino acid substitutions have been identified in murine PrP which correlate with one or the other *Sinc* allele (WESTAWAY et al. 1987). These amino acid changes are at codon 108 (leucine in s7s7 mice, phenylalanine in p7p7 mice) and codon 189 (threonine in s7s7 mice, valine in p7p7 mice). In contrast to results reported by others (BOCKMAN and KINGSBURY 1988), our studies have failed to identify WB or protease sensitivity changes which correlate with these amino acid substitutions. Recent studies have indicated that differences in WB profile associated with specific PrPs are maintained when infectivity associated with purified SAF preparations are used to initiate infection. This adds further support to the role of intrinsic properties of each scrapie strain in determining specific PrP characteristics.

4 Conclusions

Our findings suggest that the basic precursor subunit of SAF, Sp33–35, is similar among SAF associated with different scrapie strains in mice, but that variations in the post-translational processing, conformation and/or assembly of Sp33–35 into aggregates or fibrils can lead to differential protease sensitivity and/or alteration in WB profile. These patterns may be a direct result of differences at the molecular level affecting PrP generation or may be an indirect consequence of physiological differences in disease parameters generated by different scrapie strains in the infected animal. Physiological differences may result from regional and cellular variation in agent distribution as well as changes in the cellular environment, such as pH or osmolarity, in which Sp33–35 is generated. Studies examining the various post-translational modifications known to occur on Sp33–35 including *N*-linked glycosylation (BOLTON et al. 1985; HARAGUCHI et al. 1989), phosphatidylinositol-glycolipid attachment to the C-terminal portion of the protein (STAHL et al. 1987), cleavage of signal peptide (HOPE et al. 1986), intramolecular disulfide bond (TURK et al. 1988) and a modified arg at codon 25 (HOPE et al. 1988) have failed to yield differences in modification between Sp33–35 and Cp33–35. Differences seen in WB profile must therefore involve other parameters. Although three distinct WB profiles of PrPs associated with three different scrapie strains have been observed in mice of the same inbred strain, such profiles are not able to distinguish among all scrapie strains. Even within inbred mouse strains, several scrapie strains (e.g. ME7 and 22L) which possess distinct biological properites (table 2) present WB profiles which are indistinguishable. Furthermore, only a single WB profile for PrPs has been observed in hamsters despite wide variations in the biological properties associated with infections by different scrapie strains (Table 3). Putative conformational alterations in PrP precursor proteins leading to differences in characteristics of PrP, may be limited in scope. In addition, current methodologies may not be capable of detecting more subtle changes. Our results, however, clearly indicate that certain characteristics of SAF and their component PrPs are a function of the specific scrapie strains with which they are associated. Such studies provide further evidence for the existence of distinct scrapie strains and represent additional parameters, other than biological properities, by which such strains can be distinguished.

Evidence has been presented for biological and structural diversity among scrapie strains. The distinct characteristics of these strains result from a direct interaction of host and agent but are independent of host genotype in that several different scrapie strains can be distinguished after repeated passage within a single host strain. While the host coding sequence of PrP appears to play a role in the species barrier phenomenon, it is clearly not the sole determinant of scrapie strain characteristics. The importance of PrP in the scrapie infectious process is well documented. Conversion of Cp33–35 to Sp33–35 is the hallmark of these diseases. Sp33–35 is closely associated with infectivity and appears to be a major component of the agent. Mutations in the primary coding sequence of PrP are related to disease susceptibility (GOLDGABER et al. 1989; HSIAO et al. 1989) and can influence incubation periods (CARLSON et al. 1986; HUNTER et al. 1987; WESTAWAY et al. 1987). However, it appears unlikely that the diversity among scrapie strains can be explained simply by modifications to PrP. A large number of different modifications and permutations would be required to permit the biological diversity seen among scrapie strains. As stated earlier, studies have failed to identify modification differences between Sp33–35 and Cp33–35. It is more likely that PrP serves to protect and target the scrapie informational molecule. This molecule would be unique to each scrapie strain and would regulate the metabolic and molecular events of that distinct biological process. The existence of scrapie strains is an established phenomenon; the strain-specific informational molecule remains to be identified.

Acknowledgements: We thank L. Olsen and J. Dekolf for the typing of this manuscript and R. Weed and M. Nascimento for photographic assistance.

References

Bellinger–Kawahara C, Diener TO, McKinley MP, Groth DF, Smith DR, Prusiner SB (1987a) Purified scrapie prions resist inactivation by procedures that hydrolyze, modify or shear nucleic acids. Virology 160: 271–274

Bellinger–Kawahara C, Cleaver JE, Diener TO, Prusiner SB (1987b) Purified scrapie prions resist inactivation by UV Irradiation. J Virol 61: 159–166

Bendheim PE, Potempska A, Kascsak RJ, Bolton DC (1988) Purification and partial characterization of the normal cellular homologue of the scrapie agent protein. J Infect Dis 158: 1198–1208

Bockman JM, Kingsbury DT (1988) Immunological analysis of host and agent effects on Creutzfeldt–Jakob disease and scrapie prion proteins. J Virol 62: 3120–3127

Bolten DC, Bendheim PE (1988) A modified host protein model of scrapie. Ciba Foud Symp 135: 164–177

Bolton DC, McKinley MP, Prusiner SB (1982) Identification of a protein that purifies with the scrapie prion. Science 218: 1309–1311

Bolton DC, Meyer RK, Prusiner SB (1985) Scrapie PrP 27-30 is a sialoglycoprotein. J Virol 53: 596–606

Bolton DC, Bendheim PE, Marmorstein AD, Potempska A (1987) Isolation and structural studies of the intact scrapie agent protein. Arch Biochem Biophys 258: 579–590

Borras T, Gibbs CJ Jr (1986) Molecular hybridization studies with scrapie brain nucleic acids I. Search for specific DNA sequences. Arch Virol 88: 67–78

Borras T, Merendino JJ Jr, Gibbs CJ Jr (1986) Molecular hybridization studies with scrapie brain nucleic acids II. Differential expression in scrapie hamster brain. Arch Virol 88: 79–90

Bruce ME (1985) Agent replication dynamics in a long incubation period model of mouse; scrapie. J Gen Virol 66: 2517–2522

Bruce ME, Dickinson AG (1979) Biological stability of different classes of scrapie agent. In: Prusiner SB, Hadlow WJ (eds) Slow transmissible diseases of the nervous system, vol 2, Academic, New York, pp 71–86

Bruce ME, Dickinson AG, Fraser H (1976) Cerebral amyloidosis in scrapie in the mouse: effect of agent strain and mouse genotype. Neuropathol Appl neurobiol 2: 471–478

Carlson GA, Kingsbury DT, Goodman PA, Coleman S, Marshall ST, DeArmond S, Westaway D, Prusiner S (1986) Linkage of prion protein and scrapie incubation time genes. Cell 46: 503–511

Carp RI, Callahan SM (1986) Scrapie incubation periods and end-point titers in mouse strains differeing at the H-2d locus. Intervirology 26: 85–92

Carp RI, Callahan SM, Sersen EA, Moretz RC (1984) Preclinical changes in weight of scrapie-infected mice as a function of scrapie agent-mouse strain combination. Intervirology 21: 61–69

Carp RI, Merz PA, Moretz RC, Somerville RA, Callahan SM, Wisniewski HM (1985a) Biological properties of scrapie: an uncoventional slow virus. In: Maramorosch K, McKelvey Jr JJ (eds) Subviral pathogens of plants and animals: viroids and prions. Academic, New York, pp 425–463

Carp RI, Merz PA, Kascsak RJ, Merz GS, Wisniewski HM (1985b) Nature of the scrapie agent: Current status of facts and hypotheses. J Gen Virol 66: 1357–1368

Carp RI, Moretz RC, Natelli M, Dickinson AG (1987) Genetic control of scrapie: incubation period and plaque formation in I mice. J Gen Virol 68: 401–407

Carp RI, Kim YS, Callahan SM (1989a) Scrapie-induced alterations in glucose tolerance in mice. J Gen Virol 70: 827–835

Carp RI, Kim YS, Kascsak RJ, Merz PA, Rubenstein R (1989b) Classic gentics of scrapie. In: Iqbal K, Wisniewski HM, Winblso B (eds) Alzheimer's disease and related disorders. Liss, New York, pp 567–582

Carp RI, Kascsak RJ, Wisniewski HM, Merz PA, Rubenstein R, Bendheim P, Bolton D (1989c) The nature of the unconventional slow infection agents remains a puzzle. Alzheimer Dis Assoc Disord 3 (1/2): 79–99

Carp RI, Kim YS, Callahan SM (1990) Pancreatic lesions and hypoglycemia-hyperinsulinemia in scrapie-injected hamsters. J Infect Dis 161: 462–466

Czub M, Braig HR, Diringer H (1986) Pathogenesis of scrapie: Study of the temporal development of clinical symptoms, of infectivity titres and scrapie associated fibrils in brains of hamsters infected intraperitoneally. J Gen Virol 67: 2005–2009

Czub M, Braig HR, Diringer H (1988) Replication of scrapie agent in hamsters infected intracerebrally confirms the pathogenesis of an amyloid-inducing virosis. J Gen Virol 69: 1753–1756

Dickinson AG, Fraser H (1979) An assessment of the genetics of scrapie in sheep and mice. In: Prusiner SB, Hadlow WJ (eds) Slow transmissible diseases of the nervous system, vol 1. Academic, New York, pp 367–406

Dickinson AG, Meikle VMH (1969) A comparison of some biological characteristics of the mouse-passaged scrapie agents, 22A and ME7. Genet Res 13: 213–225

Dickinson AG, Outram GW (1979) The scrapie replication-site hypothesis and its implication for pathogenesis. In: Prusiner SB, Hadlow WJ (eds) Slow transmissible diseases of the nervous system, vol 2. Academic, New York, pp 13–31

Dickinson AG, Bruce ME, Fraser H, Kimberlin RH (1984) Scrapie strain differences; the implication of stability and mutation. In: Tateishi J (ed) Proceedings of workshop on slow transmissible diseases, Japanese Ministry of Health and Welfare, Tokyo, p 105

Diringer H, Hilmert H, Simon D, Werner E, Ehlers B (1983) Towards purification of the scrapie agent. Eur J Biochem 134: 555–560

Duguid JR, Rohwer RG, Seed B (1988) Isolation of cDNAs of scrapie-modulated RNAs by subtractive hybridization of a cDNA library. Proc Natl Acad Sci USA 85: 5738–5742

Fraser H (1979) Neuropathology of scrapie: the precision of the lesions and their diversity. In: Prusiner SB, Hadlow WJ (eds) Slow transmissible diseases of the nervous system, vol 1. Academic, New York, pp 387–406

Fraser H, Bruce ME (1973) Argyrophilic plaques in mice inoculated with scrapie from particular sources. Lancet i: 617–618

Fraser H, Dickinson AG (1973) Scrapie in mice: agent-strain differences in the distribution and intensity of grey matter vacuolation J Comp Pathol 83: 29–40

Goldgaber D, Goldfarbh LG, Brown P, Asher DM, Brown WT, Len S, Teener JW, Feinstone SM, Rubenstien R, Kascsak RJ, Boellard JW, Gajdusek DC (1989) PrP gene mutations in familial Creutzfeldt–jakob disease and Gerstmann–Straussler syndme. J Exp Neurol 106: 204–206

Haraguchi T, Fisher S, Olofsson T, Endo D, Groth A, Tarentino A, Borchelt DR, Teplow D, Hood L, Burlingame A, Lycke E, Kobata A, Pursiner SB (1989) Asparagine-linked glycosylation of the scrapie and cellular prion proteins. Arch Biochem Biophys 274: 1–13

Hope J, Morton LJD, Farquhar CF, Multhaup G, Beyreuther K, Kimberlin RH (1986) The major polypeptide of scrapie-associated fibrils (SAF) has the same size, charge distribution and N-terminal protein sequence as predicted for the normal brain protein (PrP). EMBO 5: 2591–2597

Hope J, Multhaup G, Reekie LJD, Kimberlin RH, Beyreuther K (1988) Molecular pathology of scrapie associated fibril protein (PrP) in mouse brain affected by the ME7 strain of scrapie. Eur J Biochem 172: 271–277

Hsiao K Baker HF, Crow TJ, Poulter M, Owen F, Terwilliger JD, Westaway D, Ott J, Prusiner SB (1989) Linkage of a prion protein missense variant to Gerstmann-Straussler syndrome. Nature 338: 342–345

Hunter N, Hope J, McConnell I, Dickinson A (1987) Linkage of the scrapie-associated fibril protein (PrP) gene and *Sinc* using congenic mice and restriction fragment length polymorphism analysis. J Gen Virol 68: 2711–2715

Kascsak RJ, Rubenstein R, Merz PA, Carp RI, Wisniewski HM, Diringer H (1985) Biochemical differences among scrapie-associated fibrils support the biological diversity of scrapie agents. J Gen Virol 66: 1715–1722

Kascsak RJ, Rubenstein R, Merz PA, Carp RI, Robakis NK, Wisniewski HM, Diringer H (1986) Immunological comparison of scrapie associated fibrils isolated from animals infected with four different scrapie strains. J Virol 59: 676–683

Kascsak RJ, Rubenstein R, Merz PA, Tonna-DeMasi M, Fersko R, Carp RI, Wisniewski HM, Diringer H (1987) Mouse polyclonal and monoclonal antibody to SAF (PrP) protein. J Virol 61: 3688–3693

Kim YS, Carp RI, Callahan SM, Wisniewski HM (1987a) Scrapie-induced obesity in mice. J Infect Dis 156: 402–405

Kim YS, Carp RI, Callahan SM, Wisniewski HM (1987b) Clinical course of three scrapie strains in mice injected stereotaxically in different brain regions. J Gen Virol 68: 695–702

Kim YS, Carp RI, Callahan SM, Wisniewski HM (1988) Adrenal involvement in scrapie-induced obesity. Proc Soc Exp Biol Med 189: 21–27

Kimberlin RH (1982) Scrapie agent: prions or virinos? Nature 297: 107–108

Kimberlin RH, Walker CA (1977) Characteristics of a short incubation model of scrapie in the golden hamster. J Gen Virol 34: 295–304

Kimberlin RH, Cole S, Walker CA (1987) Temporary and permanent modifications to a single strain of mouse scrapie on transmission to rats and hamsters. J Gen Virol 68: 1875–1881

Kimberlin RH, Walker CA, Fraser H (1989) The genomic identity of different strains of mouse scrapie is expressed in hamsters and preserved on reisolation in mice. J Gen Virol 70: 2017–2025
Kingsbury DT, Kasper KC, Stites DP, Watson JD, Hogan RN, Prusiner SB (1983) Genetic control of scrapie and Creutzfeldt–Jakob disease in mice. J Immunol 131: 491–496
Manuelidis L, Murdoch G, Manuelidis EE (1988) Potential involvement of retroviral elements in human dementias. Ciba Found Symp 135: 117–129
McKinley MP, Bolton DC, Prusiner SB (1983) A protease resistant protein is a structural component of the scrapie prion. Cell 35: 57–62
Merz PA, Rohwer RG, Kascsak R, Wisniewski HM, Somerville RA, Gibbs CJ Jr, Gajdusek DC (1984) Infection-specific particle from the unconventional slow virus diseases. Science 225: 435–440
Merz PA, Kascsak RJ, Rubenstein R, Carp RI Wisniewski HM (1987) Antisera to scrapie-associated fibril protein and prion protein decorate scrapie-associated fibrils. J Virol 61: 42–49
Meyer RK, McKinley MP, Bowman KA, Braunfeld MB, Barry RA, Prusiner SB (1986) Separation and properties of cellular and scrapie prion proteins. Proc Natl Acad Sci USA 83: 2310–2314
Oesch B, Westaway D, Walchli M, McKinely MP, Kent SBH, Aebersold R, Barry RA, Tempst P, Teplow DB, Hood LE, Prusiner SB, Weissmann C (1985) A cellular gene encodes scrapie PrP 27–30 protein. Cell 40: 735–746
Outram GW (1972) Changes in drinking and feeding habits of mice with experimental scrapie. J Comp Pathol 82: 415–427
Prusiner SB (1982) Novel proteinaceous infectious particles cause scrapie. Science 216: 136–144
Prusiner SB (1987) Prions and neurodegenerative diseases. N Engl J Med 317: 1571–1598
Prusiner SB, Gabizon R, McKinley MP (1987) On the biology of prions. Acta neuropathol (Berl) 72: 299–314
Rohwer RG (1984) Scrapie infectious agent is virus-like in size and susceptibility to inactivation. Nature 308: 658–662
Rubenstein R, Kascsak RJ, Merz PA, Papini MC, Carp RI, Robakis NK, Wisniewski HM (1986) Detection of scrapie-associated fibril (SAF) proteins using anti-SAF antibody in non-purified tissue preparations. J Gen Virol 67: 671–681
Rubenstein R, Merz PA, Kascsak RJ, Carp RI, Scalici CL, Fama CL, Wisniewski HM (1987) Detection of scrapie associated fibrils (SAF) and SAF proteins from scrapie affected sheep. J Infect Dis 156: 36–42
Scott M, Foster D, Mirenda C, Serban D, Coufal F, Walchli M, Torchia M, Groth D, Carlson G, DeArmond SJ, Westaway D, Prusiner SB (1989) Transgenic mice expressing hamster prion protein produce species-specific scrapie infectivity and amyloid plaques. Cell 59: 847–857
Stahl N, Borchelt DR, Hsiao K, Prusiner SB (1987) Scrapie prion protein contains a phosphatidylinositol glycolipid. Cell 51: 229–240
Turk E, Teplow DB, Hood LE, Prusiner SB (1988) Purification and properties of the cellular and scrapie hamster prion proteins. Eur J Biochem 176: 21–30
Westaway D, Goodman PA, Mirenda CA, McKinley MP, Carlson GA, Prusiner SB (1987) Distinct prion proteins in short and long scrapie incubation period mice. Cell 51: 651–662
Wietgrefe S, Zupancic M, Haase A, Chesebro B, Race R, Frey W II, Rustan T, Friedman RL (1985) Cloning of a gene whose expression is increased in scrapie and in senile plaques in human brain. Science 230: 1177–1179

Genetics of Prion Diversity and Host Susceptibility*

G. A. CARLSON

1 Introduction

Although all prion diseases, including familial Creutzfeldt–Jakob disease (CJD) and Gerstmann–Sträussler syndrome (GSS), are transmissible by inoculation, several features distinguish them from conventional infectious diseases. Scrapie, CJD and GSS are afebrile and proceed in the apparent absence of an immune response against the agent. The scrapie agent was long known to be highly resistant to procedures that destroy or modify nucleic acids, while being susceptible to procedures that hydrolyse, modify or denature proteins. A fresh perspective on scrapie and related disorders was provided by PRUSINER (1982), who hypothesized that the scrapie agent was fundamentally different from other infectious particles and coined the term prion to distinguish the agent from viruses and viroids. Using biochemical procedures, including limited proteolysis, to enrich scrapie-infected hamster brain fraction for infectivity, a single protein with apparent molecular weight of 27000–30000 was discovered and designated PrP 27–30 (BOLTON et al. 1982). Reverse genetics was used to clone the structural gene encoding PrP 27–30 (OESCH et al. 1985); the gene was found to be chromosomal, rather than agent-encoded, and is present in all mammals that

* This work was supported by grants NS22785 and NS14609 (S.B. Prusiner, program director) from the National Institutes of Health
McLauglin Research Institute, 1625 3rd Avenue North, Great Falls, MT 59401, USA

Current Topics in Microbiology and Immunology, Vol. 172

have been tested (BASLER et al. 1986; WESTAWAY et al. 1989). PrP mRNA is expressed at similar levels in the brains of infected and uninfected animals, but scrapie-infected individuals have two isoforms of PrP compared with the single isoform in noninfected animals (OESCH et al. 1985). In normal animals, a proteinase-sensitive protein with apparent molecular weight of 33000–35000 is found and termed PrP^{C}. In addition to PrP^{C}, scrapie-infected animals have a prion disease-specific isoform of the protein, designated PrP^{Sc}, of the same apparent molecular weight as PrP^{C}; limited proteinase K digestion of PrP^{Sc} yields PrP 27–30. The chemical or structural differences between PrP^{C} and PrP^{Sc}, which most likely arise post-translationally, are unknown. As detailed in recent review articles (PRUSINER 1987, 1989; WESTAWAY et al. 1989), a large body of biochemical, pathological and immunological data indicates that PrP^{Sc} is a major and functional component of the infectious prion. It should be noted that although most of the major recent advances in our understanding of prion diseases and the role of PrP in the disease process came from the efforts of Prusiner and his colleagues, much of this work has been confirmed or extended independently (DIENER 1987). Taken together, the data support the concepts that (1) PrP^{Sc} is a necessary component of the prion and (2) PrP is required for prion replication. Genetic analysis of host-agent interaction reinforces these concepts.

2 Prion Protein Genes

The organization and structure of the hamster PrP gene has been elucidated and has important implications for the nature of prion diseases. The entire open reading frame (ORF) or protein-coding region is contained within a single exon (BASLER et al. 1986). Ironless mammalian coding sequences also have been described for the muscarinic, 5-hydoxytryptamine and β-adrenergic receptors and for a family of three potassium channel genes (CHANDY et al. 1990). Interestingly, the diversity of potassium channel and muscarinic acetylcholine receptor genes in *Drosophila* is generated by alternative splicing involving multiple exons. In contrast, the organization of the corresponding genes in vertebrates precludes alternative RNA splicing, implying a distinct mechanism for generation of diversity. Although hybridization of hamster PrP prones to *Drosophila* DNA has been observed (WESTAWAY and PRUSINER 1986), the existence of a PrP homologue in invertebrates has not been proven (WESTAWAY et al. 1989). Nonetheless, the organization of the mammalian PrP gene provides no possibility for alternative splicing within the ORF, adding to other evidence (BASLER et al. 1986; OESCH et al. 1985) that PrP^{C} and PrP^{Sc} have same amino acid sequence.

A 5′-untranslated (5′-UT) exon is located ~10 kb from the ORF-containing exon in the hamster. The structure of the mouse PrP gene is similar, though an additional 5′-UT exon has been identified in the I/LnJ mouse (WESTAWAY et al.

1991). The hamster PrP gene promoter shows multiple transcription initiation sites, lacks a TATA box, contains and AP-1 binding consensus sequence and is GC rich (BASLER et al. 1986; SALBAUM et al. 1988). These features are similar to those exhibited by promoters of so-called housekeeping genes, though this term does not imply that the gene is expressed constitutively. The highest PrP mRNA levels are found in brain, with the greatest expression in neurons (KRETZSCHMAR et al. 1986), and PrP mRNAs increase in the neonatal hamster at different rates in various regions of the brain (MCKINLEY et al. 1987). The promoters of PrP and of the precursor protein (APP) for the A_4 or β-amyloid deposited in Alzheimer's disease display many similarities, and synthesis of both mRNAs is increased by intraventricular injections of nerve growth factor (SALBAUM et al. 1988; MOBLEY et al. 1988). The frequencies of CpG versus GpC dinucleotides in the promoter regions of PrP and APP are also considerably higher than the average of mammalian DNA; CpG is the target for DNA methylation which is known to regulate gene expression (BIRD 1986). The nature of the PrP promoter and its regulation of gene expression may impact on interpretation of incubation time results from transgenic mice constructed with cosmid clones containing PrP with its own promoter.

PrP genes from humans, Syrian hamsters, Chinese hamsters, Armenian hamsters, mice, rats and sheep have been sequenced and all predict highly conserved proteins, suggesting an important physiological function (WESTAWAY et al. 1989). OESCH et al. (this volume) discuss the functional and evolutionary aspects of prion protein sequence in detail.

3 Prion Incubation Time and Susceptibility Genes

Genetic analysis of natural and experimental scrapie infections in sheep established that host genes exert a profound influence on disease susceptibility and course (DICKINSON and FRASER 1979; PARRY 1983). Following transmission of scrapie to mice (CHANDLER 1961), the mammal of choice for genetic analysis, DICKINSON and MACKAY (1964) identified a partially inbred stain of mice derived from random bred stocks at the Moredun Institute that had a greatly prolonged scrapie incubation period, the interval between inoculation and illness. Subsequently, this influence on scrapie incubation period was shown to be the effect of a single gene designated *Sinc* (DICKINSON et al. 1968). VM/Dk and the related IM/Dk mouse strains had the $Sinc^{p7}$ allele and after inoculation with the ME7 scrapie isolate had incubation periods of approximately 300 days; most other mouse strains had the $Sinc^{s7}$ allele and short incubation periods of 150 days or less. VM/Dk mouse strains, long the only known mice with the $Sinc^{p7}$ allele, saw only very limited use outside the scrapie labs in the UK, and *Sinc* remained unmapped for over 20 years. Many "strains" of scrapie agent were classified on the basis of their behavior in $Sinc^{s7}$ and $Sinc^{p7}$ mice (DICKINSON and MEIKLE 1971).

3.1 The Prion Protein Gene Appears to Control Scrapie Incubation Time in Mice

A survey of inbred and congenic mouse strains (KINGSBURY et al. 1983) identified I/LnJ mice as having an exceptionally long incubation period following inoculation with the Chandler scrapie isolate. I/LnJ mice have incubation times of 200–400 days, while most other mouse strains die within 160 days of inoculation (KINGSBURY et al. 1983; CARP et al. 1987; CARLSON et al. 1986). (NZW/LacJ × I/LnJ) F1 mice (incubation time = 223 ± 3 days) were backcrossed to NZW/LacJ mice (incubation time = 113 ± 2 days) and the offspring were inoculated with the Chandler scrapie isolate. Two groups, each composed of approximately half the mice, were evident, one with an incubation time of 130 ± 1 day ($n = 64$) and the other with 195 ± 2 days ($n = 66$); this result is compatible with single gene control of scrapie incubation period. Using a hamster PrP cDNA as a probe (OESCH et al. 1985), restriction fragment length polymorphisms (RFLPs) distinguishing the NZW (*Prn-p*^{a}) and I/Ln (*Prn-p*^{b}) alleles of the prion protein gene were detected, and analysis of DNA from the backcross animals indicated that the gene controlling scrapie incubation time was tightly linked to *Prn-p* (CARLSON et al. 1986). This prion incubation time gene was designated *Prn-i* because VM/Dk or IM/Dk mice were not available to test for allelism with *Sinc.* Through RFLP typing of *Sinc* congenic strains, *Sinc* was also found to be genetically linked to *Prn-p* (HUNTER et al. 1987).

Compelling evidence that *Prn-p* itself controls scrapie incubation time was provided by the nucleotide sequence of cDNA clones encompassing the entire ORFs of the NZW/LacJ and I/LnJ alleles (WESTAWAY et al. 1987). Two codons differed between the long and the short incubation time mice; in *Prn-p*^{b} I/LnJ mice the amino acid at codon 108 is changed from leucine to phenylalanine, and at codon 189 from threonine to valine. The transition at codon 189 created a diagnostic Bst*EII* site in the *b* allele of *Prn-p*. Although six haplotypes of *Prn-p* have been detected among the 60 inbred mouse strains tested based on RFLP analysis, only mice with the *Prn-p*^{b} allele differ from *Prn-p*^{a} mice at codons 108 and 189 as assessed with oligonucleotide probes and the *Bst*EII site polymorphism; no other sequence differences within the open reading frame of *Prn-p* have been detected (CARLSON et al. 1988). The seven inbred strains of mice (I/LnJ, BDP/J, P/J, JE/Le, IS/Cam, VM/Dk and IM/Dk) carrying the *Prn-p*^{b} allele differed from all other strains tested in having exceptionally long scrapie incubation times. However, several observations postponed the conclusion that control of scrapie incubation time was a pleiotropic effect of *Prn-p.*

The ancestry of the seven *Prn-p*^{b} mouse strains, together with the fact that they shared flanking restriction enzyme site polymorphisms, indicated that it was highly likely that all these strains derived from the I/Ln strain progenitor in the colony of Dr. L. Strong (CARLSON et al. 1988). In addition, during the backcrosses involved in transferring the I/Ln *Prn-p*^{b} allele onto a C57BL/6J background, segregation distortion and suppression of recombination in the *B2m–Prn-p* interval were observed (CARLSON et al. 1988); these mechanisms would favor

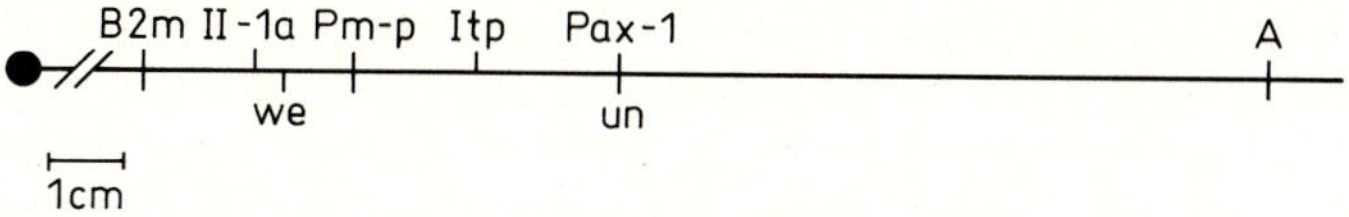

Fig. 1. Position of *Prn-p* on mouse chromosome 2 in relation to other loci. This map is derived from results compiled from several different experimental crosses (CARLSON et al. 1988 and unpublished results). With the exception of *we* and *un*, which were scored visually, all other alleles were typed by RFLPs in Southern blot analysis. The *un* mutation resulted from a point mutation in the *Pax-1* homebox gene, and RFLP typing allowed the discrimination of +/*un* and +/+ mice (BALLING et al. 1988)

linkage disequilibrium between *Prn-P* and neighboring loci. In backcross and F2 mice segregating for *Prn-p*^{a} and *Prn-p*^{b}, there were several animals discordant for incubation time phenotype and *Prn-p* genotype; if these animals represented true recombinants between *Prn-i* and *Prn-p* the two genes would be $\sim 4.8 \pm 1.3$ cM apart. The lethal nature of the scrapie bioassay precluded progeny testing, and it was possible that experimental error or cumulative effects of other genes segregating in the population were responsible for the putative recombinants.

Work with somatic cell hybrids localized the mouse *Prn-p* gene to Chr 2 and the homologous human gene, PRNP, to Chr 20pter-p12 (SPARKES et al. 1986). Mapping the mouse *Prn-p* locus in relation to other Chr 2 markers set the stage for experiments to capture *Prn-i*–*Prn-p* recombinant chromosomes should they exist. As shown in Fig. 1, *Prn-p* is located between the wellhärig (*we*- 0.9 cM proximal from *Prn-p*) and undulated (*un*- 3.9 cM distal) mutations; each of these recessive mutations is easily scored phenotypically. F1 hybrids between B6.I-*B2m*^{b} + *Prn-p*^{b} + and B10.UW (*B2m*^{a} *we Prn-p*^{a} *un*) were backcrossed to B10.UW and recombinants between *we* and *un* visually scored by separation of the wavy coat and kinky tail phenotypes. Recombinant mice were outcrossed and their

Table 1. Concordance of *Prn-p* genotype and incubation time in recombinants between *Prn-p* and flanking markers

Chromosome						Type	Number	Number of recombinant offspring	Incubation time
B2m	we		Prn		un*				
a	we		a		un	Parental	—	—	Short
b	+		b		+	Parental	—	—	Long
a	we		a	×	+	*Prn–un* recombinant	9	33	Short
b	+		b	×	un	*Prn–un* recombinant	7	19	Long
b	+	×	a		un	*we–un* recombinant	1	5	Short
a	we	×	b		+	*we–un* recombinant	3	11	Long

**un* genotype of recombinant offspring determined using Pax-1 RFLP (BALLING et al. 1988)

progeny inoculated with prions. As shown in Table 1, incubation time and *Prn-p*^{b} genotype have been completely concordant in 68 offspring carrying recombinant chromosomes from 4 *we–Prn-p* recombinant or 16 *prn-p–un* recombinant parents. These results, taken together with the results from transgenic mice to be discussed later, argue strongly that control of scrapie incubation time is a pleiotropic effect of *Prn-p*.

3.2 Linkage of Human Prion Disease to the PrP Gene

Hsiao and co-workers (1989) convincingly demonstrated linkage between the extremely rare familial disease, GSS, and PRNP. Gene sequencing in a GSS patient revealed two alleles, one with a mutation at codon 102 causing a transition from proline in the wild-type allele to leucine. This mutation creates a *DdeI* restriction site, and this plymorphism was used to demonstrate linkage between PRNP codon 102 (Leu) and disease in two families, one in the UK and the other in the USA. In contrast to the situation in inbred mice, where the distribution of the long prion incubation time allele is due to a founder effect, it is likely that the mutations in the two GSS families occurred independently. In addition to the lack of apparent genealogical relationship, a *PvuII* restriction site is present in the ORF of the GSS-linked allele of the American, nut not the British, family; this *PvuII* site is present in 90% of the general population (Wu et al. 1987). A variety of rare and nonconservative PrP sequence polymorphisms have since been described in other GSS and familial CJD pedigrees, increasing the likelihood that the PrP gene and the disease gene are allelic (Collinge et al. 1989; Goldgaber et al 1989; Owen et al. 1989). It is important to emphasize that GSS and familial CJD are transmissible to nonhuman primates (Baker et al. 1985).

In addition to genetic prion disease, there are also sporadic and infectious manifestations. Kuru is thought to have spread exclusively through a slow infectious mechanism dependent on ritualistic cannibalism (Gajdusek 1977), and some cases of CJD have been traced to iatrogenic inoculation with CJD contaminated material or instruments (Kitamoto et al. 1986; Brown 1988). However, the vast majority of CJD is sporadic, with an incidence of about 1 case per million people worldwide. If the disease were due to a conventional agent, this incidence would suggest either high prevalence of subclinical disease or an intermediate reservoir; neither has been found (Masters 1987). Although it is possible that sporadic disease results from prions that are endemic but have a very low efficiency in establishing an infection, a more intriguing, and testable, hypothesis is that infectious prions can arise through somatic mutation of the PRNP gene.

3.3 Mouse Strain and Species Barriers to Scrapie Transmission

Traditionally, the behaviors of scrapie isolates were interpreted solely in classical microbiological terms. For example, mice inoculated with sheep scrapie isolate have an extremely long incubation period, but in subsequent passages the incubation time rapidly shortens and becomes fixed—an illustration of the species barrier described by PATTISON (1965). Similarly, prions passaged through strains of mice with the *Prn-p*^{b} allele produce a shorter incubation period in *Prn-p*^{b} mice than do prions passaged through short incubation time *Prn-p*^{a} mice (CARLSON et al. 1989). BRUCE and DICKINSON (1979) had previously reported a mouse strain barrier between *Sinc*^{p7} VM and *Sinc*^{s7} C57BL mice to passage of some isolates of scrapie. A logical interpretation of these results is that the host exerts selective pressure on the agent, allowing outgrowth of resistant variants. However, given available data, an alternative explanation is equally viable.

Prn-p^{a} and *Prn-p*^{b} mice produce distinct allotypes of PrP, Prp-A and PrP-B, which differ in primary structure (WESTAWAY et al. 1987). To test whether distinguishable isolates are a transient effect of *Prn-p* genotype, the Chandler isolate of scrapie prions was serially passaged in NZW, Swiss, I/LnJ and (NZW × I/LnJ)F1 mice (CARLSON et al. 1989). A single passage through I/LnJ mice caused a dramatic reduction of the incubation time and its variance, with no further shortening in a second syngeneic passage. Conversely, this same inoculum obtained by a single passage through I/LnJ mice delayed disease onset and increased variability among NZW mice compared with inocula passaged through NZW or Swiss mice. Incubation times of *Prn-p* heterozygous mice, which express both the A and B allotypes, were constant regardless of the genotype of the previous host. Two lines of evidence implicate *Prn-p* in this phenomenon. First, Swiss and NZW mice behaved similarly; although both are *Prn-p*^{a}, they differ at many other loci. Second, in a NZW × I/Ln F2 cross high variance was only seen in mice homozygous for *Prn-p*^{b} after inoculation with prions containing Prp^{Sc}-A in spite of the fact that in this F2 cross many alleles for different genes were segregating. These results suggest that the different properties for NZW- and I/LnJ-derived prion isolates reflect an inbred mouse strain barrier, defined by the *Prn-p* genotype. Demonstration that the PrP gene also was responsible for the species barrier for transmission of scrapie came from transgenic mouse experiments.

4 Prion Protein Transgenes

Results from transgenic mice address several important issues in scrapie research (SCOTT et al. 1989). Four lines of transgenic mice were produced that harbored a cosmid insert containing the entire hamster PrP gene, including 24 kb of 5′ and 6 kb of 3′ flanking sequences. Most nontransgenic control mice inoculated with hamster prions failed to develop scrapie, and the incubation times for the few that became ill were greater than 600 days. In contrast, all four transgenic lines that expressed hamster PrP were susceptible to hamster prions, with each line having its own distinct incubation period, ranging from 47 ± 0.6 days for Tg7 to ~250 days for Tg69 (SCOTT et al. 1989; PRUSINER et al 1990). Prions produced by these transgenic mice were as infectious for hamsters as the original hamster isolate. These results indicate that the "species barrier" between mouse and hamster most probably can be explained by differences in the PrP gene sequence between the two species, which differ for 14 amino acids out of 254.

Another feature of scrapie in the transgenic mice can be attributed to the protein sequence encoded by the PrP transgene. The neuropathological features of hamster, as opposed to mouse, scrapie were evident in the transgenic mice, including the lesion profile and hamster PrP immunoreactive plaques that were morphologically identical to those in Syrian hamster scrapie (SCOTT et al. 1989). In our previous work with the Chandler mouse scrapie isolate, mice rarely had detectable plaques and consistently showed white matter involvement which was not produced in the transgenic mice inoculated with hamster prions (CARLSON et al. 1986; SCOTT et al. 1989). In earlier studies, BRUCE and co-workers (1976) came to the conclusion that the incidence of amyloid plaques in mouse scrapie is a function of both the scrapie isolate and the mouse genotype. The results from the transgenic mice indicate that the sequence of prion protein is an important factor in dictating the amyloidogenic response.

The source of the scrapie inocula determines whether the transgenic mice produce hamster or mouse infectious prions (PRUSINER et al. 1990). For example, Tg81 mice have incubation times of ~75 days with hamster prions and ~200 days with mouse prions. Brain extracts from these mice were bioassayed in both mice and hamsters, and the results showed that the prion inoculum determined which scrapie prion was produced, even though the transgenic mice were capable of producing both types of species-specific prion. Transgenic mice inoculated with hamster prions had $\sim 10^9$ ID_{50} units of hamster prions per gram of brain, while fewer than 1000 ID_{50} units of mouse prions were found. Conversely, transgenic mice inoculated with mouse prions had $\sim 10^6$ ID_{50} units of mouse prions and fewer than 100 units of hamster prions.

The quantity or expression of the transgene may also influence the course of disease, because the four lines derived from the same construct exhibit a spectrum of incubation times (SCOTT et al. 1989; PRUSINER et al. 1990). For hamster prions, incubation time was shorter in the transgenic lines with higher transgene copy number. All lines had the transgene integrated as a tandem array into a single autosomal site. The highest copy number (>50) was in Tg7 which had an incubation time for hamster prions of only 47 ± 0.6 days, compared with 75 ± 1.6 days for Tg81, which harbors >30 copies of the transgene. Tg69 and Tg71, which harbor 4–8 copies of hamster PrP, had incubation times of 250 and 172 days for hamster prions. By semi quantitative western blotting the amount of PrP expressed in brain is greatest in Tg7 followed Tg81 (SCOTT et al. 1989; PRUSINER et al. 1990), though the exact relationship between PrP gene number and protein expression remains to be determined.

Copy number also influenced the incubation time for mouse prions in the hamster PrP transgenic mice, though in the opposite direction (PRUSINER et al. 1990). Incubation times for the Chandler scrapie isolate ranged from 165 ± 5.6 days for low copy number Tg69 to 181 ± 3.2 days for high copy number Tg7. It is important to emphasize that transgene expression per se does not shorten scrapie incubation period, as shown by the prolongation of the incubation time for mouse prions in hamster PrP transgenic mice. In authentic hamsters the incubation time for hamster prions is ~75 days, while Tg69 and Tg71 have incubation times of ~250 and ~175 days in spite of having several copies of the hamster PrP transgene. Whether endogenous mouse PrP is responsible for this effect is unknown. Certainly other factors, such as position effects or insertional mutagenesis of endogenous genes, can account for variability among transgenic lines. Nonetheless, all available data point to the conclusion that the control of scrapie incubation time is a pleiotropic effect of PrP gene expression.

The impact of transgenic mice on scrapie research will be enormous. It is now feasible to determine which regions and amino acids of PrP are required for production of infectious scrapie agent, for modulation of susceptibility to disease and for control of scrapie incubation time. The use of PrP transgenes derived from GSS and familial CJD will allow determination of whether these mutations are sufficient to initiate disease. An unexpected benefit from the hamster PrP transgenic mice is the availability of a line with an incubation time of only 47 days, greatly reducing the expense of scrapie research and the time needed for many experiments.

5 Prion Genes?

Almost all available evidence suggests that prion diseases are disorders of PrP metabolism that can be initiated by inoculation of PrP^{Sc} or through mutation of the endogenous PrP gene. Militating against this idea, however, is the existence of "strains" of scrapie agent that show differences in incubation time, lesion profile, amyloidogenicity and clinical course in genetically identical hosts (DICKINSON et al. 1985; BRUCE et al. 1976; KIMBERLIN et al. 1987). An agent-specific nucleic acid genome has been invoked as the informational component of the prion accounting for these differences among scrapie isolates. Either directly or through encoded products a putative scrapie-specific nucleic acid must account for several PrP-related phenomena. First, it is extremely likely that PrP itself controls scrapie susceptibility and incubation time, which may imply direct interaction between the nucleic acid and PrP. Nucleic acid-PrP^{Sc} interaction is also demanded by copurification of infectivity and PrP^{Sc}. Second, a putative nucleic acid must somehow initiate the conversion of PrP^{C} to PrP^{Sc}; this conversion is disease-specific and not induced by other insults or neurodegenerative processes. Third, a scrapie-specific nucleic acid will also have to explain the distinct neuropathology produced by different isolates, including the presence or absence of PrP^{Sc} amyloid plaques. Because so many effects of the informational component of prions are mediated through PrP, it is prudent to consider the possibility of non-nucleic acid prion constituents rather than assuming that prion-specified behavior reflects "virulence genes" of a scrapie-specific nucleic acid. The prion protein does not exist in isolation, and its metabolism and function undoubtedly involve many factors. It is conceivable that cofactors bound to PrP or present in the crude inoculum (to our knowledge, comparison of distinct strains of purified prions has not been undertaken) skew the host's metabolic pathways to produce different manifestations of disease, thus accounting of prion-specified information. It is also conceivable that alternative conformations of PrP^{Sc} can specify information, somehow serving as a template for direct protein folding to produce more PrP^{Sc} with the same conformation. Indeed, it has been shown that PrP can assume two different forms when translated in vitro, presumably as a consequence of alternate pathways of protein folding (LOPEZ et al. 1990). Just as prions are novel and unprecedented infectious agents, the bichemical basis for "true-breeding" prion-specified behavior may also be novel. The mechanism by which the prion encodes its information is perhaps the most intriguing problem in scrapie research.

Acknowledgements. Thanks are due to Dr. Stanley B. Prusiner for providing the opportunity to work on this exciting and important problem and also to Drs. D. Westaway, M. McKinley, S. DeArmond, K. Hsiao, M. Scott, B. Oesch, N. Stahl, R. Gabizon, M. Torchia and other members of the UCSF scrapie group for collaborative studies, for sharing their diverse perspectives on prions and for helpful discussion.

References

Baker HF, Ridley RM, Crow TJ (1985) Experimental transmission of an autosomal dominant spongiform encephalopathy: does the infectious agent originate in the human genome? Br Med J 291: 299–302

Balling R, Deutsch U, Gruss P (1988) Undulated, a mutation affecting development of the mouse skeleton, has a point mutation in the paired box of Pax 1. Cell 55: 531–535

Basler K, Oesch B, Scott M, Westaway D, Wachli M, Groth DF, McKinley MP, Prusiner SB, Weissmann C (1986) Scrapie and cellular PrP isoforms are encoded by the same chromosomal gene. Cell 46: 417–428

Bird AP (1986) CpG-rich islands and the function of DNA methylation. Nature 321: 209–213

Bolton DC, McKinley MP, Prusiner SB (1982) identification of a protein that purifies with the scrapie prion. Science 218: 1309–1311

Brown P (1988) The decline and fall of Creutzfeldt–Jakob disease associated with human growth hormone therapy. Neurology 38: 1135–1137

Bruce ME, Dickinson AG, (1979) Biological stability of different classes of scrapie agent in. In: Prusiner SB, Hadlow WJ (eds) Slow transmissible diseases of the nervous system, vol 2. Academic, New York pp. 71–86

Bruce ME, Dickinson AG, Fraser H (1976) Cerebral amyloidosis in scrapie in the mouse: effect of agent strain and mouse genotype. Neuropathol Appl Neurobiol 2: 471–478

Carlson GA, Kingsbury DT, Goodman PA, Coleman S, Marshall ST, DeArmond S, Westaway D, Prusiner SB (1986) Linkage of prion protein and scrapie incubation time genes. Cell 46: 503–511

Carlson GA, Goodman PA, Lovett M, Taylor BA, Marshall ST, Peterson-Torchia ST, Westaway D, Prusiner SB (1988) Genetics and polymorphism of the mouse prion gene complex: control of scrapie incubation time. Mol Cell Biol 8: 5528–5540

Carlson GA, Westaway D, DeArmond SJ, Peterson-Torchia M, Prusiner SB (1989) Primary structure of prion protein may modify scrapie isolate properties. Proc Natl Acad Sci USA 86: 7475–7479

Carp RE, Moretz RC, Natelli M, Dickinson AG (1987) Genetic control of scrapie: incubation period and plaque formation in I mice. Virology 68: 1–7

Chandler RL (1961) Encephalopathy in the mice produced by inoculation with scrapie brain material. Lancet 1: 1378–1379

Chandy KG, Williams CB, Spencer RH, Aguilar DA, Ghanshani S, Tempel BL, Gutman GA (1990) A family of three mouse potassium channel genes with intronless coding regions. Science 247: 973–975

Collinge J, Harding AE, Owen F, Poulter M, Lofthouse R, Boughey AM, Shan T, Crow TJ (1989) Diagnosis of Gerstmann–Sträussler syndrome in familial demential with prion protein gene analysis. Lancet 2: 15–17

Dickinson AG, Fraser H (1979) An assessment of the genetics of scrapie in sheep and mice. In: Slow transmissible diseases of the nervous system, vol 1. Academic, New York, pp 367–385

Dickinson AG, MacKay JMK (1964) Genetical control of the incubation period in mice of the neurological disease, scrapie. Heredity (Edinburg) 19: 279–288

Dickinson AG, Meikle VMH (1971) Host–genotype and agent effects in scrapie incubation: change in allelic interaction with different strains of agent. Mol Gen Genet 112: 73–79

Dickinson AG, Meikle VMH, Fraser HG (1968) identification of a gene which controls the incubation period of some strains of scrapie agent in mice. J Comp Path 78: 293–299

Dickinson AG, Bruce ME, Outram GW, Kimberlin RH (1985) Scrapie strain differences: The implications of stability and mutaion. In: Tateishi J (ed) Proceedings of workshop on slow transmissible diseases. Japenese Ministry of Health and Welfare, Tokyo, pp 105–118

Diener TO (1987) PrP and the nature of the scrapie agent. Cell 49: 719–721

Gajdusek DC (1977) Unconventional viruses and the origin and disappearance of kuru. Science 197: 943–960

Goldgaber D, Goldfarb LG, Brown P, Asher DM, Brown WT, Lin S, Teener JW, Feinstone SM, Rubenstein R, Kascsak RJ, Boellaard JW, Gajdusek DC (1989) Mutations in familial Creutzfeldt–Jakob disease and Gerstmann–Sträussler–Scheinker's syndrome. Exp Neurol 106: 204–206

Hsiao K, Baker HF, Crow TJ, Poulter M, Owen F, Terwilliger JD, Westaway D, Ott J, Prusiner SB (1989) Linkage of a prion protein missense variant to Gerstmann–Sträussler syndrome. Nature 338: 342–345

Hunter N, Hope J, McConnelL I, Dickinson AG (1987) Linkage of the scrapie-associated fibril protein (PrP) gene and Sinc using congenic mice and restriction fragment length polymorphism analysis. J Gen Virol 68: 2711–2716

Kimberlin RH, Cole S, Walker CA (1987) Temporary and permanent modifications to a single strain of mouse scrapie on transmission to rats and hamsters. J Gen Virol 68: 1875–1881

Kingsbury DT, Kasper KC, Stites DP, Watson, JC, Hogan RN, Prusiner SB (1983) Genetic control of scrapie and Creutzfeldt–Jakob disease in mice. J Immunol 131: 491–496

Kitamoto T, Tateishi J, Tashima T, Takeshita I, Barry RA, DeArmond SJ, Prusiner SB (1986) Amyloid plaques in Creutzfeldt–Jakob disease stain with prion protein antibodies. Ann Neurol 20: 204–208

Kretzschmar HA, Prusiner SB, Stowring LE, DeArmond SJ (1986) Scrapie prion proteins are synthesized in neurons. Am J Pathol 122: 1–5

Lopez CD, Yost CS, Prusiner SB, Myers RM, Lingappa VR (1990) Unusual topogenic sequence directs prion protein biogenesis. Science 248: 226–229

Masters CL (1987) Epidemiology of Creutzfeldt–Jakob disease: studies on the natural mechansims of transmission. In: Prusiner SB, McKinley SB (eds) Prions: novel infections pathogens causing scrapie and Creutzfeldt–Jakob disease. Academic, New York, pp 511–522

McKinley MP, Hay B, Lingappa VR, Lieberburg I, Prusiner SB (1987) Developmental expression of prion protein gene in brain. Dev Biol 121: 105–110

Mobley WC, Nee RL, Prusiner SB, McKinley MP (1988) Nerve growth factor increases mRNA levels for the prion protein and the beta-amyloid protein precursor in developing hamster brain. Proc Natl Acad Sci USA 85: 9814–9815

Oesch B, Westaway D, Wächli M, McKinley MP, Kent SBH, Aebersold R, Barry RA, Tempst P, Teplow DB, Hood L, Prusiner SB, Weissmann C (1985) A cellular gene encodes scrapie PrP 27–30 protein. Cell 40: 735–746

Owen F, Poulter M, Lofthouse R, Collinge J, Crow TJ, Risby D, Baker HF, Ridley RM, Hsiao K, Prusiner SB (1989) Insertion in prion protein gene in familial Creutzfeldt–Jakob disease. Lancet 1: 51–52

Parry HB (1983) Scrapie disease in sheep. Academic, New York

Pattison IH (1965) Experiments with scrapie with special reference to the nature of the agent and the pathology of the disease. In: Gajdusek DC, Gibbs CJ, Alpers M (eds) Slow, latent and temperate virus infections, NINDB monograph no 2. US Government Printing Office, Washington, DC, pp 249–257

Prusiner SB (1982) Novel proteinaceous infectious particles cause scrapie. Science 216: 136–144

Prusiner SB (1987) Prions and neurodegenerative diseases. N Engl J Med 317: 1571–1581

Prusiner SB (1989) Scrapie prions. Annu Rev Microbiol 43: 345–374

Prusiner SB, Scott M, Foster D, Pan K-M, Groth D, Mirenda C, Torchia M, Yang S-L, Serban D, Carlson GA, Hoppe PC, Westaway D,' DeArmond SJ (1990) Transgenetic studies implicate interactions between homologous PrP isoforms in scrapie prion replication. Cell 63: 673–686

Salbaum JM, Weidemann A, Lemaire H-G, Masters CL, Beyreuther K (1988) The promoter of alzheimer's disease amyloid A4 precursor gene. EMBO J 7: 2807–2813

Scott M, Foster D, Mirenda C, Serban D, Coufal F, Wächli M, Torchia M, Groth D, Carlson G, DeArmond SJ, Westaway D, Prusiner SB (1989) Transgenic mice expressing hamster prion protein produce specific scrapie infectivity and amyloid plaques. Cell 59: 847–857

Sparkes RF, Simon M, Cohn VH, Fournier REK, Lem J, Klisak I, Heinzmann C, Blatt C, Lucero M, Mohandas T, DeArmond SJ, Westaway D, Prusiner SB, Weiner LP (1986) Assignment of the human and mouse prion protein genes to homologous chromosomes. Proc Natl Acad Sci USA 83: 7358–7362

Westaway D, Prusiner SB (1986) Conservation of the cellular gene encoding the scrapie prion protein. Numcleic Acid Res 14: 2035–2044

Westaway D, Goodman PA, Mirenda CA, McKinley MP, Carlson GA, Prusiner SB (1987) Distinct prion proteins in short and long scrapie incubation period mice. Cell 51: 651–662

Westaway D, Carlson GA, Prusiner SB (1989) Unravelling prion diseases through molecular genetics. Trends Neurosci 12: 221–227

Westaway D, Mirenda CA, Foster D, Zebarjadian Y, Scott M, Torchia M, Yang S-L, Serban H, DeArmond SJ, Ebeling C, Prusiner SB, Carlson GA (1991) Paradoxical shortening of scrapie incubation times by expression of prion protein transgenes derived from long incubation period mice. Neuron, in press

Wu Y, Robakis NK, Dobkins C, Devine-Gage E, Merz P, Wisniewski HM (1987) A PvuII RFLP detected in the human prion protein gene. Nucleic Acid Res 15: 3191

Natural Transmission and Genetic Control of Susceptibility of Sheep to Scrapie

N. HUNTER

1 Introduction

Scrapie has often been the subject of controversy and speculation both by farmers and by scientists. Not the least strange of numerous etiologic ideas was that put forward by ROCHE–LUBIN in 1848 who suggested that scrapie was caused by sexual excess, sunstroke after shearing, or lightning. The recent suggestion that kuru, a human scrapie-like disease, was the result of a volcanic eruption (SCHOENTAL 1990) is therefore part of a long tradition. The lightning analogy is not a bad one, however. One of the most common remarks made by farmers having to deal with a scrapie outbreak is that it appeared suddenly and apparently from nowhere; reports like this are still being made from, for instance, Sweden (ELVANDER et al. 1988), Cyprus (TOUMAZOS 1988) and Italy (MECHELLI and MANTOVANI 1988). Another common belief is that the disease was "brought in" by a recently purchased ram. Such anecdotal evidence is difficult to substantiate, especially as scrapie has such stigma attached to it that true incidence figures have been difficult to obtain. It is probably a much more common disease than would seem from official diagnosis figures (CHATELAIN et al. 1983; N. HUNTER 1991). In addition, the lack of a preclinical dignostic test means that

Institute for Animal Health, AFRC/MRC Neuropathogenesis Unit, Ogston Building, West Mains Road, Edinburgh EH9 3JF, UK animals incubating scrapie cannot be identified until they develop symptoms. The number of carrier-status animals, if indeed they exist at all, is completely unknown, despite their obvious importance in any hypothesis of natural transmission mechanisms. In this chapter the evidence for transmission and contagion in the field will be reviewed; both old and recent studies on genetics of differences in susceptibility to experimentally induced scrapie will be used to shed light on the natural disease.

2 Natural Transmission

The pathogenesis of scrapie in the experimental mouse models is very predictable over the long time courses involved if eight factors (DICKINSON 1975) are controlled. These are (in the host) the genotype and sex, the route of injection and the age at injection and (in the inoculum) the strain and dose of agent and the genotype and organ source of the inoculum. Studies of natural transmission among sheep in field conditions do not have the luxury of such control and should thus be interpreted with caution.

The average age of sheep at appearance of scrapie symptoms is 3.5 years (STAMP 1962; DICKINSON et al. 1964) with most animals affected aged between 2.5 and 4.5 years. Scrapie sheep younger than 1.5 years are extremely rare but have been reported in North India (ZLOTNIK and KATIYAR 1961), in France (JOUBERT et al. 1972) and in the Suffolk breed (FOSTER and DICKINSON 1989). The disease appears to run in families. Whether or not a lamb contracts scrapie depends more on the present or future scrapie status of its mother than on anything else. These findings may be connected, as DICKINSON et al. (1966) pointed out: an infectious disease whose normal route of transmission is maternal (ewe to lamb) would be likely to die out in the field if it affected young ewes before breeding age. A natural selection might therefore operate to make most field strains of the disease those which take longer to incubate.

The familial appearance of scrapie can be interpreted in different ways, one of these being that it is purely a genetic disease. A primary advocate of this view was PARRY (1960), who believed that scrapie was caused by a defective and recessive gene and invoked the term provirus (first used by Darlington in the 1940s) to account for the disease being both genetic and transmissible (PARRY 1962). His work has been elegantly and convincingly disputed by DICKINSON et al. (1965), amongst others, but the major problem with Parry's view for a number of people has been the undoubted existence of contagious spread of scrapie between unrelated animals (BROTHERSTON et al. 1968; HOURRIGAN et al. 1979). The "genetic and transmissible" idea has also been put forward to explain the human diseases Creutzfeldt-Jakob disease and Gerstmann–Sträussler syndrome (HSIAO et al. 1989; OWEN et al. 1990).

Familial incidence can also be the result of inherited susceptibility to an infectious and transmissible agent (ASHER et al. 1983). If this were true for scrapie, it would join a growing list of diseases that blur the traditional distinction between infectious and genetic diseases. In an environment where exposure to the infectious agent is both high and uniform, genetic factors in the host influencing the process of infection and pathogenesis become so important that the disease itself can appear to be genetic in origin. Such a scenario has been postulated for AIDS in high-incidence populations (DIAMOND 1987).

2.1 Maternal and Contagious Spread

In the study of natural transmission, it is extremely difficult to ensure that the test animals come from scrapie-free environments, are not already harbouring infection but are at the same time susceptible to disease. Observation is necessary over extremely long timescales to be certain of picking up all possible cases. Two large-scale studies go a long way towards fulfilling the exacting controls required and demonstrate that both maternal and contagious transmission can occur. In one of these, carried out in Scotland (DICKINSON et al. 1974), Scottish Blackface sheep were used as the indicator animals because no scrapie case had been reported in this breed for the previous 30 years. It was known to be possible to infect Scottish Blackface sheep by inoculation, however (GORDEN 1946). The source flock was kept on a remote hill farm and observed for 13 years during which time no case of scrapie was found in a total of over 18000 animals. Seventy-five sheep were taken away from this flock and mixed with scrapie sheep (Suffolk × Scottish Blackface originating from a high-incidence Suffolk flock at Moredun) in normal field conditions in an area not previously used for scrapie studies. The incidence of scrapie in the indicator animals was 28% in 5–6 years. This level in the source Scottish Blackface flock would have given rise to over 200 scrapie cases in this time and so was regarded as highly significant. All suspected cases were subjected to confirmation by histopathology and rejected if negative. Maternal transmission was also found to be important in this study, the scrapie status of the mother being much more important than that of the father in predicting which lambs would ultimately succumb.

HOURRIGAN et al. (1979) began a field trial (using extremely stringent controls) in 1964 at Mission, Texas, to establish the possibility of lateral scrapie transmission. Sheep judged to be scrapie-free were run in field conditions with sheep known to have been exposed to the disease. A diagnosis of scrapie was in all cases confirmed by histopathological examination of brain and/or transmission to mice. A scrapie-free label was only assigned after observation for 8 years after introduction into the programme at 2 years of age. Animals dying were examined carefully for any signs of scrapie and histopathology was carried out. Any animals over which there was some doubt either clinically or pathologically were excluded from the final report. The exposed animals brought into the field station (547 in number) were known to be from scrapie flocks but were apparently healthy. The breeds were Cheviot, Hampshire, Montadale, and Suffolk in 21 blood lines. The unexposed group consisted of the following: 31 Targhees, 31 Rambouillets (no scrapie case has been reported in the USA in either of these breeds, although Rambouillet sheep are credited with bringing scrapie into North India; ZLOTNIK and KATIYAR 1961), 33 Hampshires, 28 Suffolks and 17 goats. The previously exposed group included 333 Suffolks, and 29% of these developed scrapie. Their progeny, bred at the station, numbered 446, of which 34% later came down with scrapie. Of the unexposed indicator animals, five (3.5%) contracted scrapie and all at the later than average age of 6–8 years. Their progeny, exposed from birth to infection, came down with scrapie at the rate of 27% at the more usual age range of 2.5–5 years. The low level of lateral transmission compared to maternal transmission is quite clear from this study. Although the incidence of five animals contracting scrapie out of a total of 123 can be criticised, their late age of onset was argued to indicate that they were not exposed to scrapie before coming to the field station. However, unlike the Dickinson study, the source flocks were not followed up, and the "scrapie-free' label is more difficult to support.

A number of studies indicate that in close penning conditions scrapie can be transmitted laterally (e.g. BROTHERSTON et al. 1968; DICKINSON et al. 1974). These studies have been criticised on the grounds that the conditions are not similar enough to those in the field, where sheep will be much more spread out. Icelandic farmers might disagree with this, however, because, as reported by PALSSON (1979), it is normal practice for them to keep their sheep in close penning conditions throughout the 5–6 months of winter. Icelandic scrapie, or rida, is not treated with the secrecy common elsewhere and, being openly acknowledged, is easier to study. Palsson believed that the winter conditions were more important in the spread of rida from sheep to sheep than pasture conditions and also claimed evidence of sheep in the preclinical state being a source of infection to their flockmates.

2.2 Route of Transmission

The route of transmission of natural scrapie is not known. However, extensive studies of the spread of infection within naturally or experimentally infected animals have suggested possible pathways.

The high-incidence Suffolk sheep of Mission, Texas, have provided information on the body distribution of scrapie (HADLOW et al. 1982). Tissues taken from sheep that varied in age from newborn lambs to 8.5 years and were known to be at high risk of developing scrapie naturally were tested by mouse bioassay for the presence of infectivity. Scrapie was detected in lymphatic tissues and, interestingly, intestine in clinically normal 10- to 14-month-old lambs, although titres were generally low. Infectivity was first detected in the central nervous system in a clinically normal 25-month-old ewe, and by the time disease developed in nine animals at approximately 3.5 years of age, scrapie was present in the alimentary tract from the tonsil to the distal colon as well as in the spleen and central nervous system. Scrapie was not detected in other tissues such as heart, lung, kidney and skeletal muscle, nor was it found in high-risk sheep that remained clinically normal until 4.5–8.5 years of age.

Becuse of the apparent early involvement of the alimentary tract in this form of Suffolk sheep scrapie, an oral route of transmission was postulated; however, milk, colostrum and faeces were also tested (by mouse bioassay) by HOURRIGAN et al. (1979) and nothing was found. PATTISON et al. (1972, 1974) had already demonstrated that it was possible to produce scrapie in Herdwick sheep by feeding them placental material from scrapie-affected Swaledale ewes, but infection did not seem to be present in the foetus (HOURRIGAN et al. 1979), and so the idea of oral transmission of scrapie via pasture contaminated by infected placentae was put forward. Contamination of pasture as a source of infection is supported by some findings in Icelandic scrapie (PALSSON 1979) and in the work of GREIG (1940 a,b, 1950) and could be quite long lived because of scrapie's infamous resistance to inactivation. Skin abrasions are also possible natural entry points for scrapie. This route has not been explored experimentally, although DICKINSON (1976), quoting from the unpublished work of D.R. Wilson in the 1950s, said that it was possible to inoculate sheep successfully with scrapie using scarification.

2.3 Experimental Mouse Models

Studies of scrapie induced experimentally in mouse models using peripheral routes of inoculation are believed to mimic natural infection more closely than, for instance, intracerebral (ic) injection. Using intraperitoneal (ip), intravenous (iv) and subcutaneous (sc) routes, various studies have suggested that replication of scrapie in the mouse takes place first in the lymphoreticular system (LRS) and that fatal disease only develops if infection spreads to the central nervous system (CNS). Scrapie is found in the CNS later, first in the thoracic spinal cord and always at predictable stages in the incubation period (KIMBERLIN and WALKER 1979, 1980, 1982). These findings are quite similar to those of HADLOW et al. (1982), described above, for natural infectivity in Suffolk sheep. KIMBERLIN and WALKER (1989) also describe mouse studies using an intragastric injection route which it was hoped would compare well with the postulated natural oral route in sheep. Replication was found to start in Peyer's patches, preceding replication in the spleen, and in contrast to other routes (e.g. ip), splenectomy did not increase the incubation period (delay appearance of clinical symptoms). Infection seemed to be spreading directly to the CNS via the enteric and sympathetic nervous system. The reason for the involvement of the LRS in early stages of disease is not clear, but in one mouse model (scrapie strain 87V injected ip as a 1% brain homogenate into IM mice) there is long-term persistence of high scrapie titres in mouse spleen without either accompanying replication in the brain or development of scrapie symptoms. This may be a mechanism by which carrier-status animals can be produced (COLLIS and KIMBERLIN 1985).

Very low infecting doses may also result in the host becoming a carrier. One of the earliest studies of scrapie in mice (CHANDLER 1963) showed that incubation period was dose dependent. In other words, incubation period lengthened as the amount of infectivity in the inoculum was reduced. At very low levels of scrapie infection replication may be at such a low level that disease does not develop within the animal's normal lifespan (DICKINSON et al. 1975). Such a situation, if shown to exist in sheep (and as yet there is no proof for this), could lead to the maintenance of a low level of infection in a flock without any signs of symptoms. Some of the Suffolk sheep used in the trials described by HOURRIGAN et al. (1979) were known to come from flocks with scrapie cases and did not develop scrapie for 5 to nearly 11 years. Such sheep could be shedding scrapie infectivity for all or part of this time and be a source of infection for their flockmates.

3 Genetics of Susceptibility

Early experimental studies of transmission of scrapie using sheep often gave confusing and contradictory results. One of the studies most often quoted is that described by GORDON in 1966, in which 24 different breeds of sheep were injected ic with the source of scrapie known as SSBP/1 (described by DICKINSON 1976). The animals were observed for 2 years and the numbers developing scrapie were noted, ranging from Herdwicks at 78% susceptibility to Dorset Downs, which were all apparently resistant. Later work gave different results, with Herdwicks at 30% susceptibility (PATTISON 1966), and there were also problems with the original Dorset Downs, some of which did eventually develop SSBP/1 scrapie at much later dates (DICKINSON 1976). It became clear that there was as much variation in the "take" of scrapie inoculation between different flocks of the same breed as there was between different breeds (NUSSBAUM et al. 1975), and the need was perceived to develop lines of sheep with more predictable response to induced scrapie if any progress in studying the disease was to be made. There are now three such flocks in Britain—Cheviots, Herdwicks and Swaledales.

3.1 Development of Selected Sheep Lines

The Cheviot flock [at the Institute for Animal Health, Neuropathogenesis Unit (NPU), Edinburgh] was started in 1961 from a foundation flock of 303 ewes and 15 rams with no record of exposure to scrapie. They were split into two lines (positive and negative) on the basis of their response to sc injection with SSBP/1 (DICKINSON et al. 1968; DICKINSON 1976) and a policy of careful recording was

Table 1. Incubation periods of SSBP/1 and CH164 in Cheviot sheep of differing *Sip* genotype

Source of scriple	Route of injection	Incubation period (days ± SEM)	
		Positive line (*Sip sAsA* or *sApA*)	Negative line (*Sip pApA*)
SSBP/1	ic	197 ± 7	917 ± 90
	sc	313 ± 9	Survival
CH1641	ic	595 ± 122	360 ± 15

Data adapted from DICKINSON et al. (1968) and FOSTER and DICKINSON (1988c). ic, intracerebral; sc, subcutaneous

adopted to minimise inbreeding. The positive line develop scrapie in around 300 days whereas the negative line survive a natural lifespan. Negative-line sheep are not believed to be truly resistant as they can contract scrapie in around 1000 days following ic injection. Positive-line sheep die with 200 days of ic injection of SSBP/1 (Table 1).

From line-cross and back-cross segregation experiments it became clear that a single gene (now known as *Sip*) controls susceptibility to SSBP/1. *Sip* has two alleles, *sA* and *pA* (DICKINSON and OUTRAM 1988). The negative line (which used to be known as LIP, long incubation period) are all *Sip pApA*. The positive line (previously called SIP, short incubation period) are either *Sip sAsA* or *Sip sApA*, as *sA* is thought to be dominant. Isolates of scrapie can be classified according to their relative effects on sheep of the different *Sip* genotypes. Most isolates (A group) produce the disease in carriers of the *sA* allele faster than in *Sip pApA* sheep, but with at least one isolate (CH1641) (C group; FOSTER and DICKINSON 1988a), the ranking is not clear-cut and may even be reversed (Table 1).

Also in 1961, W.S. Gordon at Compton began a similar selection process using SSBP/1 and Herdwick sheep (NUSSBAUM et al. 1975). Positive and negative lines were produced in the same way, and susceptibility differences were shown again to be under the control of a single gene with two alleles. The allele conferring high susceptibility to SSBP/1 was dominant. There is, as yet, no formal proof that the *Sip* gene is acting in the Herdwicks as well as the Cheviots, but the results of crossing experiments suggest that this is the case (J.D. Foster personal communication). Unfortunately the numbers of sheep remaining in the Herdwick selected lines are now very low. Those that do remain (around 30 sheep in total) are at the NPU in Edinburgh.

During the course of the development of the Herdwick lines, it became apparent that resistance to experimental scrapie could be identified within one generation. A low-susceptibility Swaledale flock was started at Compton in 1973 (HOARE et al. 1977; DAVIES and KIMBERLIN 1985) by the relatively simple means of inoculating the animals sc with scrapie and breeding from the survivors. This time the sources of scrapie used were two pools of Swaledale natural scrapie

Table 2. Incubation periods of SW73 in the foundation flock of the Swaledale sheep selected line

Animals	*n*	Incubation period (days ±SEM)
Challenged	294	
Inercurrent deaths	63 (21% of total)	
Scrapie cases	196 (85% of remainder)	
Early	156 (80% of cases)	287 ± 7
Late	40 (20% of cases)	1207 ± 38
Survivors	35 (15% of remainder)	

Observation period 7 years. Data adapted from DAVIES and KIMBERLIN (1985)

brains, SW73 and SW75. Of the foundation flock, 85% of the animals developed scrapie within the 7-year observation period. Susceptibility was again determined by a single gene with two alleles, the dominant one resulting in a short incubation period, but this time the dominance appeared only to be partial. The scrapie cases fell into two distinct groups: early—80% of cases—which occurred before 500 days (Table 2), and late—after 870 days. The early group were thought to be homozygous "susceptible" and the later group heterozygous. The progeny of the surviving sheep had a greatly reduced susceptibility to experimental challenge with scrapie, and 72 of them were used to form a nucleus flock of scrapie-resistant (or negative-line) Swaledales now at the Ministry For Agriculture, Fisheries and Food Redesdale Experimental Husbandry Farm in Northumberland. Although again it has not been formally proved that the *Sip* gene operates in these Swaledales, it is thought to be extremely likely (N. Hunter, unpublished work).

What relevance does scrapie in these experimentally derived lines of sheep have to the natural disease? The sheep themselves are extremely useful in transmission and molecular genetic studies because their response to challenge is known in advance. They may also prove useful in studies of natural scrapie. There have been outbreaks of natural scrapie in the positive lines of both Cheviots and Herdwicks but not in the negative lines (DICKINSON 1974; PATTISON 1974) nor in the Swaledale nucleus flock (A.J. Chalmers, personal communication). The gene *Sip*, which controls the incubation period of SSBP/1 in the Cheviots, is also thought to affect the host response to natural scrapie, although *sA* is thought to be recessive in this case. Crossbreeding experiments described by FOSTER and DICKINSON (1988b) involved negative-line Cheviots (*Sip pApA*) and Suffolks from the flock bred for high incidence of natural scrapie (DICKINSON et al. 1974). Progeny were observed for development of scrapie both naturally and following experimental challenge with both SSBP/1 and SUF81 (a homogenate of five natural scrapie brains from the Suffolk flock). The incidence of scrapie was close to the expected ratio for the segregation of a single gene, with one dominant allele, giving good evidence that the *Sip* gene affects the host response to both experimental and natural scrapie in these two breeds of sheep. The differences in apparent dominance of the alleles of this gene between the

Cheviot, Herdwick and Swaledale experimental flocks (dominance or partial dominance of *sA*) and natural scrapie (where *sA* is thought to be recessive) have yet to be satisfactorily explained.

3.2 Biochemical and Genetic Studies of Selected Sheep Lines

The Herdwick lines have been studied in an attempt to find genetic or biochemical differences between them which could be linked to the differences in susceptibility. No correlation was found with susceptibility and phenotypes of albumin, pre-albumin, esterase, haemoglobin, transferrin, reduced glutathione and α-mannosidase or with lymphocyte antigens (COLLIS and MILLSON 1975; COLLIS et al. 1977; CULLEN et al. 1984). A linkage of OLA haplotypes was described in Ile de France sheep with natural scrapie (MILLOT et al. 1988) but this has been disputed (CULLEN 1989)—it may be due to a founder effect.

The Cheviot sheep have provided what may be the first reliable in vitro marker for susceptibility differences—restriction fragment length polymorphisms (RFLP) of the sheep PrP gene. The PrP protein is described in detail elsewhere in this volume. Suffice it to say that sheep, like rodents, goats, cows and man, have a PrP gene (GOLDMANN et al. 1990; WESTAWAY and PRUSINER 1986; ROBAKIS et al. 1986), and the abnormal scrapie-associated fibrils (SAF) made up of PrP protein can be extracted from scrapie sheep brain (HOPE et al. 1986; RUBENSTEIN et al. 1987). In mice, mutations in and around the PrP gene were found to be linked to the alleles of *Sinc* (the murine homologue of *Sip*) (CARLSON et al. 1986; HUNTER et al. 1987; WESTAWAY et al. 1987). In a study of the PrP genes in the two lines of Cheviot sheep, two RFLP (*Eco*RI and *Hin*dIII) were found which appeared to be linked to alleles of *Sip* (HUNTER et al. 1989). The *Sip sA* allele was linked to 6.8-kb *Eco*RI and 3.4-kb *Hin*dIII fragments of the PrP gene; the *Sip pA* allele to 4.4-kb *Eco*RI and 5.0-kb *Hin*dIII fragments. The original study on a small number of animals has been extended to 77 positive-line and 31 negative-line Cheviot sheep. (The number of animals in each line is variable but usually around 100). All 31 negative-line (*Sip pApA*) animals have one PrP-*Eco*RI fragment of 4.4-kb. Of the positive line (*Sip sAsA* or *sApA*), 25% (24 animals) have the 6.8-kb fragment and 65% (50 animals) have both fragments (Table 3). This strengthens the preliminary results suggesting that the 6.8-kb fragment is a marker for *Sip sA*, and the study is being extended to the other selected lines of sheep, the Swaledales and the Herdwicks, and to the natural disease.

Although the results so far are encouraging, much more knowledge is needed about both the PrP and *Sip* genes before regarding the apparent genetic linkage as a foolproof indicator of a sheep's response to challenge with scrapie. All field strains of scrapie may not be the same as can be seen from the quite different responses of the Cheviot lines to SSBP/1 and CH1641 (Table 1). An animal with low susceptibility to one type of scrapie may well develop disease when exposed to a different type, for example on being sold into an area with a different endemic strain.

Table 3. *PrP* gene restriction fragment length polymorphism within the Neuropathogenesis Unit Cheviot flock

Sheep line (*Sip* genotype)	*Eco*RI fragment size					
	6.8 kb		Both		4.4 kb	
	n	%	*n*	%	*n*	%
Positive line ($n = 77$) (*Sip sAsA* or *sApA*)	27	35	50	65	0	
Negative line ($n = 31$) (*Sip pApA*)	0		0		31	100

GOLDMANN et al. (1990) have demonstrated an amino acid polymorphism of the PrP protein itself and this may be linked to alleles of *Sip*. The sheep therefore joins an increasing list of mammals in which such linkage has been found. These include mice (WESTAWAY et al. 1987), hamsters (LOWENSTEIN et al. 1990) and man: familial Creutzfeldt–Jakob disease (OWEN et al. 1989, 1990) and Gerstmann–Sträussler syndrome (HSIAO et al. 1989; COLLINGE et al. 1989). The transfer of the hamster PrP gene into mice renders the transgenic mouse more susceptible to hamster-passaged scrapie (M. SCOTT et al. 1989), and so there is a great deal of support for the idea that the PrP gene may itself be the gene which controls incubation period or survival time. However, there are one or two reports of recombinants in the mouse between PrP and *Sinc* (CARLSON et al. 1989; RACE et al. 1990), and many may wonder how such apparently minor changes in the structure of a single protein can mean the difference between life and death on exposure to scrapie-like agents.

3.3 Comparison with Other Diseases

There are not many studies of genetically controlled differences in reaction to disease in the non-experimental situation with which to compare scrapie. However, it is almost invariably taken for granted that during the spread of a natural infection, in animal or human communities, the outcome will vary from individual to individual (DICKINSON and MACKAY 1967). The variability can be the result of many factors, some of which have been discussed above in relation to scrapie and which are often ignored in the experimental situation, when the desired result is to produce disease in the laboratory animal as quickly and as uniformly as possible. Scrapie, often believed to be an unusual and mysterious entity, really has much in common with other infectious diseases.

An example of a natural disease outbreak showing genetic differences in susceptibility was ectromelia (mousepox) in an experimental mouse colony. Although many mouse strains were exposed to the disease, some were very much more susceptible than others in terms of disease symptoms and death.

C57BL mice, which had only mild skin lesions, were carriers of mousepox and sources of infection to other animals (DICKINSON and MACKAY 1967). Resistance of C57BL mice to mousepox appeared to be inherited as a single, dominant, autosomal factor (SCHELL 1960). In experimental studies of this disease, the response of the mice to footpad inoculation (considered a "natural" route) showed that, although C57BL mice were very much less susceptible than "stock" mice, it *was* possible to produce disease in the animals if a high enough dose was used. Susceptibility differences were not maintained if other routes of injection were used; for instance, the ic route produced disease equally well in C57BL mice and in stock mice. If the relatively fast ic route, a very "unnatural" one, had been the only one used, the genetic effect on disease control would not have been noticed (SCHELL 1960).

Resistance of mice to ip injection of street rabies virus (SRV) is genetically controlled by the concurrent presence of each of two segregating genes (LODMELL 1983). SJL/J mice are resistant, have high serum titres of neutralising antibody, and SRV seldom invades the brain. A/WySnJ mice are susceptible. The factors responsible for the differing susceptibilities of these mouse strains are thought to be controlled by the immune response (LODMELL et al. 1989).

Mouse hepatitis virus (MHV) produces infection ranging from persistent to acute, and genetic evidence has suggested that a single host mutation controls the development of a semiresistant state. Lysis of cells exposed to MHV is much reduced in mutants, and the resulting persistent infection is correlated with an apparent cell resistance to virus-induced cell fusion (DAYA et al. 1989).

The phenotype may not always reflect the genotype (BANG 1965), especially in the uncontrolled conditions of a natural outbreak of disease. In 1952, in a paper read before the National Academy of Sciences, SABIN said "The vast majority of viruses which attack the nervous system of human beings and animals produce recognisable disease and death in only a small proportion of infected individuals." Nowadays he would have to reference this pronouncement! However, it is clear from even a cursory reading of books and journals concerned with infectious diseases that the presence of virus in an organism is not always indicated by disease symptoms (e.g. NETTLETON 1987; LACKNER et al. 1989; J.L. SCOTT et al. 1990). There is a view which says that a virus which causes an inapparent infection has evolved to a favourable equilibrium with its host: examples are the enteroviruses which multiply in the gut and poliovirus, which produces no symptoms in over 90% of infections. The largely inapparent infection can be thought of as the result of interaction between the virus and its natural host, lethal disease only coming from the infection of a new host (yellow fever virus infects Old World monkeys without producing disease but is often a fatal infection for man and some New World monkeys) (PRIMROSE and DIMMOCK 1980).

In relation to scrapie, the persistence of some viral infections which cause disease only rarely is of great interest. One example is Epstein-Barr virus (EBV), to which large numbers of people have antibodies resulting from glandular fever (mononucleosis) infections of varying severity, the majority so mild that they go unnoticed by the individuals concerned. Some go on to develop Burkitt's lymphoma, but not all (ZUR HAUSEN 1980).

Around 70% of people worldwide (except for a few small populations) have antibodies to the papovaviruses BKV and JC, mostly acquired by the age of 14 (SHAH et al. 1973; PADGETT and WALKER 1973; MANTIJARVI et al. 1973; PADGETT 1980). The primary infection is usually subclinical but is followed by lifelong persistence of the virus. To maintain such prevalence the viruses must be transmitted easily, but the routes of transmission are uncertain and no animal reservoir has been found. Very rarely these viruses have been found to be associated with disease (often in immune-suppressed people) although, because they are so ubiquitous, it is very difficult to prove a causal relationship. For instance, JC virus, if it reaches the oligodendrocytes in the brain, will cause progressive multifocal leukoencephalopathy (PML) a demyelinating disease which is fatal (PADGETT 1980; SMALL et al. 1986). Mice transgenic for JC T-antigen have myelin deficiencies (SMALL et al. 1986). BKV and JC also have been shown to produce tumours in neural tissue when inoculated into newborn hamsters (PADGETT et al. 1977; VARAKIS et al. 1978), and JC will cause tumours in owl monkeys (LONDON et al. 1978).

Despite its very unusual physical and chemical stability, the biology of scrapie is not very much out of the ordinary as long as the effects of route of infection and amount of infectivity are fully appreciated. Carrier status (or persistent low level infection) and genetic susceptibility could explain the "lightning" effect (scrapie disease appearing from nowhere) complained of by farmers and shepherds. The imported ram, often blamed for subsequent outbreaks, might change the dominant genotype of a flock within only a few years, as one ram can be used to serve 30–50 ewes per year. He may bring in infection himself, but this would not necessarily have a sudden effect on the flock unless it was an unusual strain to which they were susceptible but had not previously been exposed. He may, however, bring in a susceptible genotype, which, inherited by his offspring, would make them more susceptible to prevailing scrapie strains present in their mothers at a low level or in the pasture. In order for disease to develop at all, a certain threshold dose is needed, even if the animal is of a susceptible genotype. A slowly increasing load of infection in a flock could remain unseen for generations before reaching the critical threshold and suddenly producing disease.

All of these theories concerning scrapie and similar diseases depend on the existence of carrier status animals and their availability as sources of infection. With all the recent worries about bovine spongiform encephalopathy and other apparently foodborne "slow viral" infections, it has become more important than ever to establish whether or not carrier status exists. Random bioassays would be time-consuming and perhaps uninformative. CATHALA et al. (1985) reported on a small study of 32 ewes selected at random from the slaughterhouse in an area of France with endemic scrapie. Seven tissues were taken from each sheep and inoculated into mice which were, for 2 years, fed, cleaned and watched for signs of transmission. None were observed. Primary transmissions from scrapie sheep do not always "take" in mice (FRASER 1983), and if any of the tested sheep had been preclinically infected with scrapie, the titre might have been too low to show up within 2 years. A positive result is naturally of interest, but a negative result tells us nothing.

Although the potential of the genetic tests for degrees of susceptibility of mammals to the scrapie-like diseases might be of short-term help in management of infections, a preclinical diagnostic test would make this a lot easier in sheep and cows and would make it possible to study these diseases in humans.

References

Asher DM, Masters CL, Gajdusek DC, Gibbs CJ (1983) Familial spongiform encephalopathies. In: Kety SS, Rowland LP, Sidman RL, Matthysse SW (eds) Genetics of neurological and psychiatric disorders. Raven, New York, pp 273–291

Bang FB (1965) Genetic and cellular factors in resistance to virus diseases. In: Gajdusek DC, Gibbs CJA, Alpers M (eds) Slow, latent and temperate virus infections. US Department of Health Education and Welfare, Washington DC, pp 389–392 (Health service publication no 1378) (National Institute of Neurological Diseases and Blindness monograph no 2)

Brotherston JG, Renwick CC, Stamp JT, Zlotnik I (1968) Spread of scrapie by contact to goats and sheep. J Comp Pathol 78: 9–17

Carlson GA, Kingsbury DT, Goodman PA, Coleman S, Marshall ST, DeArmond S, Westaway D, Prusiner SB (1986) Linkage of prion protein and scrapie incubation time genes. Cell 46: 503–511

Carlson GA, Goodman PA, Lovett M, Taylor BA, Marshall ST, Petersa-Forchia M, Westaway D, Prusiner SB (1989) Genetics and polymorphism of the mouse prion gene complex: control of scrapie incubation time. Mol Cell Biol 8: 5528–5540

Cathala F, Brown P, Gray F, Sulima M, Chatelain J, Gibbs CJ (1985) Failure to detect scrapie virus in sheep at slaughter in a highly endemic region of France. Eur J Epidemiol 1: 90–93

Chandler RL (1963) Experimental scrapie in the mouse. Res Vet Sci 4: 276–285

Chatelain J, DeLasnerie-Laupretre N, Cathala F (1983) Scrapie in France: some possible predisposing factors in the naturally-acquired disease of sheep. Vet Microbiol 8: 511–515

Collinge J, Harding AE, Owen F, Poulter M, Lofthouse R, Boughey AM, Shah T, Crowe TJ (1989) Diagnosis of Gerstmann–Sträussler syndrome in familial dementia with prion protein gene analysis. Lancet i: 15–17

Collis SC, Kimberlin RH (1985) Long-term persistence of scrapie infection in mouse spleens in the absence of clinical disease. FEMS Microbiol Lett 29: 111–114

Collis SC, Millson GC (1975) Transferrin polymorphism in Herdwick sheep. Anim Blood Groups Biochem Genet 6: 117–120

Collis SC, Millson GC, Kimberlin RH (1977) Genetic markers in Herdwick sheep: no correlation with susceptibility or resistance to experimental scrapie. Anim Blood Groups Biochem Genet. 8: 79–83

Cullen PR (1989) Scrapie and the sheep MHC: claims of linkage refuted. Immunogenetics 29: 414–416

Cullen PR, Brownlie J, Kimberlin RH (1984) sheep lymphocyte antigens and scrapie. J Comp Pathol 94: 405–415

Davies DC, Kimberlin RH (1985) Selection of Swaledale sheep of reduced susceptibility to experimental scrapie. Vet Rec 116: 211–214

Daya M, Wong F, Cervin M, Evans G, Vennema H, Spaan W, Anderson RV (1989) Mutation of host cell determinants which descriminate between lytic and persistent mouse hepatitis virus infection results in a fusion-resistant phenotype. J Gen Virol 70: 3335–3346

Diamond JM (1987) AIDS: infectious, genetic or both? Nature 328: 199–200

Dickinson AG (1974) Natural genetic infection, 'spontaneous generation' and scrapie. Nature 252: 179–180

Dickinson AG (1975) Host-pathogen interactions in scrapie. Genetics 79: 387–395

Dickinson AG (1976) Scrapie in sheep and goats. In: Kimberlin RH (ed) Slow virus diseases of animals and man. North-Holland, Amsterdam, pp 209–241

Dickinson AG, MacKay JMK (1967) Virus hosts and genetic studies. In: Maramorosch K, Koprowski H (eds) Methods in virology, vol. 1 Academic, New York, pp19–61
Dickinson AG, Outram GW (1988) Genetic aspects of unconventional virus infections: the basis of the virino hypothesis. In: Bock G, Morsh J (eds) Novel infectious agents and the central nervous system. Wiley, Chichester, pp 63–83 (CIBA Found symp 135)
Dickinson AG, Young GB, Stamp JT, Renwick CC (1964) A note on the distribution of scrapie in sheep of different ages. Anim Prod 6: 375–377
Dickinson AG, Young GB, Stamp JT, Renwick CC (1965) An analysis of natural scrapie in Suffolk sheep. Heredity (Edinburgh) 20: 485–503
Dickinson AG, Young GB, Renwick CC (1966) Scrapie: experiments involving maternal transmission in sheep. Report of scrapie seminar, 1964 ARS 91–53. US Department of Agriculture, 244–248, Washington DC
Dickinson AG, Stamp JT, Renwick CC, Rennie JC (1968) Some factors controlling the incidence of scrapie in Cheviot sheep injected with a Cheviot passaged scrapie agent. J Comp Pathol 78: 313–321
Dickinson AG, Stamp JT, Renwick CC (1974) Maternal and lateral transmission of scrapie in sheep. J Comp Pathol 84: 19–25
Dickinson AG, Fraser H, Outram GW (1975) Scrapie incubation time can exceed natural lifespan. Nature 256: 732–733
Elvander M, Engvali A, Klingeborn B (1988) Scrapie in sheep in Sweden. Acta Vet Scand 29: 509–510
Foster JD, Dickinson AG (1988a) The unusual properties of CH1641, a sheep passaged isolate of scrapie. Vet Rec. 123: 5–8
Foster JD, Dickinson AG (1988b) Genetic control of scrapie in cheviot and Suffolk sheep. Vet Rec 123: 159
Foster JD, Dickinson AG (1989) Age at death from natural scrapie in a flock of Suffolk sheep. Vet Rec 125: 415–417
Fraser H (1983) A survey of primary transmission of Icelandic scrapie (rida) to mice. In: Court LA, Cathala F (eds) Virus non conventionnels et affection du system nerveux central. Masson, Paris, pp 34–46
Goldmann W, Hunter N, Foster JD Salbaum JM, Beyreuther K, Hope J (1990) Two alleles of a neural protein gene linked to scrapie in sheep. Proc Natl Acad Sci USA 87: 2476–2480
Gordon WS (1946) Louping-ill, tick-borne fever and scrapie. Vet Rec 58: 516–520
Gordon WS (1966) Variation in susceptibility of sheep to scrapie and genetic implications. In: Report of scrapie seminar, 1964 ARS 91–53. US Department of Agriculture 53–67, Washington DC
Greig JR (1940a) Scrapie. Trans Highl Agric Soc Scot 52: 71–90
Greig JR (1940b) Observations on the transmission of the disease by mediate contact. Vet J 96: 203–206
Greig JR (1950) Scrapie in sheep. J Comp Pathol 60: 263–266
Hadlow WT, Kennedy RC, Race RE (1982) Natural infection of Suffolk sheep with scrapie virus. J Infect Dis 146: 657–664
Hoare M, Davies DC, Pattison IH (1977) Experimental production of scrapie–resistant Swaledale sheep. Vet Rec 101: 482–484
Hope J, Morton LJD, Farquhar CF, Multhaup G, Beyreuther K, Kimberlin RH (1986) The major poltypeptide of scrapie-associated fibrils (SAF) has the same size, charge distribution and N-terminal sequence as predicted for the normal brain protein (PrP). EMBO J 5: 2591–2597
Hourrigan J, Klingsporm A, Clark WW, deCamp M (1979) Epidemiology of scrapie in the United States. In: Prusiner SB, Hadlow WJ (eds) Slow transmissible diseases of the nervous system, vol 1. Academic, New York, pp. 331–356
Hsiao K, Baker HF, Crow TJ, Poulter M, Owen F, Terwilliger JD, Westaway D, Ott J, Prusiner SB (1989) Linkage of a prion protein missense variant to Gerstmann–Sträussler syndrome. Nature 338: 342–345
Hunter N, Hope J, McConnell I, Dickinson AG (1987) Linkage of the scrapie–associated fibril protein (PrP) gene and *Sinc* using congenic mice and restriction fragment polymorphism analysis. J Gen Virol 68: 2711–2716
Hunter N, Foster JD, Dickinson AG, Hope J (1989) Linkage of the gene for the scrapie-associated fibril protein (PrP) to the *Sip* gene in Cheviot sheep. Vet Rec 124: 364–366
Hunter N, Foster JD, Benson G, Hope J (1991) Restriction fragment length polymorphisms of the scrapie-associated fibril protein (PrP) gene and their association with susceptibility to natural scrapie in British sheep. J Gen Virol 72: (in press)

Joubert L, Lapras M, Gastelli J, Prare M, Laurent D (1972) Un foyer de tremblante du mouton en Provence. Bull Soc Sci Vet Med Comp 74: 165–184

Kimberlin RH, Walker CA (1979) Pathogenesis of mouse scrapie: dynamics of agent replication in spleen, spinal cord and brain after infection by different routes. J Comp Pathol 89: 551–562

Kiimberlin RH, Walker CA (1980) Pathogenesis of mouse scrapie: evidence for neural spread of infection the CNS. J Gen Virol 51: 183–187

Kimberlin RH, Walker CA (1982) Pathogenesis of mouse scrapie: patterns of agent replication in different parts of the CNS following intraperitional infection. J R Soc Med 75: 618–624

Kiimberlin RH, Walker CA (1989) Pathogenesis of scrapie in mice after intragastric infection. Virus Res 12: 213–220

Lackner AA, Marx, PA, Lerche NW, Gardner UB, Kluge JD, Spinner A, Kwang H-S, Lowenstine LJ (1989) Asymptomatic infection of the central nervous system by the macaque immunosuppressive type D retrovirus-SRV-1. J Gen Virol 70: 1641–1651

Lodmell DL (1983) Genetic control of resistance to street rabies virus in mice. J Exp Med 157: 451–460

Lodmell DL, Wiedbrauk DL, Ewalt LC (1989) Interferon induced within the central nervous system during infection is inconsequential as a mechanism responsible for murine resistance to street rabies virus. J Gen Virol 70: 473–478

London WT, Houff SA, Madden DL, Fuccillo DA, Gravell M, Wallen WC, Palmer AE, Sever JL, Padgett BL, Walker DL, ZuRhein GM, Ohashi T (1978) Brain tumors in owl monkeys inoculated with a human polyoma virus (JC virus). Science 201: 1246–1249

Lowenstein DH, Butler DA, Westaway D, McKinley MP, DeArmond SJ, Prusiner SB (1990) Three hamster species with different scrapie incubation times and neuropathological features encode distinct prion proteins. Mol Cell Biol 10: 1153–1163

Mantijarvi RA, Meurman OH, Vihma L, Berlund B (1973) A human papovavirus (BK): biological properties and seroepidemiology. Ann Clin Res 5: 283

Mechelli L, Mantovani AL (1988) Scrapie in sheep in central and southern Italy. Vet Res Commun 12: 165–167

Millot P, Chatelain J, Dautheville C, Salmon D, Cathala F (1988) Sheep major histocompatibility (OLA) complex: linkage between a scrapie susceptibility/resistance locus and the OLA complex in Ile-de-France sheep progenies. Immunogenetics 27: 1–11

Nettleton PF (1987) Pathogenesis and epidemiology of Border disease. Ann Rec Vet 18: 147–155

Nussbaum RE, Henderson WM, Pattison IH, Elcock NV, Davies DC (1975) The establishment of sheep flocks of predictable susceptibility to experimental scrapie. Res Vet. Sci 18: 49–58

Owen F, Poulter M, Lofthouse R, Collinge J, Crow TJ, Risby D, Baker HF, Ridley RM, Hsiao K, Prusiner SB (1989) Insertion in prion protein gene in familial Creutzfeldt–Jakob disease. Lancet i: 51–52

Owen F, Poulter M, Shah T, Collinge J, Lofthouse J, Baker H, Ridley R, McVey J, Crow TJ (1990) An in-frame insertion in the prion protein gene in familial Creutzfeldt–Jakob disease. Mol Brain Res 7: 273–276

Padgett BL (1980) Human papovaviruses. In: Tooce J (ed) Molecular biology of tumor viruses, 2nd ed 2: DNA tumor viruses. Cold Spring Harbor Laboratory, New York, pp 339–370

Padgett BL, and Walker DL (1973) Prevalence of antibodies in human sera against JC virus, an isolate from a case of PML. J Infect Dis 127: 467

Padgett BL, Walker DL, ZuRhein GM, Varakis JN (1977) Differential neurooncogenicity of strains of JC virus, a human polyoma virus, in newborn Syrian hamsters. Cancer Res 37: 718–720

Palsson PA (1979) Rida (scrapie) in Iceland its epidemiology. In: Prusiner SB, Hadlow WJ (eds) Slow transmissible diseases of the nervous system, vol 1. Academic, New York, pp 357–366

Parry HB (1960) Scrapie: a transmissible hereditary disease of sheep. Nature 185: 441–443

Parry HB (1962) Scrapie: a transmissible and hereditary disease of sheep. Heredity 17: 75–105

Pattison IH (1966) The relative susceptibility of sheep, goats and mice to two types of the goat scrapie agent. Res Vet Sci 7: 207–212

Pattison IH (1974) Scrapie in sheep selectively bred for high susceptibility. Nature 248: 594

Pattison IH, Hoare MN, Jebbett JN, Watson WA (1972) Spread of scrapie to sheep and goats by oral dosing with foetal membranes from scrapie affected sheep. Vet Rec 90: 465–468

Pattison IH, Hoare MN, Jebbett JN, Watson WA (1974) Further observations on the production of scrapie in sheep by oral dosing with foetal membranes from scrapie-affected sheep. Br Vet J 130: 65–67

Primrose SB, Dimmock NJ (1980) Introduction to modern virology, 2nd ed, Blackwell, Oxford

Race RE, Graham K, Ernst D, Caughey B, Chesebro B (1990) Analysis of linkage between scrapie incubation period and the prion protein gene in mice. J Gen Virol 71: 493–497

Robakis NK, Devine-Gage EA, Jenkins EC, Kascsak RJ, Brown WT, Krawczun MS, Silverman WP (1986) Localization of a human gene homologous to the PrP gene on the p arm of chromosome 20 and detection of PrP-related antigens in normal human brain. Biochem Biophys Res Commun 140: 758–765

Roche-Lubin (1848) Memoire practique sur la maladie des betes a laine connue sous les noms de prurigo, lombaire, maladie convulsive, trembleuse, tremblant, etc. Rec Med Vet Ser 3 (5): 698–714

Rubenstein R, Merz PA, Kascsak RJ, Carp RI, Scalici CL, Fama CL, Wisniewski HM (1987) Detection of scrapie associated fibrils (SAF) and SAF proteins from scrapie-affected sheep. J Infect Dis 156: 36–42

Sabin AB (1952) Nature of inherited resistance to viruses affecting the nervous system. Proc Natl Acad Sci USA 38: 540–546

Schell K (1960) Studies on the innate resistance of mice to infection with musepox. II. Route of inoculation and resistance: and some observations on the inheritance of resistance. Aust J Exp Biol 38: 289–300

Schoental R (1990) Kuru, mycotoxins and volcanic eruptions. J App Toxicol 10: 153

Scott JL, Blanchard-Channell M, Scibienski RJ, Scott ML (1990) Interaction of bluetongue virus with bovine lymphocytes. J Gen. Virol 71: 363–368

Scott M, Foster D, Mirenda C, Serban D, Confal F, Walchli M, Torchis M, Groth D, Carlson G, DeArmond SJ, Westaway D, Prusiner SB (1989) Transgenic mice expressing hamster prion protein produce species-specific scrapie infectivity and amyloid plaques. Cell 59: 847–857

Shah KV, Daniel RW, Warszawski R (1973) High prevalence of antibodies to BK virus and SV40 related papovaviruses. J Natl Cancer Inst 54: 945

Small JA, Scangos GA, Cork L, Jay G, Khoury G (1986) The early region of Human papovavirus JC induces dysmyelination in transgenic mice. Cell 46: 13–18

Stamp JT (1962) Scrapie: a transmissible disease of sheep. Vet. Rec 74: 357–362

Toumazos P (1988) First report of ovine scrapie in Cyprus. Br Vet J 1144: 98–99

Varakis J, ZuRhein GM, Padgett BL, Walker DL (1978) Induction of peripheral neuroblastomas in Syrian hamsters after injection as neonates with JC virus, a human polyoma virus. Cancer Res: 1718–1722

Westaway D, Prusiner SB (1986) Conservation of the cellular gene encoding the scrapie prion protein. Nucleic Acids Res 14: 2035–2045

Westaway D, Goodman PA, Mirenda CA, McKinley MP, Carlson GA, Prusiner SB (1987) Distinct prion proteins in short and long scrapie incubation period mice. Cell 51: 651–662

Zlotnik I, Katiyar RD (1961) The occurrence of scrapie disease in sheep of the remote Himalayan foothills. Vet Rec 73: 543–544

Zur Hausen H (1980) Oncogenic herpes viruses. In: Toote J (ed) Molecular biology of tumor viruses, 2nd ed, pt 2. Cold Spring Harbor, New York, pp 747–795

The Scrapie Agent In Vitro

R. RACE

1 Introduction

Precise biochemical characterization of the scrapie agent has not been attained. This is thought by many investigators to be due to the agent's tenacious association with membranes (HUNTER et al. 1968; HUNTER 1979) as well as other characteristics considered unusual for conventional viruses (ALPER 1985; BELLINGER-KAWAHARA et al. 1987; DEES et al. 1985; GIBBONS and HUNTER 1967; KASPER et al. 1982; ROHWER 1984; BROWN et al. 1982). As a result, scientists have sought tissue cultures supporting scrapie replication as an alternate source of scrapie agent. It has been reasoned that such cultures would be free of at least some of the tissue-derived by-products encountered in animal tissue sources and might provide the agent in a substrate that would allow determination of its structure. In addition, infected cultures could provide a system whereby the biology of agent-cell interactions could be studied in vitro. Such cultures could also conceivably lead to the development of an in vitro system for rapid agent titration.

The first culture in which scrapie agent replicated continuously over a large number of in vitro passages involved a cell line designated SMB (CLARKE and HAIG 1970a). Prior to the establishment of the SMB culture, several studies were

National Institutes of Health, National Institute of Allergy and Infections Diseases, Laboratory of Persistent Viral Diseases, Rocky Mountain Laboratories, Hamilton, MT 59840, USA

Current Topics in Microbiology and Immunology, Vol. 172

done in which explant cultures were established from the brains of infected mice and their growth and morphological characteristics compared to similar cultures derived from normal mice. Using explant techniques, FIELD and WINDSOR (1965) reported that brain explants derived from scrapie-infected mice grew more rapidly than those derived from normal mice and resulted in the rapid liquefaction of the chicken-plasma clots in which they were grown. Several other investigators also described increased growth potential in cultures established from scrapie-infected mice compared to those established from normal mice (GUSTAFSON and KANITZ 1965; HAIG and PATTISON 1967; CASPARY and BELL 1971; BUENING and GUSTAFSON 1971). However, only in the SMB culture was agent serially and continuously passaged beyond the initial explant culture.

Because of our desire to biochemically characterize the scrapie agent, we considered it essential that a suitable in vitro culture system capable of replicating scrapie agent be utilized. The SMB cell culture (CLARKE and HAIG 1970) was available, but because little information regarding its characterization had been published, and because the amount of agent available from it seemed low, especially after prolonged passage in vitro, we felt it necessary to develop additional scrapie-infected cultures. To do so we initiated several studies aimed at establishing cell cultures in which high titers of scrapie agent might be obtained. Our initial attempt sought to take advantage of the fact that scrapie agent reaches high titer in mouse spleen within a few weeks after infection as well as the availability of several splenotropic neoplastic cell lines which can be maintained in continuous passage in vitro or maintained in vivo in appropriate mouse strains. We reasoned that scrapie agent replicating in the spleens of infected mice might infect splenotropic tumor cells inoculated into mice already infected with scrapie. The tumors could subsequently be explanted and their in vitro passages resumed, the tumor cells presumably infected with scrapie agent. Several splenotropic tumor cell lines representing various lineages were

Table 1. Attempt to transfer scrapie agent from scrapie-affected mice to splenotropic tumors in vivo

Tumor designation	Cell lineage	Earliest in vitro passage assayed	Result of mouse inoculation
5402	Macrophage	3	Neg.
7320	Macrophage	2	Neg.
2800	Erythroid	1	Neg.
1593	B-lymphocyte (late)	2	Neg.
BCA	B-lymphocyte (early)	7	Neg.
A13	T-lymphocyte	2	Neg.
101	Erythroid	1	Neg.
LST	Ia^+	5	Neg.

Mice were inoculated initially with $10^5 LD_{50}$ of the Chandler strain of scrapie agent (EKLUND et al. 1967) and 45–60 days later they were inoculated with cell suspensions of the tumors indicated above. They were allowed to live until tumor growth threatened their lives, usually 7–28 days after tumor inoculation. They were then killed by cervical dislocation and the tumors explanted and passaged continuously in vitro. At various passage levels cell aliquots were freeze thawed, sonicated, and inoculated into RML Swiss mice which were then observed at least weekly for clinical signs of scrapie. The experiment was discontinued and surviving assay mice killed after 1 year. Tumor lines were obtained from Dr. Bruce Chesebro, Laboratory of Presistent Viral Diseases, Hamilton, Montana

inoculated intravenously into mice which had been infected 45–60 days earlier with the Chandler isolate of scrapie agent (Table 1). Approximately 10–20 days later tumors were recovered from the mice and in vitro passaging of the tumor cells was resumed. After various numbers of in vitro passages, cell aliquots of respective tumors were lysed and assayed for the presence of scrapie agent by bioassay in susceptible mouse recipients. Unfortunately, none of the cultures established in this way gave evidence of scrapie replication. We did not attempt to infect any of these cell lines by exposing them directly to scrapie agent in vitro.

We also attempted to establish scrapie-infected cultures by fusing spleen cells from scrapie-infected mice with NS-1 cells (P3-Ns-1-Ag4/1) (CHESEBRO et al. 1981), using conventional techniques used to establish hybridoma cultures secreting specified monoclonal antibodies. We reasoned that scrapie agent might be incorporated into the immortalized fused cell and continue to be replicated in them. Unfortunately, no evidence of scrapie agent persistence was found in any of over 100 clones that were analyzed.

Other approaches to establish scrapie-infected cultures have, however, been successful, and we, as well as several other investigators, have recently reported on cultures in which scrapie agent does replicate continuously. Cells from several sources and a variety of experimental approaches have been used to establish scrapie-infected cultures.

2 Adaptation of Scrapie Agent to In Vitro Growth

As mentioned before, an explant technique was used to establish the first in vitro culture in which scrapie agent replicated continuously (CLARKE and HAIG 1970). ASHER et al. (1979) also established a scrapie-infected culture using an explant technique, but did so by transforming the mouse brain-derived culture with SV40. However, this culture lost infectivity after 12 in vitro passages, whereas the SMB culture established by CLARKE and HAIG (1970) remained infectious for over 150 in vitro passages. Several successful adaptations of scrapie agent to in vitro culture have been reported since these early explant cultures were described. All of the recent adaptations have utilized a simple approach involving the exposure of cell monolayers or cell suspensions to homogenates of brain derived from scrapie-affected animals. In some instances relatively crude brain homogenates, diluted to various levels, were added to monolayers of cells (ROIKHEL et al. 1987; MARKOVITS et al. 1983: RUBENSTEIN et al. 1984; CLARKE 1979), while in other situations highly purified (BUTLER et al. 1988) or at least partially purified preparations were used (RACE et al. 1987). Some investigators sought to improve the efficiency of the infectious process by pretreating cells with chemicals which would enhance phagocytosis (CHEREDNICHENKO et al. 1985b) or which would promote membrane permeability (CLARKE and MILLSON 1976) or cell fusion (CLARKE 1979). It is now clear, however, that no special pretreatment of otherwise susceptible cells is required (MARKOVITS et al. 1983; RACE et al. 1987; ROIKHEL et al. 1987;

Butler et al. 1988), except for the necessary pretreatment of rat pheochromocytoma cells (PC-12 cells) with nerve growth factor several weeks prior to infection (Rubenstein et al. 1984). All of the cell lines that have been productively infected with scrapie agent are of central nervous system (CNS) origin except for mouse L fibroblasts (Clarke and Millson 1976). Cherednichenko et al. (1985) infected L23 cells, a line of unspecified origin. Cells that have been infected include cells derived from rats (Rubenstein et al. 1984; Roikhel et al. 1987) or lines derived from mice (Markovits et al. 1981; Race et al. 1987; Butler et al. 1988). Interestingly, at least three of the cell lines which have been successfully infected (Markovits et al. 1981; Race et al. 1987; Butler et al. 1988) have a common origin but different passage histories.

During the course of our studies, we independently sought to infect cell lines equivalent to each of these latter three cell lines. They included NIE-115 cells similar to those used by Markovits et al. (1981) which we obtained from Dr. J. Portis, Rocky Mountain Laboratories, Hamilton, MT; N2a cells similar to those used by Butler et al. (1988) which we obtained from Dr. Kingsbury, Georgetown University, Washington, D.C., and cells designated C-1300 (American Type Tissue Culture Collection). We successfully infected all of these cell lines with mouse-derived scrapie agent using a protocol which we described before (Race et al. 1987). If cultures representing this cell lineage are excluded from those successfully infected, the number of positive cultures and number of susceptible cells becomes quite restricted. This particular cell lineage, therefore, appears to be especially susceptible to scrapie infection if mouse-adapted agent is used to infect them. This conclusion is clearly apparent in view of the large number of cell lines which could not be infected over a period of many years by several investigators (Race et al. 1987; Markovits et al. 1983; Butler et al. 1988; Elleman, 1984; Yanagihara et al. 1980).

3 Infection of Cell Lines In Vitro May Be Species Restricted

There is evidence to suggest that the transmission of agents of the subacute spongiform encephalopathy group is species restricted. In fact, the species trophism of various isolates of scrapie agent has served as a basis for classifying strains of scrapie agent (Kimberlin et al. 1975; Kimberlin and Walker 1978; Bruce and Dickinson 1979; Gibbs et al. 1979). Crossing of various species barriers either has not occurred or has involved prolonged periods of adaptation in which primary passage from one species to another results in unusually long incubation periods followed by progressively shorter incubation periods on subsequent passages. Except for the infection of rat cells using mouse brain homogenates (Rubenstein et al. 1984; Roikhel et al. 1987), the infection of tissue culture cells has exhibited evidence of species specificity. In our efforts to develop in vitro cell lines in which scrapie agent replicates, we attempted to infect

Table 2. Infection of tissue culture cell lines using mouse or hamster-derived scrapie agent

Cell designation	Species of origin	Tissues of origin or cell type	Result when the origin of the infecting material was	
			Mouse	Hamster
N2a	Mouse	Neuroblastoma	Pos.	Neg.
C-1300	Mouse	Neuroblastoma	Pos.	Neg.
NIE-115	Mouse	Neuroblastoma	Pos.	Neg.
C-127	Mouse	Mammary carcinoma	Neg.	Neg.
BHK	Hamster	Kidney	Neg.	Neg.
CCL 61	Hamster	Ovary	Neg.	Neg.
HTB10	Human	Neuroblastoma	Neg.	Neg.
CCL 127	Human	Neuroblastoma	Neg.	Neg.
HTB 14	Human	Glioblastoma	Neg.	Neg.

N2a cells were obtained from Dr. David Kingsbury, Georgetown University, Washington, D.C. and from Dr. Donald Lodmell, Laboratory of Persistent Viral Diseases, Hamilton, Montana. C-1300 (CCL 147), BHK, CCL 61, HTB 10, CCL 127 and HTB 14 cells were obtained from the American Type Culture Collection, Rockville, Maryland. C-127 cells were obtained from Dr. Bruce Chesebro, Laboratory of Persistent Viral Diseases, Hamilton, Montana. NIE-115 cells were obtained from Dr. John Portis, Laboratory of Persistent Viral Diseases, Hamilton, Montana.

Attempts to infect various cell lines were done as described (RACE et al. 1987). Results are based on whether scrapie was diagnosed in recipient animals following inoculation of aliquots of cells after sufficient passages in vitro to rule out potential detection of scrapie agent in the inoculum used to infect the cells. All lines were bioassayed in both mice and hamsters

several existing cell lines with mouse and/or hamster-derived scrapie agent. We were especially interested in obtaining a hamster-infected culture because the incubation period of scrapie is shorter in hamsters than in any other susceptible laboratory animal and because higher agent titers are reached in hamster brain than in brain or other tissues from other species (MARSH and KIMBERLIN 1975; BROWN et al. 1981). Cell monolayers representing several cell culture lines and various species were exposed to partially purified homogenates of brain derived from clinically sick mice or clinically sick hamsters. Although some of the mouse-derived lines became infected with mouse-derived scrapie agent, none became infected using hamster-derived scrapie agent (Table 2). Cells derived from hamsters or humans were not infected with either the mouse- or the hamster-derived scrapie agent. We were somewhat surprised that none of the cultures representing cells of hamster derivation became infected with hamster agent. However, none of the hamster cell lines available to us at the time were derived from the CNS. With the exception of mouse L cells, all of the cell lines infected with mouse scrapie have been of CNS origin. Other investigators have reported that productive in vitro infections were not obtained when hamster glioma cells or hamster neuroblastoma cells were exposed to mouse agent (BUTLER et al. 1988). It would be interesting to know if these hamster CNS-derived cells were susceptible to hamster-derived scrapie agent. Recently, TARABOULOS et al. (1990) used an explant technique to establish a continuously dividing cell culture from the brain of a scrapie-infected hamster. The primary culture established in this way did not contain proteinase K(PK)-resistant prion protein (PrP). However, subsequent exposure of these cells to purified hamster agent did lead to cultures persistently infected with hamster-derived scrapie agent and the demonstration of PK-resistant PrP in them (TARABOULOS et al. 1990). These results, taken collectively, suggest that not only the species from which a cell line has been derived but also its tissue origin may be very important in determining its potential to become infected when exposed to scrapie agent. It is likely that efficient adaptation of scrapie agent to tissue culture has the minimum requirement that the infectious agent, tissue culture target cell, and assay species be as close to homologous as possible.

4 Scrapie Agent Titers In Vitro

There is now no doubt that scrapie agent will replicate continuously in vitro. There has been concern, however, that the amount of agent which can be realistically recovered from tissue culture sources may not be sufficient for biochemical analyses. Various investigators have reported from one to four mouse 50% lethal doses (LD_{50}) per cell (RUBENSTEIN et al. 1990), to one LD_{50} per 10000 cells (ASHER et al. 1979). Generally, though one mouse LD_{50} has required approximately 100 cells (CLARKE and HAIG 1970; RACE et al. 1987; BUTLER et al. 1988). The highest reported titer, four LD_{50} per cell, has been obtained in scrapie-infected PC-12 cells (RUBENSTEIN et al. 1990). PC-12 cells are terminally differentiated by the addition of nerve growth factor (NGF) and then exposed to scrapie agent. Cultures are not split or serially passaged once they have been established. Titers in the PC-12 cultures increase approximately 1000-fold during their 70 days in culture, indicating that the amount of agent per cell can increase substantially during the culture period. In contrast to the scrapie-infected PC-12 cultures are the neuro 2a type cultures (MARKOVITS et al. 1981; RACE et al. 1987; BUTLER et al. 1988). These cells require continual serial in vitro passage, which likely prevents significant accumulation of agent within individual cells and, thus, the relatively low titers of scrapie agent per cell compared to the PC-12 cells. However, they offer other advantages which will be discussed later. LD_{50} determinations for all of the scrapie-infected cell lines for which this information is available were done by inoculating mice with dilutions of cell homogenates and, therefore, reflect LD_{50} values for a given number of cell equivalents. A significant question with regard to all of the positive cultures is whether the infectivity that has been detected represents uniform infection of most or all of the cells of a given culture or rather relatively greater infection of very few cells. To obtain this information, we utilized limited dilution cloning to determine the frequency of infected cells in various cultures which we developed. Briefly, this involved establishing cultures from single cells, allowing the cells to expand and finally analyzing the cultures by infecting mice with cell aliquots of the expanded cultures. Using this technique, we found that in the earliest cultures we developed, <2% of the cells contained scrapie agent. However, by recloning from the few scrapie-positive cultures we found, it was possible to increase the frequency of positive cells to ~88% (RACE et al. 1987). The very low frequency of positive cells in our initial cultures may explain ithe relatively quick loss of infectivity (after only 10–15 in vitro passages) which we observed in them, and may also explain the transient infectivity of cultures developed by others (MARKOVITS et al. 1981; ASHER et al. 1979). When cells from cultures containing 88% scrapie positive cells were titered in mice, we found that one mouse LD_{50} required approximately 100 cell equivalents. This result suggested that bioassay techniques may significantly underestimate the amount of agent contained in the cultures analyzed. Immune responses by the mouse toward either cell-associated antigens or components of the scrapie agent may result in the inactivation of a significant amount of the agent which was inoculated. Indeed, it has been reported that the addition of mouse peritoneal macrophages significantly reduced the yield of infectivity from scrapie agent-infected PC-12 cultures (RUBENSTEIN et al. 1984).

5 Scrapie Agent Survival In Vitro

It has been mentioned that in the earliest cultures which we developed, as well as some of those developed by others, infectivity was transient, that is, agent persisted in the cultures long enough to prove that agent had replicated but was lost soon thereafter. Cultures which we have established from single cells infected with scrapie agent have maintained relatively high levels of agent over significantly more in vitro passages. Stable levels of infectivity persisting for 50 or more in vitro passages have been observed consistently. Unfortunately, however, persistence of agent has not been an absolute characteristic of our cultures. Some cultures have lost agent, and at unpredictable times after their initiation. We have not yet identified factors which account for the loss of infectivity, though virtually any sort of stress seems to lead to the loss of agent (Race, unpublished). In some instances, though, seemingly identical cultures derived as clones from a common parent culture have exhibited differences in agent persistence. BORCHELT et al. (1990) observed a similar situation in cultures recovered from stocks which had been held in liquid nitrogen but reported stable infectivity in one line maintained continuously in vitro for 18 months.

It is clear that scrapie-positive cultures can differ in their in vitro passage requirements. For example, the scrapie-infected cultures which we have utilized are routinely passaged twice each week at a 1:20 dilution, whereas those of others are passaged once every 7–10 days at a 1:5 or 1:10 dilution (BORCHELT et al. 1990). An understanding of mechanisms of agent persistence in vitro could be relevant in vivo, and these mechanisms, when identified, could provide valuable information regarding agent-cell interactions in scrapie.

6 Detection of Proteinase K-Resistant PrP in Cell Cultures

Until recently the lack of absolutely stable levels of infectivity within given cultures necessitated that cells from individual cultures be assayed in mice to verify the continued presence of scrapie agent. This situation imposed obvious limitations on the use of scrapie-infected cultures. This problem has been overcome recently with the detection of PK-resistant PrP in scrapie-infected cell cultures (BUTLER et al. 1988; CAUGHEY et al. 1989). The detection of PK-resistant PrP and the presence of scrapie agent have been shown to be correlated in mouse (BOLTON et al. 1984; KASCSAK et al. 1985) and hamster brain (BOLTON et al. 1982; MCKINLEY et al. 1983) and recently in a limited number of scrapie-infected cell cultures (BUTLER et al. 1988; CAUGHEY et al. 1990). To extend this observation and further substantiate the association between the detection of PK-resistant PrP and the presence of scrapie agent, we analyzed several cell cultures which represented clones derived from a single scrapie-infected parent culture. An aliquot of each culture was inoculated into mice to detect scrapie agent while another aliquot of each culture was analyzed by immunoblotting for the presence of PK-resistant PrP. Cultures which gave PK-resistant PrP bands on immunoblots always caused scrapie in mice, and cultures which failed to kill mice never gave PK-resistant PrP bands on immunoblots. However, a few clones caused scrapie when inoculated into mice but failed to give PK-resistant PrP bands on immunoblots. So, although positive demonstration of PK-resistant PrP on immunoblots was always associated with infectivity, infectivity was not necessarily associated with a positive immunoblot. This result very likely reflects greater sensitivity of the bioassay for infectivity than of PrP detection by immunoblot. Detection of PK-resistant PrP in cell culture does appear to offer an excellent method of monitoring at least the presence of infectivity in given samples provided a certain, as yet undetermined, threshold amount of agent is present.

Based on this result, we felt that it might be possible to develop a quantitive assay where the amount of PK-resistant PrP detected by immunoblot in a given sample could be compared to the amount present in a standard preparation that has previously determined PrP-infectivity correlations. Before such an assay could be developed, however, we felt it necessary to determine the influence of PK on cell-associated scrapie agent, because PK is utilized in procedures for the purification of PK-resistant PrP. Reliable correlations would be impossible if the PK destroys cell-associated agent. To make this determination we exposed cell lysates to varying amounts of PK for various lengths of time. PK-resistant PrP was reduced by 95% when exposed to PK at a concentration of 500 μg/ml for 6 h at 37 °C. However, exposure of cell lysates to 100 μg PK for 1 h at 37 °C resulted in little or no detectable loss of scrapie agent. The use of PK, therefore, seems appropriate in protocols designed to isolate PK-resistant PrP and should not compromise efforts to establish assays correlating detection of PK-resistant PrP and agent infectivity titers.

7 Morphological Changes in Scrapie-Infected Cell Cultures

Although cytolytic changes have not been observed in scrapie-infected cell cultures, various other observations have been made. For instance, one group of investigators described increased catecholamine levels and decreased free serotonin and noradrenaline levels in scrapie-positive cultures compared to uninfected ones (MARKOVITS et al. 1983). They also described increased cell proliferation rates and various morphologic differences between cells exposed to homogenates containing scrapie agent and cells not exposed. Unfortunately, it was not clear that the changes described were necessarily due only to the scrapie agent rather than to other factors present in the brain homogenate used to infect the cells, especially since these cultures had very little to no agent remaining after only a few in vitro passages (MARKOVITS et al. 1981). CHEREDNICHENKO et al. (1985) described several chromosome alterations in scrapie-infected cultures, including chromosome rearrangements and changes in chromosome morphology such that three new marker chromosomes were identified in infected cultures and another lost. Most of these changes occurred over the course of 95 in vitro passages. This group also described other changes such as cell vacuolization, pyknotic degenerated nuclei and foamy cytoplasm in cells of infected cultures, and although these cultures were certainly infected with scrapie agent at the time of the analyses it was not clear whether appropriate controls were included in the experiments. Uninfected cells, unless they were exposed to brain homogenates prepared in a similar way, would not be satisfactory. ROIKHEL et al. (1987) described vacuolization and hypertrophy of the Golgi apparatus in infected cells compared to uninfected cells which had been exposed to normal brain homogenate. Several investigators have described an increased rate of cell proliferation in scrapie-infected cultures (ROIKHEL et al. 1987; MARKOVITS et al. 1983) while others have described the opposite effect (CHEREDNICHENKO et al. 1985a; BORCHELT et al. 1990). The significance of these observations is unclear, and all need to be confirmed.

Over a period of several years, we have observed a number of scrapie-infected clones which gave distinct and unique morphologic phenotypes that were unlike the uninfected cells from which they were derived. In every instance, however, we identified uninfected clones which were identical to infected ones. In other instances, infected clones have reverted to a non-infected state with no change in morphologic appearance. Indeed, we have also been able to alter the morphologic appearance of various clones by varying such parameters as cell density, source of fetal calf serum and temperature of incubation. It seems probable that the morphology of various cultures results from the interaction of many parameters. Scrapie agent in the initial preparation seems to be related to initial morphologic alterations in some cultures, but morphologic changes, once induced, do not seem dependent on the continued presence of scrapie agent (Race, unpublished). Our results in this regard point to the fact that investigators should be especially careful to substantiate that changes in the cellular morphology of various scrapie-infected cultures are truly scrapie agent dependent and/or induced. Clearly, much remains to be learned about the interactions of the scrapie agent with its in vitro cellular host(s).

8 Creutzfeldt–Jakob Agent in Cultured Cells

Several attempts have been made over the past decade to establish the causative agent of Creutzfeld–Jakob disease (CJD) in cell culture. Early approaches utilized explant techniques and showed that agent derived from either human or animal brain could persist in primary cultures for as long as 9 months but was quickly lost when serial passages were initiated (ASHER et al. 1979; YANAGIHARA et al. 1980).

Others have reported that explant cultures derived from CJD-infected mice or hamsters lose contact inhibition and become immortalized. One such hamster culture, as well as a culture derived from a CJD-infected guinea pig, were reported to contain CJD agent after sufficient in vitro passages had elapsed to exclude the detection of agent present in the original brain (MANUELIDIS et al. 1987).

Several attempts to establish CJD-infected cultures by exposing various cell lines to brain homogenates derived from CJD-infected humans or animals have also been made. The earliest attempts were not successful (ASHER et al. 1979; YANAGIHARA et al. 1980). Recently, however, similar efforts using the same neuro 2a cells which we (RACE et al. 1987) and others (BUTLER et al. 1988) successfully infected with scrapie agent has led to the establishment of tissue cultures persistently infected with the agent of mouse-adapted CJD (KINGSBURY et al. 1984; BUTLER et al. 1988).

9 Limitations and Advantages of Scrapie-Infected Cultures

The greatest potential limitation to the use of any of the scrapie-infected cultures that have been described so far relates to the relatively low amount of agent that can be easily obtained from cultures compared to the amount which can be obtained from the brains of scrapie-infected mice or hamsters. The highest titer that has been reported from a tissue culture source is four LD_{50} per cell. At this level, one 150-cm^2 flask of cells provides approximately the same amount of scrapie agent as the brain of one scrapie-infected mouse. However, to achieve titers this high requires that cells remain in culture for approximately 70 days (RUBENSTEIN et al. 1990). The need for cells to be in culture for such a long period of time would, it seems, present certain logistical problems in planning of experiments unless cultures are continually and routinely established at various intervals. Furthermore, scrapie-infected PC-12 cells require the constant presence of NGF, which is currently extremely expensive if purchased from commercial suppliers. To obtain large numbers of cells could be impractical considering the cost. Nevertheless, if four LD_{50} per cell turns out to be a reliable measure of the amount of agent in these cells, then they should be particularly useful for structural and biochemical analyses of the scrapie agent. Scrapie-infected PC-12 cells may also be particularly well suited for studies involving the influences of various neurotrophic factors on scrapie agent replication.

The neuro 2a type cultures require approximately 100 cells per LD_{50} or approximately 400 150-cm^2-flask equivalents of cells to equal the amount of infectivity that can be obtained from a single scrapie-positive mouse brain (RACE et al. 1987; BUTLER et al. 1988). They also present the potential problem of agent loss from established cultures and inconsistent agent titers, as the frequency of infected cells changes at various passages within various cultures. The neuro 2a cells offer certain advantages over the PC-12 cells. Because they can be passaged continuously in vitro, cells can be available for use at any time. They can be monitored for the presence of scrapie agent by analyzing them for the presence of PK-resistant PrP. Infected cells can be preserved by freezing and recovered whenever they are needed. They can be expanded quickly to whatever numbers are required. Furthermore, they do not require the addition of NGF or other additives and, in contrast to the PC-12 cells, are easily handled in the laboratory.

Several other scrapie-infected culture systems which have been reported have not been sufficiently well characterized in terms of the quantity of agent present to allow comparison with those discussed above (MARKOVITS et al. 1983; CHEREDNICHENKO et al. 1985a; ROIKHEL et al. 1987). Scrapie-infected cells of any derivation should be useful in studying the species barrier restrictions observed for scrapie isolates in vivo and are, in fact, being used in several studies of this type (Caughey, unpublished).

From a practical perspective, tissue cultures, with the possible exception of the PC-12 cells, cannot at this time begin to provide the amounts of agent that can be obtained from animal tissues. Nevertheless, culture-derived agent can be titered over a range of several $\log_{10}$ of infectivity. Infected cultures are, therefore, satisfactory, if not ideal, for a variety of experimental procedures. To date they have been used in several studies of prion protein biosynthesis (STAHL et al. 1987; RACE et al. 1988; CAUGHEY et al. 1989, 1990; BORCHELT et al. 1990). The neuro 2a cell lines are being used extensively in studies aimed at defining mechanisms which may account for the post-translational change of PrP to PK-resistant PrP. Especially exciting are the opportunities now available to manipulate the PrP gene on a molecular level using scrapie-infected or -susceptible cells. There is little doubt that infected cells, even with the low scrapie titers currently associated with them, will play a progressively more important role in studies designed to enhance our understanding of agent-cell interactions in this unique disease, as well as studies which may lead to an understanding of the structural nature of the infectious agent.

References

Alper T (1985) Scrapie agent unlike viruses in size and susceptibility to inactivation by ionizing or ultraviolet radiation. Nature 317: 750

Asher DM, Yanagihara RT, Rodgers NG, Gibbs CJ Jr, Gajdusek DC (1979) Studies of the viruses of spongiform encephalopathies in cell cultures. In: Prusiner SB, Hadlow WJ, (eds) Slow transmissible diseases of the nervous system, vol 2, Academic, New York. pp 235–242

Bellinger-Kawahara C, Cleaver JE, Diener TO, Prusiner SB (1987) Purified scrapie prions resist inactivation by UV irradiation. J Virol 61: 159–166

Bolton DC, McKinley MP, Prusiner SB (1982) Identification of a protein that purifies with the scrapie prion. Science 218: 1309–1311

Bolton DC, McKinley MP, Prusiner SB (1984) Molecular characteristics of the major scrapie prion protein. Biochemistry 23: 5898–5906

Borchelt DR, Scott M, Taraboulos A, Stahl N, Prusiner SB (1990) Scrapie and cellular prion proteins differ in the kinetics of synthesis and topology in cultured cells. J C Biol 110: 743–752

Brown P, Rohwer RG, Moreau-Dubois M-C, Green EM, Gajdusek DC (1981) Use of the golden hamster in the study of scrapie virus. Adv Exp Med Biol 134: 365–373

Brown P, Rohwer RG, Green EM, Gajdusek DC (1982) Effect of chemicals, heat and histopathological processing on high infectivity hamster-adapted scrapie virus. J Infect Dis 145: 683–687

Bruce M, Dickinson AG (1979) Biological stability of different classes of scrapie agent. In: Prusiner SB, Hadlow WJ, (eds) Slow transmissible diseases of the nervous systems, Academic, New York, pp 71–86

Buening G M, Gustafson DP (1971) Growth characteristics of scrapie agent-infected mouse brain cell cultures, Am J Vet Res 32: 953–958

Butler DA, Scott M, Bockman JM, Borchelt DR, Taraboulos A, Hsiao KK, Kingsbury DT, Prusiner SB (1988) Scrapie-infected murine neuroblastoma cells produce protease-resistant prion proteins. J Virol 62: 1558–1564

Caspary EA, Bell TM (1971) Growth Potential of scrapie mouse brain in vitro. Nature 229: 269–270

Caughey B, Race RE, Ernst D, Buchmeier MJ, Chesebro B (1989) Prion protein (PrP) biosynthesis in scrapie-infected and uninfected neuroblastoma cells. J Virol 63:175–181

Caughey B, Neary K, Buller R, Ernst D, Perry L, Chesebro B, Race R (1990) Normal and scrapie-associated forms of prion protein differ in their sensitivities to phospholipase and proteases in intact neuroblastoma cells. J Virol 64: 1093–1101

Cherednichenko YN, Mikhailova GR, Rajcani J, Zhdanov VM (1985a) In vitro studies with the scrapie agent. Acta Virol 29: 285–293

Cherednichenko YN, Fadeeva LL, Sologub VK (1985b) Antigenic changes in the cells latently infected with the scrapie agent. Acta Virol 29: 515

Chesebro B, Wehrly K, Cloyd M, Britt W, Portis J, Collins J, Nishio J (1981) Characterization of mouse monoclonal antibodies specific for Friend murine leukemia virus-induced erythroleukemia cells: Friend-specific and FMR-specific antigens. Virology 112: 131–144

Clarke MC (1979) Infection of cell cultures with scrapie agent. In: Prusiner SB, Hadlow WJ (eds) Slow transmissible diseases of the nervous system, vol 2. Academic, New York. pp 225–232

Clarke MC, Haig DA (1970a) Multiplication of scrapie agent in cell culture. Res Vet Sci 11: 500–501

Clarke MC, Haig DA (1970b) Evidence for the multiplication of scrapie agent in cell culture. Nature 225: 100–101

Clarke MC, Millson GC (1976) Infection of a cell line of mouse L fibroblasts with scrapie agent. Nature 261: 144–145

Dees C, Wade WF, German TL, Marsh RF (1985) Inactivation of the scrapie agent by ultraviolet irradiation in the presence of chlorpromazine. J Gen Virol 66: 845–849

Eklund CM, Kennedy RC, Hadlow WJ (1967) Pathogenesis of scrapie virus infection in the mouse. J Infect Dis 117: 15–22

Elleman CJ (1984) Attempts to establish the scrapie agent in cell lines. Vet Res Commun 8: 309–316

Field EJ, Windsor GD (1965) Cultural Characters of scrapie mouse brain. Res Vet Sci 6: 130–132

Gibbons RA, Hunter GD (1967) Nature of the scrapie agent. Nature 215: 1041–1043

Gibbs CJ Jr, Gajdusek DC, Amyx H (1979) Strain variation in the viruses of Creutzfeldt-Jakob disease and kuru. In: Prusiner SB, Hadlow WJ (eds) Slow transmissible diseases of the nervous system, vol 2. Academic, New York, pp 87–110

Gustafson DP, Kanitz CL (1965) Evidence of the presence of scrapie in cell cultures of brain. In: Gajdusek DC, Gibbs CJ, Alpers M, (eds) Slow latent and temperate virus infections. US Department of Health, Education and Welfare, Bethesda, pp 221–236

Haig DA, Pattison IH (1967) In vitro growth of pieces of brain from scrapie-affected mice. J Pathol Bacteriol 93: 724–727

Hunter GD (1979) The enigma of the scrapie agent: biochemical approaches and the involvement of membranes and nucleic acids. In: Prusiner SB, Hadlow WJ, (eds) Slow transmissible diseases of the nervous system, vol 2. Academic, New York, pp 365–385

Hunter GD, Kimberlin RH, Gibbons RA (1968) Scrapie: a modified membrane hypothesis. J Theor Biol 20: 355–357

Kascsak RJ, Rubenstein R, Merz PA, Carp RI, Wisniewski HM, Diringer H (1985) Biochemical differences among scrapie-associated fibrils support the biological diversity of scrapie agents. J Gen Virol 66: 1715–1722

Kasper KC, Stites DP, Bowman KA, Panitch H, Prusiner SB (1982) Immunological studies of scrapie infection. J Neuroimmunol 3: 187–201

Kimberlin RH, Walker CA (1978) Evidence that the transmission of one source of scrapie agent to hamsters involves separation of agent strains from a mixture. J Gen Virol 39: 487–496

Kimberlin RH, Walker CA, Millson GC (1975) Interspecies transmission of scrapie-like diseases. Lancet ii: 1309–1310

Kingsbury DT, Smeltzer D, Bockman J (1984) Prurification and properties of the V. Fu. Isolate of the agent of Creutzfeldt–Jakob disease. 6th International Congress on Virology, 7 Sept 1984, Sendai, p 70 Abstr (no W 47–6)

Manuelidis EE, Fritch WW, Kim JH, Manuelidis L (1987) Immortality of cell cultures derived from brains of mice and hamsters infected with Creutzfeldt–Jakob disease agent. Proc Natl Acad Sci USA 84: 871–875

Markovits P, Dormont D, Delpech B, Court L, Latarjet R (1981) Essais de propagtion in vitro de l'agent scrapie dans des cellules nerveuses de souris. C R Acad Sci (Paris) 293: 413–417

Markovits P, Dautheville C, Dormont D, Dianoux L, Latarjet R (1983) in vitro propagation of the scrapie agent. Transformation of mouse glia and neuroblastoma cells after infection with the mouse-adapted scrapie strain c-506. Acta Neuropathol (Berl) 60: 75–80

Marsh RF, Kimberlin RH (1975) Comparison of scrapie and transmissible mink encephalopathy in hamsters. II. Clinical signs, pathology, and pathogenesis. J Infect Dis 131: 104–110

McKinley MP, Bolton DC, Prusiner SB (1983) A protease-resistant protein is a structural component of the scrapie prion. Cell 35: 57–62

Race RE, Fadness LH, Chesebro B (1987) Characterization of scrapie infection in mouse neuroblastoma cells. J Gen Virol 68: 1391–1399

Race RE, Caughey B, Graham K, Ernst D, Chesebro B (1988) Analyses of frequency of infection, specific infectivity, and prion protein biosynthesis in scrapie-infected neuroblastoma cell clones. J Virol 62: 2845–2849

Rohwer RG (1984) Scrapie infectious agent is virus-like in size and susceptibility to inactivation. Nature 308: 658–661

Roikhel VM, Fokina GI, Lisak VM, Kondakova LI, Korclev MB, Pogodina VV (1987) Persistence of the scrapie agent in glial cells from rat gasserian ganglion. Acta Virol 31: 36–42

Rubenstein R, Carp RI, Callahan SM (1984) In vitro replication of scrapie agent in a neuronal model: infection of PC12 cells. J Gen Virol 65: 2191–2198

Rubenstein R, Scalici CL, Papini MC, Callahan SM, Carp RI (1990) Further characterization of scrapie replication in PC12 cells. J Gen Virol 70: (in press)

Stahl N, Borchelt DR, Hsiao K, Prusiner SB (1987) Scrapie prion protein contains a phosphatidylinositol glycolipid. Cell 51: 229–240

Taraboulos A, Serban D, Prusiner SB (1990) Scrapie prion proteins accumulate in the cytoplasm of persistently infected cultured cells. J Cell Biol (in press)

Yanagihara RT, Asher DM, Gibbs CJ Jr, Gajdusek DC (1980) Attempts to establish cell cultures infected with the viruses of subacute spongiform encephalopathies (40974). Proc Soc Exp Biol Med 165: 298–305

The Scrapie Agent: "A Virus by Any Other Name"

R. G. ROHWER

1 Introduction

1.1 The Case for a Virus

Scrapie is a fatal, central nervous system disease of sheep and goats caused by a serially transmissible agent (CUILLE and CHELLE 1936) that passes bacterial filters (CUILLE and CHELLE 1938; WILSON et al. 1950). Serial transmissions at high dilutions of inoculum demonstrate that the agent is self-propagating. Filterability establishes that it is subcellular. Other estimates of size (see below) based on sedimentation velocity, filtration, exclusion chromatography and sensitivity to ionizing radiation place the scrapie agent with the smaller viruses.

At least 20 distinct strains of mouse-adapted scrapie are recognized (DICKINSON et al. 1984). In general the phenotypes of these strains are conserved

University of North Carolina, Chapel Hill, North Carolina, USA

Current Topics in Microbiology and Immunology, Vol. 172

Table 1. Virucidal treatments that kill ≥95% of scrapie infectivity

Heat (100°C)	Chloroform-methanol 2:1
Heat (121°C)	Ether
Phenol extraction	Guanidinum hydrogen bromide
Sodium hypochlorite	Lithium chloride 12*M*
Sodium metaperiodate	Potassium permanganate
Chlorine dioxide	Sodium dodecylsulphate
Lysol	Sodium lauryl sarcosinate
Iodine	Triton X-100
Formaldehyde	NaOH 0.1 *N*
Acetone	Guanidinum Thiocyanate 0.2*M*
2-Chloroethanol	Guanidinium hydrochloride 0.2*M*
	Potassium thiocyanate 6*M*

From MILLSON et al. 1976; PRUSINER et al. 1981a; BROWN et al. 1982; ROHWER 1984a, b

during passage through different inbred mouse lines or during passage through alternative species (BRUCE and DICKINSON 1987; KIMBERLIN et al. 1989). Only rarely does a strain phenotype change (BRUCE and DICKINSON 1980, 1987; KIMBERLIN et al. 1989). Strain variation and mutation imply the existence of agent-encoded genetic information.

Scrapie infectivity, like that of many other viruses, is sensitive to the treatments listed in Table 1.

Until recently there was far less evidence that non-A, non-B hepatitis was viral, and yet few doubted it (HOLLINGER 1990a). Acquired immunodeficiency syndrome was quickly recognized to be a viral disease largely on the basis of epidemiological evidence of its transmissibility (CATES 1990). Hepatitis B and Borna disease were also considered viral on epidemiological grounds long before there was any physical evidence for a virus (HOLLINGER 1990b; LUDWIG et al. 1988).

1.2 The Case Against a Virus

In contrast, the nature of the scrapie agent has been, and remains, extremely controversial in spite of the compelling indicators of a viral etiology enumerated above. Disputed is the interpretation and significance of several key physical and biological properties of the agent. First, the inactivation rate constant for ionizing radiation is the largest yet measured for a virus, and this parameter has been used to calculate a target size of 50 000–150 000 Da. Second, even though the vast majority (95%) of the infectivity is sensitive to virucidal treatments (Table 1), small refractory subpopulations survive a number of these same treatments under conditions that are sterilizing for many viruses. These include 60-min. exposures to 100°C or 121°C (ROHWER 1984a), formalin fixation and paraffin embedding (BROWN et al. 1982), and dilution into common chemical inactivants based on hypochlorites, iodophors, aldehydes, phenolics, peroxides, and quaternary amines (ROHWER 1984b; BROWN et al. 1986). Finally, the infection is noninflammatory and no agent-specific antibody has been identified (BROWN 1990).

Many have felt that these properties are incompatible with what has come to be meant by a virus, i.e., an autonomous nucleic acid-based genome packaged for extracellular survival and host-to-host, and cell-to-cell, transmission in a particle constructed from virus-encoded proteins and other macromolecules of virus or host origin. In the case of the scrapie agent, it is maintained that its target size is too small, its resistance to inactivation is too great, and its apparent lack of immunogenicity is too unlikely.

These arguments have been bolstered by the tenacious association of infectivity with a brain amyloid, the scrapie-associated fibril (SAF) (P.A. MERZ et al. 1981; PRUSINER et al. 1983; BRAIG and DIRINGER 1985), also called the prion rod (PRUSINER et al. 1983; ROHWER 1984c), an insoluble, protease-resistant condensation of the host (CHESEBRO et al. 1985; BASLER et al. 1986) PrP protein. This association has been interpreted by supporters of the prion hypothesis as establishing the conjunction of infectivity and the PrP protein. They propose that the protein itself, or some small aggregate of it, is the agent (PRUSINER 1989). Supporting this position is the observation that infectivity is protease sensitive and nuclease resistant (PRUSINER 1982). Several anomalous measurements by filration (PRUSINER et al. 1982; COEROLI-VIGOUROUX et al. 1983; POCCHIARI et al. 1988) and exclusion chromatography (PRUSINER 1982) have also figured prominently in support of the notion of subviral size.

At issue are: (1) the credibility of the target size calculation; (2) the credibility of the anomalous filtration and exclusion chromatography experiments indicating subviral size; (3) the significance of small subpopulations of inactivation-resistant infectivity to an assessment of the structure of the majority population; (4) the significance of our inability as investigators to demonstrate an antigen, or, should there in fact be no immune response to the agent, the significance of a low immunogenicity to a virus model; (5) the interpretation of nuclease resistance, and (6) the integrity of the association between infectivity and PrP. These will be addressed in turn.

2 Points at Issue

2.1 The Target Size Calculation

Inactivation of viruses by ionizing radiation has been used to estimate their size (LEA 1955). When radioactive particles pass through a dispersion of viruses, the smaller the virus, the less likely it is to be hit by a particle and the more resistant it will be to inactivation. The inactivation rate constant, D_{37}, is the dose depositing, on average, one inactivating event per unit of infectivity. Thus, the smaller the virus the greater will be its inactivation rate constant. (Since dose is distributed randomly, the actual population survival at the inactivating dose is given by the

P_0 term of the Poisson distribution and is $1/e = 0.37$. Thus the rate constant is often designated D_{37}.) Nucleic acid accounts for most, if not all, of the radiosensitivity of viruses (HRADECNA 1966; GINOZA and NORMAN 1957; WATSON 1950; LYTLE and GINOZA 1969; LATARJET and FREDERICQ 1955). Hence, the greater the D_{37}, the smaller the viral genome.

Scrapie infectivity has the largest inactivation rate constant yet measured for a virus. This value has seemed so surprising to some, that the measurement has now been repeated 13 times. D_{37} values ranged from 4.2×10^6 to 4.9×10^6 rads for determinations on dried specimens or on suspensions frozen at $-78\,°C$ (ALPER et al. 1966; ALPER and HAIG 1968; GIBBS et al. 1978; FIELD et al. 1969). Somewhat lower values of 2.5×10^6 to 4.2×10^6 rads were obtained for suspensions frozen at $-135\,°C$ (BELLINGER-KAWAHARA et al. 1988).

The D_{37} value is expressed in units of dose/inactivation where dose is given in units of energy/mass. In its simplest form, the mass of the "target" is computed directly from the D_{37} by solving for mass:

$$MW = kE/D_{37}$$

where *MW* is the molecular weight in daltons, *k* is a constant needed to convert all values to the same units, and *E* is the energy expended in a single inactivation event. Attempts have been made to compute the value of *E* from first principles (LEA 1955). In practice, *E* is estimated from the molecular weight and D_{37} of molecules of known size and varies from 60 to 100 eV (ALPER 1987).

Calculated target sizes for scrapie infectivity have ranged from 50 000 to 150 000 Da of nucleic acid, equivalent to 150–500 nucleotides. These estimates of subviral size have been the single most important factor in both instigating and perpetuating the idea of an "unconventional agent".

But how good are these calculations? Validation requires comparison of calculated values with known molecular weights of independently characterized standards. The genomic molecular weights of most well-characterized viruses, which includes most of those for which D_{37} values have been determined in the past, are now known with great accuracy. More than 150 have been completely sequenced (GenBank 1990) and as many others have been restriction-mapped. Virus nucleic acid molecular weights are plotted against D_{37} in Fig. 1 (ROHWER 1984b). The dotted line is computed from the target theory as represented in the equation above with $E = 66$ (ALPER 1987). Various refinements of the theory have the effect of modifying the value of *E* and shifting the line up or down slightly while retaining the same slope (LEA 1955; ALPER 1987). However, since the target theory considers viruses, enzymes and other macromolecules as a single group there is always but one line.

In contrast, the overwhelming evidence is that single-stranded (ss) and double-stranded (ds) viruses comprise distinct groups with respect to their sensitivities to ionizing radiation (TESSMAN et al. 1957; HOTZ and MULLER 1961; SCHAMBRA and HUTCHINSON 1964; GINOZA and MILLER 1965; GINOZA 1968). Separate regressions of the empirically determined values for ss and ds viruses bear this out (the solid lines in Fig. 1). Considering our current knowledge of the

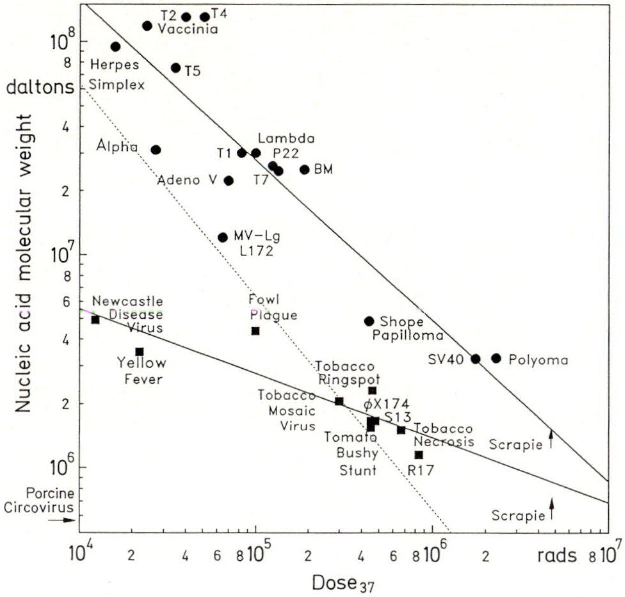

Fig. 1. Radiosensitivity of viruses as a function of nucleic acid molecular weight. *Circles* and *squares* represent viruses with double-stranded and single-stranded nucleic acids respectively. The *solid lines* are separate regressions of these two classes. The *dotted line* is that calculated from the target theory. *Arrows* indicate the positions of the scrapie intercepts of the ss and ds regression lines. The molecular weight of the ss DNA of the porcine circovirus (Tischer et al. 1982) is indicated at the ordinate. Data are as referenced in ROHWER (1984b) except that nucleic acid molecular weights have been updated with sequence-derived values (acid form) where available (GenBank 1990) to include: adenovirus V ≡ adenovirus II, fowl plague virus ≡ influenza virus A, herpes simplex virus type 1, lambda, Newcastle disease virus ≡ parainfluenza virus 3, φX174, polyoma virus, R17 ≡ MS2, S13, Shope papilloma virus, SV 40, T7, tobacco mosaic virus, tomato bushy stunt virus, vaccinia virus and yellow fever virus. The D_{37} value for vaccinia is from MCCREA (1960), who first recognized the now well established biphasic nature of vaccinia virus inactivation (DECKER et al. 1969) and reported the initial component. The D_{37} value for Newcastle disease virus (WOESE and POLLARD 1954) was determined on a dry specimen as preferred by Alper (see text). (From ROHWER 1984b)

differences in structure and biology of the ss and ds viruses, this finding may not seem surprising. Nevertheless, it can not be predicted from the target theory. Nor can the target calculation account for the considerable scatter within each strandedness class. The scatter and outliers may reflect procedural variations in the measurement of D_{37} or contributions of structures other than nucleic acids to the inactivation process (WATSON 1950; LATARJET and FREDERICQ 1955; HRADECNA 1966; LYTLE and GINOZA 1969) or both, but whatever the cause, they represent practical limitations on the analysis that are recognized by the standard curves but can not be utilized in the target calculation.

When the D_{37} for scrapie is compared to the extrapolated lines in Fig. 1, one obtains values of 1.5×10^6 Da or 0.90×10^6 Da depending upon whether or not the genome is ds or ss. The smallest known autonomously replicating virus, the porcine circovirus, has an ss DNA genome of 0.58×10^6 Da (TISCHER et al. 1982).

Were D_{37} values available for the circoviruses, parvoviruses, hepadnaviruses, and hepatitis delta agent, the D_{37} for the scrapie virus might not appear so unique and might well be bracketed in Fig. 1 by other values.

Direct comparison with standards was not an approach that was available in the 1930s and 1940s to the originators of the target theory (LEA 1955). Viruses were recognized to be subcellular, but consistent estimates of size based upon filtration, diffusion, sedimentation, and electron microscopy existed for only a few of them. It was recognized that there was a radiosensitive component of viruses that was distinct from the complete particle, but it was another decade before the genomic role of nucleic acids in virus replication was widely appreciated (HERSHEY and CHASE 1952; FRAENKEL-CONRAT 1956). Meanwhile, the target calculation was used to corroborate the small sizes reported by other methods.

However, a standard curve is always preferable when the means exists to construct one. It can, in one implementation, validate the method, minimize the number of assumptions required to estimate the result, incorporate any inherent variability in the measurements or the presumed relationship, and shortcut otherwise complex and imprecise data reductions. It seems likely that, had the early investigators had access to the abundant structural information that we now possess, they would also have favored an empirical approach for analyzing unknowns. To do otherwise is analogous to computing molecular weights from first principles on the basis of electrophoretic mobility or S_{20W} values.

The use of standard curves for the analysis of D_{37} data has been disputed by Alper (ALPER 1985, 1987; ROHWER 1986), the author of the original target size calculation. While widely considered equivalent by many investigators (TERZI 1961; GINOZA 1963; LATARJET 1979), the data in Fig. 1 include D_{37} values obtained from dry specimens, broth suspensions, and/or frozen samples. Alper conducted her experiment under oxygen on a dry specimen of scrapie brain homogenate and argues that comparison with broth or frozen samples would invalidate the result. In fact, when the data obtained from dry samples, frozen samples, or broth suspensions are segregated, one obtains essentially the same result from each standard curve (R.G. ROHWER , unpublished). For example, the D_{37} values for scrapie infectivity obtained from dry and frozen specimens (−78°C) are indistinguishable (ALPER et al. 1966; ALPER and HAIG 1968; FIELD et al. 1969; GIBBS et al. 1978).

Relatively few viruses tolerate dehydration. Nevertheless, Fig. 2 gives the standard curves for dry specimens under oxygen constructed from those determinations that do exist. Figure 3 demonstrates that, even if one considers only the measurements that were made by Alper, one obtains a consistent result. The intercept of the ds line is 1.5×10^6, of the ss line, 0.40×10^6. While extrapolations over several decades are seldom valid, it is nevertheless clear that these data do not support the target calculation. Furthermore, the dearth of D_{37} data for dried preparations of small ds viruses indicates that the validity of the target computation for this group has never in fact been tested under these conditions.

Alper, acknowledging the need to validate the target calculation, has also constructed a standard curve relating virus nucleic acid molecular weights to D_{37}. The D_{37} data employed were obtained in her laboratory and are the same as those presented in Fig. 3. In contrast to Fig. 3, her standard curve shows an excellent correlation of molecular weight with the line predicted by the target calculation (ALPER 1987). However, this is due to the use of outdated molecular weight estimates dating from 1971 or earlier. Particularly misleading is her value of 4–8 × 10^7 Da for the molecular weight of the yellow fever virus (YFV) RNA estimated from the core diameter of a YFV particle in an electron micrograph (BERGOLD and WEIBEL 1962). The sequence-determined molecular weight, 3.7 × 10^6 Da, is more than tenfold lower (RICE et al. 1985). This error has the effect of moving the YFV point onto the line for the ds viruses and, in her limited data set, obscuring the differences in sensitivity between the ss and ds viruses. When

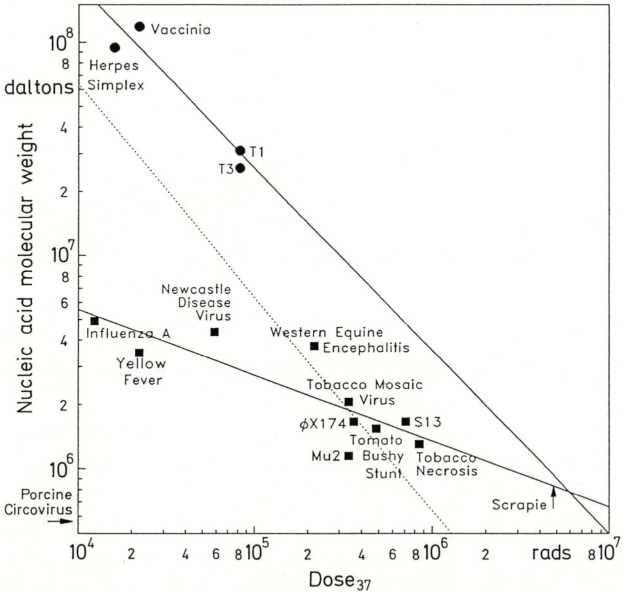

Fig. 2. Radiosensitivity of viruses irradiated dry in oxygen as a function of nucleic acid molecular weight. *Circles* and *squares* represent viruses with double-stranded and single-stranded nucleic acids respectively. The *solid lines* are separate regressions of these two classes. The *dotted line* is that calculated from the target theory. The *arrow* indicates the position of the scrapie intercepts of the ss and ds regression lines. The molecular weight of the ss DNA of the porcine circovirus (TISCHER et al. 1982) is indicated at the ordinate. D_{37} data: herpes simplex virus type 1, T1, T3, yellow fever virus, Mu2 (Alper & Haig 1968); vaccinia, as in Fig. 1; influenza virus A (Kroeger and Kempf 1959); Western equine encephalitis virus (Hruskova 1969); tobacco mosaic virus, tomato bushy stunt virus, tobacco necrosis virus (Lea and Smith 1940, 1942); S13 (Lea and Salaman 1946); ΦX174 (GINOZA 1963). Molecular weight data: herpes simplex virus type 1, influenza virus A, Western equine encephalitis virus ≡ Sindbis virus, Mu2 ≡ MS2, Newcastle disease virus ≡ parainfluenza virus 3, ΦX174, S13, tobacco mosaic virus, tomato bushy stunt virus, vaccinia virus, yellow fever virus (GenBank 1990); T3 (BAILEY and MCALLISTER 1980); T1 (Ramsay and Ritchie 1980); tobacco necrosis virus as in Fig. 1

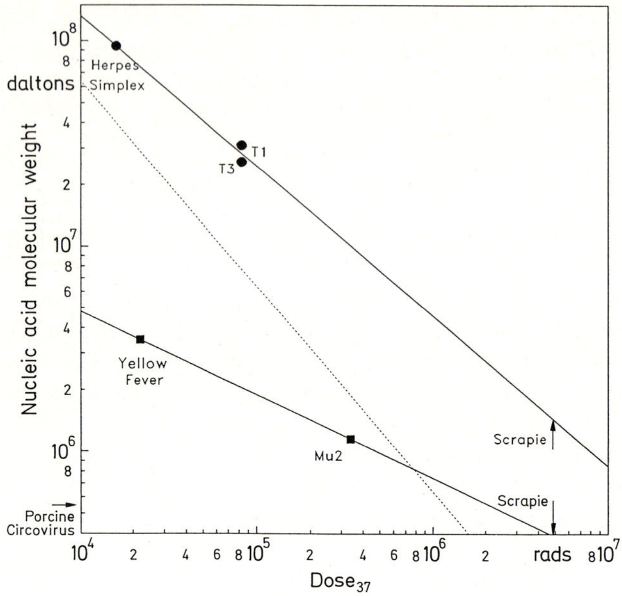

Fig. 3. Viruses irradiated dry under oxygen: Alper's data only. *Circles* and *squares* represent viruses with double-stranded and single-stranded nucleic acids respectively. The *solid lines* are separate regressions of these two classes. The *dotted line* is that calculated from the target theory. The *arrows* indicate the positions of the scrapie intercepts of the ss and ds regression lines. The molecular weight of the ss DNA of the porcine circovirus (TISCHER et al. 1982) is indicated at the ordinate. All D_{37} data are from ALPER et al. (1966) and ALPER and HAIG (1968). Molecular weight values are as in Fig. 2

the same D_{37} data are plotted with accurate molecular weights one obtains the pattern shown in Fig. 3. It is curious that, while tacitly admitting to the greater utility and predictive power of the standard curve by using it to validate the target calculation, Alper nevertheless abandons it for her analysis of the scrapie agent nucleic acid molecular weight.

It has been argued (ALPER 1987) that empirical comparisons with virus standards prejudice the case for viruses. Rather, the point is that the D_{37} for scrapie infectivity is consistent with a viral agent. It does not prove that the agent is in fact a virus, and it implies nothing about whatever else it might be. It does indeed greatly strengthen the case for a virus etiology, but only because, with the exception of the target size calculation, there is little reason to expect that scrapie is not caused by a virus.

It would be a mistake to consider the molecular weight estimates obtained from the regression intercepts in Fig. 1–3 as particularly accurate. Few of the viruses in Fig. 1 fall exactly on the regression lines, and there is no reason to expect that the scrapie virus would be any more likely to do so. Nevertheless, it is probable that the true molecular weight would not deviate any further from the line than do the values for the standard viruses.

The small target size calculated for the scrapie agent is central to the prion hypothesis (PRUSINER 1982, 1989; BELLINGER-KAWAHARA et al. 1988). Yet, if the molecular mass of the agent were in fact only 50 000–150 000 daltons, a paradox results if we consider, as is proposed, that SAF or prion rods are merely large aggregates of a much smaller unit of infectivity, the as yet hypothetical prion (PRUSINER et al. 1983). In infected brain homogenates, the scrapie isoform of PrP, PrP^{Sc}, is largely sequestered into SAF or prion rods and significant infectivity copurifies with these structures (DIRINGER et al. 1983a; MCKINLEY et al. 1983a; PRUSINER et al. 1984). Since the prion itself is postulated to be some small multimer of this same isoform of the host PrP protein, the SAF must themselves be large aggregates of prions (PRUSINER et al. 1983). If this were indeed the state of the infectivity in brain homogenates, then the assumption of simple first-order kinetics in the derivation of the scrapie agent's inactivation rate constant, the starting point for both the target analysis and the standard curve analysis, does not apply (ROHWER and GAJDUSEK 1980; ROHWER 1983).

Aggregation provides both intrinsic resistance to inactivation by virtue of the functional redundancy of the infectivity-bearing structures, and vastly increases the opportunities for multiplicity reactivation should an aggregate of damaged viruses enter a cell (KIM and SHARP 1967a, b, 1968; ROHWER and GAJDUSEK 1980; ROHWER 1984b). In this case the appearance of first-order kinetics can be given by any of several distributions of aggregates of heterogeneous size (ROHWER and GAJDUSEK 1980). The effect of this error in interpretation is always to overestimate the D_{37} and thereby underestimate the size of the intrinsic unit of infectivity (ROHWER and GAJDUSEK 1980). Thus the prion can not be both small and aggregated unless it is exceptionally radiosensitive.

The only determination of D_{37} in which an attempt was made to control aggregation state suggests that infectivity is not associated with PrP (BELLINGER-KAWAHARA et al. 1988). Within the sensitivity of the method, inactivation of SAF (i.e., prion rods), sonicated SAF, and emulsified brain homogenate all gave similar values of D_{37}. If infectivity resides in SAF or prion rods as multimers of PrP^{Sc}, then the infectivity titer per gram of PrP^{Sc} protein in the dispersions should have been greater than that in the parent brain homogenate, and the dispersions should have shown a greater sensitivity to inactivation than intact SAF. Either the dispersions were ineffective in disaggregating SAF, or SAF are not composed of large aggregates of infectious agent.

If, as is suggested in the prion hypothesis, the scrapie agent does not contain a genomic nucleic acid, then (PRUSINER 1982, 1989), in the absence of some other standard for comparison, one could not deduce its physical size either by direct comparison to virus standards, for which the majority of the radiosensitivity under direct effect is due to nucleic acid (WATSON 1950; LATARJET and FREDERICQ 1955; GINOZA and NORMAN 1957; HRADECNA 1966; LYTLE and GINOZA 1969), or by means of the target theory, for which an arbitrary assumption would have to be made about its inactivation efficiency and the relation of its radiosensitive components to its overall structure (LATARJET et al. 1959). For example, in the case of viruses, the radiosensitive target and the size calculated is that of the genome, not the entire particle (GINOZA and NORMAN 1957; HRADECNA 1966). Thus there is no a priori reason based upon the D_{37} to expect that the prion, as postulated, would be small.

2.1.1 Ultraviolet Inactivations

Scrapie is also resistant to inactivation by 254-nm ultraviolet (UV) light (ALPER et al. 1967, 1978; HAIG et al. 1969, 1977 LATARJET et al. 1970; BELLINGER-KAWAHARA et al. 1987a). However, it is not generally possible to correlate UV inactivation kinetics with size due to the variable contributions of potential interferences from UV-absorbing chromophores in the medium or in direct association with the agent, agent concentration effects, the pyrimidine concentration and distribution in the nucleic acid, the repair efficiency of the host, and multiplicity reactivation phenomena.

Suspensions of the scrapie agent show greater sensitivities to inactivation at ultraviolet wavelengths below the 254-nm maximum typical of some viruses. However, a similar inactivation spectrum is observed for tobacco mosaic virus (KLECZKOWSKI and MCLAREN 1967), for the protein synthetic function of ribosomes (EKERT et al. 1970), and, to a lesser extent, for potato X virus (KLECZKOWSKI and GOVIER 1969). The lack of other examples may only reflect the limited number of viruses for which an inactivation spectrum has been determined rather than the actual prevalence of this presumed anomaly in nature.

In summary: In contrast to the original target size calculation, direct comparison of the inactivation rate constant of the scrapie agent with those of other viruses places scrapie with the smaller viruses. The circoviruses and hepatitis delta agent have genomes even smaller than that estimated for the scrapie agent, and these and other small viruses might well show inactivation kinetics similar to those for scrapie if measured.

2.2 Other Measurements of Agent Size

There have been numerous attempts to establish agent size by filtration, sedimentation velocity, exclusion chromatography, and gel electrophoresis. Filtration estimates of particle size range between 30 and 50 nm (STAMP et al. 1959; EKLUND et al. 1963; HAIG and CLARKE 1965; GIBBS 1967; KIMBERLIN et al. 1971). Exceptions include two reports of passage through 100000-Da molecular filters (PRUSINER et al. 1982, 1983; POCCHIARI et al. 1988) and one report of passage through 7-, 5- and 2-nm; filters (COEROLI-VIGOUROUX et al. 1983). However, none of the filtration experiments conducted to date have included internal controls for filter integrity or tests to verify pore size before and after filtration. Thus, while failure to pass infectivity is an interpretable result, passage of infectivity is not, since one cannot know without a challenge test whether or not the filter was intact or whether or not its pore size was as rated. Furthermore, ultrafilters are not absolute. For example, in the experiment of POCCHIARI et al. (1988), filtrate concentrations were less than 1% of retentate for a 100,000-Da filter, well within the manufacturer's expectations for a challenge by particles of $\sim 2 \times 10^6$ daltons. Similar recoveries might have been expected for the porcine circovirus with a particle mass of $\sim 2.6 \times 10^6$ daltons (estimated from TISCHER et al. 1982). Filtrate recoveries were also $<1\%$ of input infectivity in two of the experiments of COEROLI-VIGOUROUX et al. (1983). In a third, recoveries from the 5-nm and 2-nm filters exceeded that of the 7-nm filter, suggesting a leak. Prusiner and colleagues claim 100% recoveries in filtrates from 15-nm Nucleopore filters (PRUSINER et al. 1982) and passage through 100,000-Da molecular filters (PRUSINER et al. 1983). Full details have never been provided for either of these experiments, but high recoveries in ultra-filtrates are often indicative of leaks or ruptures.

Several different matrices for exclusion chromatography have been investigated for their ability to size the agent. Infectivity apparently sticks to porous glass, as, even in the presence of 0.5% SDS, it elutes in a broad featureless peak extending from the void volume to molecular weights of 350 Da and less (PRUSINER et al. 1980d, 1981a). Nevertheless, a significant portion of the infectivity elutes in the void volume, indicating that it does not bind to the column and therefore must be larger than the 24-nm exclusion limit. Fractionation on Bio-Gel A 50m showed less evidence of binding. The bulk of the recovered infectivity was in the void volume, indicating particle masses in excess of 5×10^6 Da (KIMBERLIN et al. 1971).

At variance with these results is a report by PRUSINER (1982) of an HPLC fractionation on a Toya Soda TSK 4000 column in which the infectivity peak migrated between bovine serum albumin (66000-Da) and ovalbumin (44000-Da) markers. Although this result figured prominently as evidence for subviral size in the original proposal of the prion hypothesis, the experiment has never been completely reported. A likely explanation for this anomaly has been provided by DIRINGER and KIMBERLIN (1983), who showed that the detergent in which the sample was applied also elutes at the position of the infectivity. They interpret this to mean that the infectivity, which otherwise would be excluded from the matrix, absorbs to column until it is swept out by the detergent peak traveling as much smaller micelles.

For both filtration and exclusion the relationship of the measured size to the actual mass of the unit particle is not known. All that can be said is that the result is consistent with viral dimensions. However, a subviral size is also easily accommodated by postulating that the infectivity has either self-aggregated or is complexed with larger nonessential structures.

Ironically, neither the filtration nor the exclusion chromatography results indicating subviral size are consistent with the postulated structure of the prion as it exists in extracellular preparations (PRUSINER et al. 1983). If PrP^{Sc} is aggregated into fibrils, then it is these fibrils, rather than the unit prion, that determine the dimensions of the infectivity-bearing particle. SAF are 10–20 nm in diameter and 100 to >1000 nm long (P.A. MERZ et al. 1981, 1983; PRUSINER et al.

1983; ROHWER 1984c). For comparison, tobacco mosaic virus is 18 nm × 300 nm (FRAENKEL-CONRAT and Wagner 1974). As a consequence, infectivity in tissue homogenates should fractionate like viruses whether or not it is composed of prions. [The detergent treatments used in these protocols were insufficient to disrupt the fibrillar structure of the SAF or prion rods (PRUSINER et al. 1982).]

Scrapie sediments with heterogeneous size, presumably reflecting its propensity to associate with cell debris. However, the minimum S_{20W} value for the infectivity is approximately 40S (PRUSINER et al. 1978a, b, 1980a, c; DIRINGER and KIMBERLIN 1983; DIRINGER et al. 1983a; SKLAVIADIS et al. 1989). In comparison, the small, spherical, porcine circovirus sediments at 52S (TISCHER et al. 1982). Irregularly shaped particles of the same mass sediment even more slowly. For example, the 40S ribosomal subunit has an RNA mass identical to that of the circovirus DNA.

Scrapie infectivity has been recovered from agarose-acrylamide electrophoresis gels comigrating with viroid RNA and low-molecular-weight DNA restriction fragments (MALONE et al. 1979). At the time of this discovery, it was considered proof of the subviral size of the scrapie agent (PRUSINER et al. 1980a, b, d). However, it was subsequently shown that viruses readily penetrate these gels, that fractionation of the small spherical viruses is on the basis of charge rather than molecular weight and that bacteriophage MS2 migrates almost as far as scrapie infectivity under the same conditions (ROHWER 1984b). Due to variable and often poor recoveries of input scrapie (but not bacteriophage) infectivity, this approach has not proved useful for agent purification (PRUSINER et al. 1980a, b, d).

2.3 Resistance to Inactivation

Two concerns have motivated inactivation studies on the scrapie agent, a desire for structural information about the particle and a practical interest in disinfection. These two interests have conflicting data requirements, and confusion over the legitimate conclusions that can be drawn from these data has contributed greatly to the impression of unconventionality attached to the spongiform agents.

Valid structural conclusions about the typical virus particle in a population can only be drawn from the behavior of the majority population of the virus. The majority behavior is revealed in the initial rate of inactivation. This value must be obtained from a kinetic measurement of the inactivation executed in sufficient detail to permit extrapolation of the inactivation curve to zero exposure.

Deviations in the kinetics that are observed at higher levels of inactivation represent the behavior of only a small fraction of the starting population. This behavior is not representative and therefore is not relevant to a discussion of virus structure. Stated another way, 90% of the population is killed during the first log of inactivation, 99% during the second. Should the rate of inactivation slow at that point, the new rate would represent the behavior of < 1% of the original viruses.

Inactivation-resistant subpopulations are a familiar occurrence in virology (BLOCK 1983). They constitute the major complication in water purification efforts (SHARP et al. 1975; BERG 1967; JOHNSON 1975) and the greatest threat to the safety of inactivated vaccines (SALK and GORI 1960). Often the size of the inactivation-resistant fraction is inversely correlated with the purity of the virus preparation. In these instances the resistance of the minority population is a property of the suspension rather than of the virus. Attempts to deduce virus structure on the basis of such behavior would be entirely misleading (ROHWER and GAJDUSEK 1980; ROHWER 1983, 1984b).

These principles are evident in Fig. 4, which shows the inactivation of scrapie infectivity and four bacteriophages in 0.525% sodium hypochlorite (ROHWER 1984b). All of the viruses are rapidly inactivated to ⩽0.1% survival levels upon exposure to the hypochlorite solution. However, the scrapie infectivity and two of the bacteriophages that have also been suspended in brain homogenate (solid symbols) resist further inactivation. The same bacteriophages when suspended in buffer (open symbols) show no resistance. Brain homogenate has no effect on the inactivation kinetics of the other two bacteriophages.

Scrapie infectivity can be recovered after 60 min of exposure to 121 °C (GORDON 1960; PATTISON and MILLSON 1961; EKLUND et al. 1963; HUNTER and MILLSON 1964; ZLOTNIK and RENNIE 1967; ADAMS 1970; MOULD and DAWSON 1970;

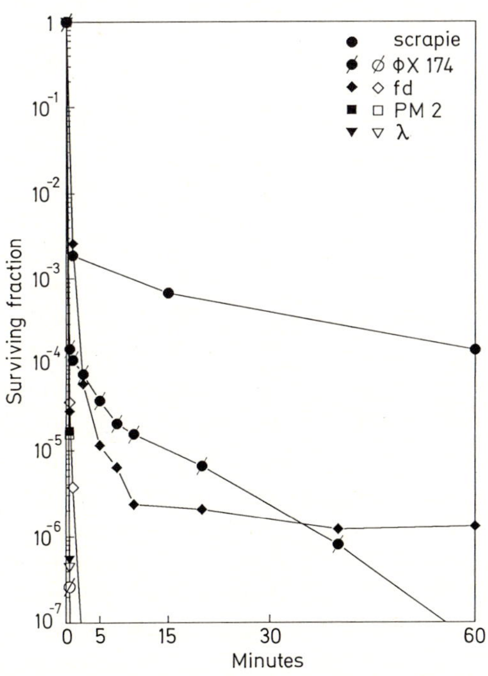

Fig. 4. Inactivation of scarapie infectivity and four bacteriophages by 0.525% sodium hypochlorite. *Solid symbols* represent inactivations in 10% brain homogenate; *open symbols* represent inactivations in PBS. (From ROHWER 1984b)

BROWN et al. 1982). This behavior has seemed one of the most remarkably at variance with a viral structure. Figure 5 presents the kinetics of such an inactivation (ROHWER 1984a). One could hardly claim that the scrapie agent is insensitive to exposure to 121°C when 99.99999% of the population is killed during the 60-s transit from 4°C to that temperature. On the other hand, it is the case that the occasional animal inoculated with the undiluted 60-min sample will, after an extended period, develop scrapie.

This experiment illustrates the conflicting data requirements for structural inferences and for disinfection. Clearly scrapie infectivity is sensitive to 121°C. On the other hand, a 1-h exposure to this temperature may be insufficient for sterilization of the agent. Structural information about the typical virus particle must be drawn from the behavior of the typical virus as revealed by the majority population. In contrast, sterilization requires the inactivation of the most inactivation-resistant member of the entire population. In Fig. 5 that is one individual out of 10 million. It would be of great interest to know what confers resistance on that individual. It is also of great practical importance due to the serious agricultural and public health problems that have resulted from this resistance. Nevertheless, whatever the mechanism, it is likely to be largely irrelevant to first-order structural considerations of the majority population. Stated another way, it would be foolish to base purification schemes on this rare behavior. On the other hand it is crucial to any consideration of disinfection.

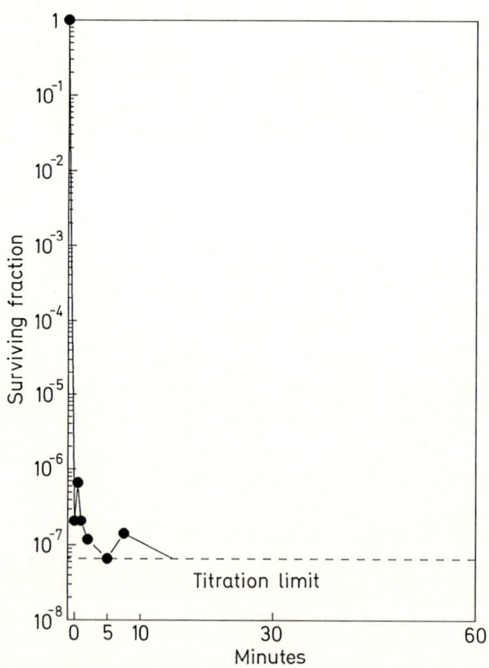

Fig. 5. Inactivation of scrapie infectivity by exposure to 121°C. (From Rohwer 1984a)

2.4 Lack of Immunogenicity

The scrapie-infected host does not mount a readily identifiable immune response to the infection, nor has any disease-specific alteration of host immune system functionality been detected as a consequence of infection (see BROWN 1990 for a recent review). On the other hand, besides the central nervous system, the only other tissues that harbor significant titers of scrapie infectivity are immune tissues, suggesting ample opportunities for interactions and modulations that could lead to suppression of anti-scrapie responses (LAVELLE 1973; MANUELIDIS et a. 1978; CARP and CALLAHAN 1981; HADLOW et al. 1982; KURODA et al. 1983; G.S. MERZ et al. 1987). Such interactions have been described for other viral systems (NASH 1985; BROWNE et al. 1990; KOTWAL et al. 1990; HSU et al. 1990) It is also of potential significance that activated lymphocytes express the PrP protein on their surfaces (CASHMAN et al. 1990).

Viruses also vary greatly in their intrinsic antigenicity (HUSO et al. 1988), and immunogenicity can depend strongly on the circumstances of the infection. Lymphocytic choriomeningitis virus (LCMV) infection in utero or of neonatal mice results in complete tolerance of the host to the viral proteins while infection of the adult results in a massive humoral response to the agent (LEHMANN-GRUBE et al. 1983; OLDSTONE et al. 1985). The history of hepatitis B and hantavirus immunity demonstrates that even high-titer humoral antibodies can be elusive when the corresponding antigen has not been identified (HOLLINGER 1990a; LEE et al. 1978; LEE 1982). Not all viruses elicit neutralizing antibodies (MAJER 1972; MASSEY and SCHOCHETMAN 1981; HUSO et al. 1988). Without an identifiable antigen, neutralization is the only sure test for a scrapie-specific antibody (GIBBS 1967). (Loss of infectivity has been demonstrated in response to anti-PrP antibody (GABIZON et al. 1988, 1989). However, since infectivity associates with SAF or prion rods this may be a secondary consequence of aggregation or precipitation of those structures. It is likely that antibodies against plasma membrane antigens might give a similar result.)

The apparent lack of infection-specific immunogenicity can not be cited as evidence against a virus origin for scrapie disease without jeopardizing the prion (PRUSINER 1989) and virino (DICKINSON and OUTRAM 1988; KIMBERLIN and HOPE 1987) models as well. Prions are postulated to be composed largely or completely of the scrapie isoform of PrP. The virino is encapsidated by it. Yet the scrapie isoform of PrP readily elicits hyperimmunity when injected into rabbits or mice (DIRINGER et al. 1984; BENDHEIM et al. 1984; BARRY et al. 1985; CHO 1986; KASCSAK et al. 1987) and its biophysical properties confirm its divergence from the normal protein present in uninfected animals (BOLTON et al. 1982; MC KINLEY et al. 1983a). It would be remarkable if SAF or the post-translationally modified PrP^{Sc} from which it is composed did not present new, "foreign", epitopes to the host. The failure of the host to recognize SAF is at least as provocative as is its apparent failure to respond to the agent.

Whatever the explanation, the apparent lack of host immunity to the scrapie infection is better considered an anomaly than a fatal flaw to a virus etiology.

2.5 Nuclease Resistance

Among many other functions of the viral capsid is the protection it affords the viral genome as it transits hostile extracellular and extraorganismic spaces on its journeys from cell to cell and host to host. For viruses that infect the gut, the capsid must be capable of withstanding low pH, proteolysis, detergent extraction (by bile salts), and other digestive processes. The selective pressures shaping other viruses produce different spectrums of resistance. However, for all viruses protection of the nucleic acid core constitutes an irreducible minimum requirement of capsid function.

The resistance of scrapie infectivity to nucleolytic attack (HUNTER 1965; MILLSON et al. 1976; PRUSINER et al. 1980b, d; DIENER et al. 1982; CHO 1983; BELLINGER-KAWAHARA et al. 1987b) is often cited as evidence that the agent lacks a nucleic acid genome and is not, therefore, a virus (PRUSINER 1982, 1988; DIENER et al. 1982; BELLINGER-KAWAHARA et al. 1987b). In fact, insensitivity to nucleases is a property of viruses that distinguishes them from their nucleic acids.

Scrapie infectivity is inactivated by proteinase but only under extreme conditions (HUNTER and MILLSON 1967; HUNTER et al. 1969; MARSH and HANSON 1969; MILLSON et al. 1976; CHO 1980, 1983; PRUSINER et al. 1981b; LAX et al. 1983). This is again suggestive of a virus. Nevertheless, the combination of protease sensitivity with nuclease insensitivity inspired the original description of the prion (PRUSINER 1982), while in fact this combination of relative protease insensitivity and total nuclease insensitivity is even more suggestive of a virus than either attribute independently.

The failure to demonstrate scrapie-specific nucleic acids in infectious preparations of prion rods (SAF) has been taken as evidence that the agent must, at best, contain only small amounts (4–45 nucleotides by one estimate, BELLINGER-KAWAHARA et al. 1987a) of subgenomic nucleic acid (DIENER et al. 1982; MCKINLEY et al. 1983b; BELLINGER-KAWAHARA et al. 1987b, 1988). Moreover, since the major mass constituent of SAF or prion rods is the PrP^{Sc} protein, it is reasoned that the agent must be comprised largely of this protein (MCKINLEY et al. 1983a; PRUSINER et al. 1983; BOLTON et al. 1984; MEYER et al. 1986).

This argument depends critically upon the specific infectivity of the PrP^{Sc} preparation with respect to PrP^{Sc}. The most optimistic estimates are from 10000 to 100000 PrP^{Sc} per unit of scrapie infectivity (PRUSINER et al. 1982, 1983; see discussion accompanying OESCH et al. 1988 for similar estimates from other groups). By comparison small icosahedral viruses contain 60 copies of capsid protein per virion. A small isometric virus with a specific infectivity close to one infectious dose per particle would account for only $60/10^4$ to $60/10^5$ of the protein in a prion rod (SAF) preparation and would be unlikely to be recognized by either acrylamide gel electrophoresis, electron microscopy, or most other molecular comparisons without the assistance of agent-specific molecular probes.

If scrapie is caused by a virus then the infectivity contaminating PrP preparations must contribute nucleic acid. Lack of detectable nucleic acids depends upon the method of detection, and two independent laboratories have now both demonstrated nucleic acids in preparations of prion rods after digestions by DNase I and micrococcal nuclease, and Zn^{2+} hydrolysis (OESCH et al. 1988; AIKEN et al. 1990). This is in agreement with other laboratories reporting significant nucleic acid contamination of SAF or prion rod preparations (DIRINGER et al. 1983b; P.A. MERZ 1988; NARANG et al. 1988; SKLAVIADIS et al. 1989, 1990). Where identified, these nucleic acids have been of host origin (except for four as yet uncharacterized clones reported by OESCH et al. 1988).

Nonspecific association of host nucleic acids with PrP might be expected due to the strong basic character of the PrP protein (calculated pI for amino acids 23–254 is 10.88). Host nucleic acid contamination would possibly overwhelm the contributions of a small virus. A circovirus contaminating the preparation at a titer of 10^8/ml would contribute only 0.1 pg/ml of ss DNA. A similarly sized virus could be excluded from consideration as the scrapie agent only if infectivity could be purified to a specific infectivity of $>10^8$ units per 0.1 pg nucleic acid.

Differential screening and subtractive hybridization have been employed in several attempts to identify a scrapie-specific mRNA from scrapie-infected tissues (WIETGREFE et al. 1985; DUGUID et al. 1988, 1989). While these approaches have proved useful in identifying modulated mRNAs from several host genes that had not previously been suspected in the pathology of the disease, they have not to date revealed nonhost sequences. This failure has been cited as proof that there is no agent-specific nucleic acid (BROWN et al. 1990a, b; PRUSINER et al. 1990; GAJDUSEK and GIBBS 1990; GAJDUSEK 1990), although never by the authors of the experiments. In fact, this methodology is in its infancy. Only abundant message species have been screened by the methods employed, and those screenings have not been exhaustive. Subtraction is a methodology that promises improvements and with development may eventually reveal the scrapie nucleic acid. In the meantime it should not be overlooked as a means of gaining access to new and unsuspected features of the pathology.

2.6 The Integrity of the Infectivity/PrP Association

If the scrapie agent is composed of the PrP^{Sc} protein then it should not be possible to separate infectivity from PrP^{Sc}, and so far there has not been an incontestable demonstration of the physical separation of the two species. Nevertheless, there is compelling evidence that they are separate entities.

2.6.1 Strain Variation and Mutation

Without question the most important phenomena at variance with the prion hypothesis are strain variation and mutation.

Lack of a credible replication mechanism has been a continuing weakness of the prion hypothesis (PRUSINER 1988). It is currently proposed that the post-translationally modified scrapie isoform of the PrP protein, PrP^{Sc}, is able to induce this same modification in the cellular isoform of PrP, PrP^C, either within a given host or when inoculated into a new host. The propagation of this change constitutes infection (BASLER et al. 1986; STAHL et al. 1987; GAJDUSEK 1990). Since there is only a single copy of the host PrP gene, and no introns within the PrP open reading frame, there should be only a single PrP translation product (CHESEBRO et al. 1985; BASLER et al. 1986; ROBAKIS et al. 1986). Therefore, one might expect that every host species would exhibit a single distinctive spongiform encephalopathy based upon its particular PrP protein.

In contrast, 20 distinct strains of mouse scrapie can be reliably differentiated by their patterns of incubation times in different inbred mouse lines, by the type and distribution of histopathological lesions that they produce (plaque producers vs non-plaque producers, white matter vs gray matter involvement, cerebellar vs noncerebellar, etc.), and by differences in their abilities to infect alternative host species (DICKINSON and FRASER 1975, 1979; DICKINSON et al. 1984). Furthermore, new mouse strains continue to be isolated and the total number of strains possible is not known (FRASER et al. 1988; BARLOW and MIDDLETON 1990; TATEISHI et al. 1990). Other hosts also exhibit a variety of strain phenotypes (KIMBERLIN and WALKER 1978; DICKINSON and FRASER 1979; BUYUKMIHCI et al. 1987; KIMBERLIN et al. 1989).

Scrapie strains retain their phenotypes when passaged through different inbred mouse lines or during passage through alternative species (BRUCE and DICKINSON 1987; KIMBERLIN et al. 1989). For example, the mouse 22A strain gives a long incubation time in the C57Bl mouse, a short incubation time in the VM mouse, and produces distinct patterns of vacuolar lesions in the brains of each strain. The mouse ME7 strain gives the opposite pattern of incubation times and produces different lesion profiles (DICKINSON and MEIKLE 1969; DICKINSON and OUTRAM 1979; FRASER 1979). Both strains when passaged through hamsters and back to mice give their characteristic patterns of mouse disease (KIMBERLIN et al. 1989). Neither strain when propagated in the hamster behaves like the extensively utilized hamster 263K strain, nor do they themselves appear similar in the hamster (KIMBERLIN and WALKER 1977). For example, strain 263K has an incubation time of 60 days, whereas 22A stabilized at 151 days and ME7 at 251 days after several hamster passages (Table 2). Thus strain variation is reposited

Table 2. Incubation times of scrapie strains 263K, 22A, and ME7 in mouse and hamster

Scrapie strain	VM mouse p7p7[a]	C57Bl mouse s7s7[a]	Golden Syrian hamster
263K	–[b]	–[b]	60 days
22A	200 days	463 days	151 days
ME7	345 days	180 days	251 days

[a] VM and C57 Bl mice are homozygous for the p7 and s7 alleles respectively of a gene affecting scrapie incubation time in the mouse

[b] Hamster strain 263K does not infect the mouse

in the agent, not in the host (although host factors also influence the course of the infection, DICKINSON et al. 1968; DICKINSON and FRASER 1975; and see below).

To account for strain variation by post-translational modification of the PrP protein requires at least 20 and possibly a much larger number of distinct post-translational modifications of the PrP protein, each capable of specifically inducing further modifications of only that type. Each of these modified forms must be able to induce a concomitant modification in the hamster with sufficient fidelity that it can be passaged back to the mouse. Such changes must be subtle as it has not as yet been possible to detect any differences in the covalent modifications to the scrapie vs the cellular isoform of the PrP protein within a given host animal in spite of intensive efforts to do so (PRUSINER et al. 1983; HOPE et al. 1986; MEYER et al. 1986; Sklaviadis et al. 1986; BOLTON et al. 1987; BOLTON and BENDHEIM 1988; TURK et al. 1988; FRASER et al. 1989; CAUGHEY et al. 1990; SOMERVILLE and RITCHIE 1990). This failure is not just a matter of being unable to detect changes, as differences are readily demonstrated between PrP proteins isolated from different inbred mouse lines (KASCSAK et al. 1985, 1986).

Occasionally strain phenotype changes spontaneously during passage (BRUCE and DICKINSON 1979, 1987; KIMBERLIN et al. 1987, 1989). It is also possible to select for changes by serial passage through a new host species (DICKINSON and FRASER 1979; GIBBS et al. 1979; KIMBERLIN et al. 1987, 1989). These phenomena suggest mutation and adaptation through selection of mutants.

Strain variation and mutation are properties of viruses. In contrast, it is difficult to imagine how a single copy gene with a single open reading frame (CHESEBRO et al. 1985; BASLER et al. 1986; ROBAKIS et al. 1986) can give rise to strain variation and how this strain identity can be maintained during passage through alternative hosts. Mechanisms have been proposed (BOLTON and BENDHEIM 1988; PRUSINER 1988; WEISSMANN 1989; WILLS 1989). However, the contortions required to circumvent the single copy constraint, while interesting intellectual exercises, have nevertheless produced ever more unlikely explanations. As the number of recognizable strains rises (new strains have been recently isolated from bovine spongiform encephalopathy and Creutzfeldt–Jakob disease; FRASER et al. 1988; BARLOW and MIDDLETON 1990; TATEISHI et al. 1990), the mechanisms postulated begin to require all of the plasticity and resilience of nucleic acid encoding systems and formally converge on a mutable genomic system. It would be remarkable indeed for such a complex system, if it existed, to have involved so little of biology as to have completely escaped notice until now. Certainly such ad hoc hypotheses should not be favored over the established viral paradigm for transmissible diseases.

2.6.2 Separations of PrP^{Sc} and Infectivity

Several laboratories have reported progress in fractionating infectivity away from PrP^{Sc} (DIRINGER et al. 1983a; BRAIG and DIRINGER 1985; AIKEN et al. 1989; SKLAVIADIS et al. 1989). Thus highly purified mitoplasts prepared from scrapie-infected brains still harbor significant infectivity, $10^{8.5}$ IU/ml, while exhibiting no

PrP signal in a western blot. In comparison, unfractionated brain at the same concentration of infectivity produces a strong signal (AIKEN et al. 1989).

While it is apparent that the specific infectivity of this fraction with respect to PrP is greater than that of the starting brain homogenate, one can not exclude the possibility that the remaining $10^{8.5}$ IU/ml of infectivity is due to PrP. Proof of separateness requires reductions in PrP contamination to levels that can no longer account for infectivity in terms of the prion model. Due to the small size postulated for the prion this may be difficult. For example, as little as 0.06 pg/ml of PrP could account for the infectivity in the mitoplast preparation if the agent were a tetramer of PrP. Such small quantities would not have been detected by the methods used. Nevertheless this result does establish that, if infectivity resides in PrP^{Sc}, then all PrP^{Sc}-bearing structures are not equally infectious.

Fractionation by sedimentation velocity has yielded similarly equivocal results. A portion of the infectivity is separable from the bulk of the PrP^{Sc} in centrifugations conducted according to several protocols (DIRINGER et al. 1983a; SKLAVIADIS et al. 1989, 1990). Nevertheless, there always remains readily detectable PrP^{Sc} in association with the infectivity. As was the case for mitoplasts, the specific infectivity of this fraction with respect to PrP^{Sc} has increased compared to whole brain. Thus once again, if infectivity resides in PrP^{Sc}, then all PrP^{Sc}-bearing structures can not be equally infectious.

In the case of mitoplasts, differences in aggregation state could account for differences in specific infectivity between mitoplast and other fractions (ROHWER and GAJDUSEK 1980). One would predict that mitoplast-associated prions would be less aggregated than those in whole brain homogenates, resulting in less PrP^{Sc} mass per unit of infectivity. As a consequence, infectivity associated with mitoplasts should be physically smaller than that found in the cytoplasm. However, this explanation is inconsistent with the sedimentation results, as the fractions with high specific infectivity sediment faster than those of lower specific infectivity and thus must be larger.

As an alternative explanation for the differences in the specific infectivities of PrP^{Sc} populations, the prion infection might, as is common in virology (and may or may not be common in prionology), produce excess particulate antigen as a by-product of replication. In this case infectivity would have to be selectively compartmentalized in mitoplasts and, from the sedimentation data, the infectious form of PrP^{Sc} sequestered there would have to be larger than the uninfective forms of PrP^{Sc} found extramitochondrially. However, the larger size of the infectious form could not derive from aggregation because that would reduce the specific infectivity. Instead, the infectious form of PrP^{Sc} but not the uninfectious PrP^{Sc} antigen must attach to some larger structure during sedimentation. Since PrP^{Sc} comprises both high specific infectivity forms and low specific infectivity forms, some other feature besides PrP^{Sc} itself would have to distinguish the two states. This example only serves to demonstrate that, while there is seldom a smooth fit between observation and model, it is generally possible, via ad hoc adjustments, to keep the prion "hypothesis" alive if so inclined.

A simple explanation for this fractionation behavior is that infectivity and PrP^{Sc} are separate entities. Indeed, the sedimentation behavior is as expected for a small virus. Nevertheless, without rigorous quantitation, neither the mitoplast nor the sedimentation results can be considered proof that infectivity is separable from PrP^{Sc}. However, both results constrain the prion hypothesis in new ways. First, there must be at least two classes of PrP^{Sc}, one that is associated with infectivity and one that is not. Second, the infectivity-bearing particle is either of virus dimensions or specifically attaches to something that confers virus-like size during sedimentation velocity.

2.6.3 Proportionality of PrP^{Sc} and Infectivity

If infectivity and PrP^{Sc} are identical, then procedures that destroy one should also destroy the other with identical kinetics. This is purported to be the case and is cited as the principal evidence that "the prion contains only one major protein, namely PrP^{Sc}" (McKinley et al. 1983a). In the experiment in question, reductions in infectivity titer due to protease digestions appear to parallel destruction of PrP^{Sc}. However, this conclusion is an artifact of the data presentation. Plotting the logarithm of time conceals the disparate behaviors of the two curves (Fig. 6a, b). The same data plotted with a linear time axis reveal distinctly different inactivation kinetics for infectivity and PrP^{Sc}, suggesting separate entities (Fig. 6c, d). The PrP^{Sc} digestion at 100 ug/ml of proteinase K shows evidence of cooperativity consistent with the disruption of multimeric structures. Neither infectivity curve displays this behavior.

It is also claimed that, for a large number of sucrose gradient fractions from multiple gradients and for the protease digestions discussed above, PrP^{Sc} concentrations were "directly proportional" to infectivity titers, thereby indicating that PrP^{Sc} and infectivity were one and the same. This was demonstrated by plotting log PrP^{Sc} concentration versus log titer (see Fig. 2A–C and 5 in McKinley et al. 1983a). Again the log log format obscures the nature of the relationship between infectivity and titer. The "visually fitted" lines are linear only in transformation. The equations describing them are in fact power functions of the form:

$$(\mathrm{PrP^{Sc}\ concentration}) = b(\mathrm{infectivity\ titer})^a$$

where a and b are constants. This relationship is "directly proportional" (McKinley et al. 1983a; Prusiner 1989) only if $a = 1$. For equations of the form:

$$y = bx^a$$

y_1/y_2 is given by:

$$\frac{y_1}{y_2} = \frac{bx_1^a}{bx_2^a} = \left(\frac{x_1}{x_2}\right)^a$$

Table 3 gives the values of a and b for each of the correlations as plotted by McKinley et al. (1983a). Only the data in their Fig. 2A approach linearity. In their

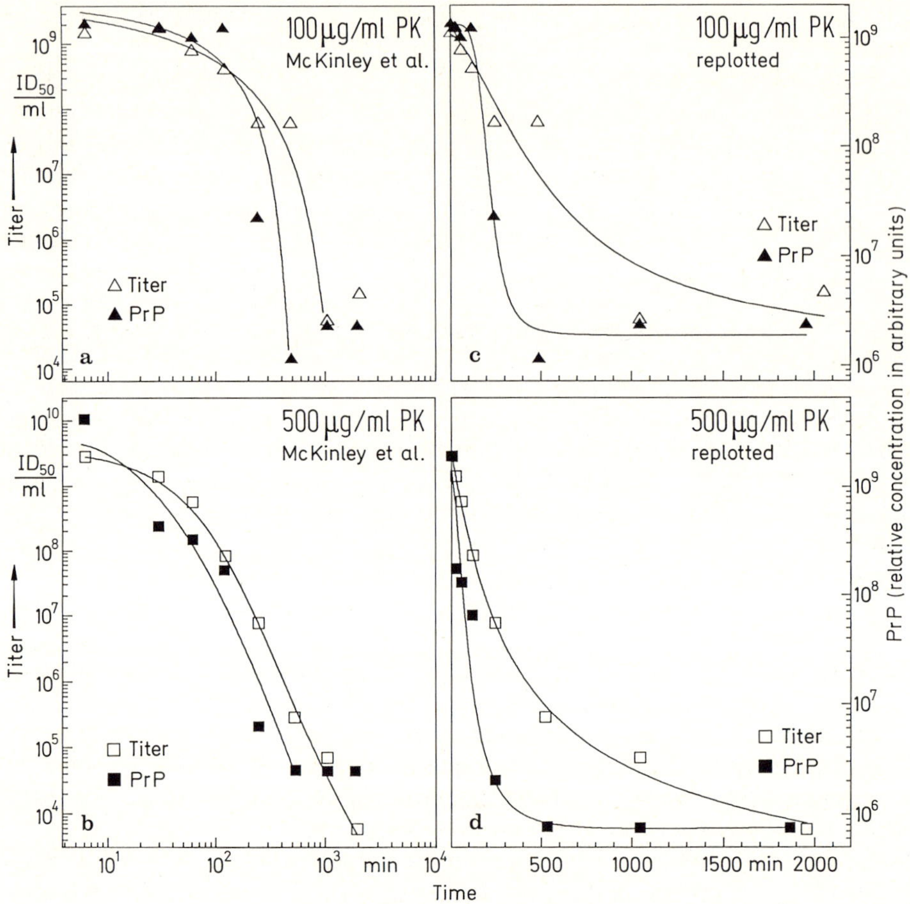

Fig. 6 a–d. A comparison of the kinetics of inactivation of scrapie infectivity and PrP degradation by proteinase K. SAF/prion rods were incubated with proteinase K at 100 µg/ml, (**a, c**) and 500 µg/ml (**b, d**). Data and lines in **a** and **b** are as plotted by McKinley et al. (1983). The identical data are replotted on a linear time axis in **c** and **d** to reveal the differences in kinetic behavior of PrP and infectivity. The PrP data are in arbitrary units of relative concentration. To facilitate comparison, in **d** the PrP data have been scaled by a linear transfromation [1.052 log(PrP concentration) − 0.8427] so as to bring the first and last points into conjuntion with the first and last points of the infectivity data. (**a, b** from McKinley et al. 1983)

Table 3. The relationship of PrP^{Sc} concentration and titer in the correlations plotted in McKinley et al. (1983a)

Figure	Constants in the expression: $PrP^{Sc} = b\ (\text{titer})^a$	
	a	*b*
2A	0.83	0.32
2B	1.88	6.5×10^{-10}
2C	1.77	5.1×10^{-9}
5	0.66	56

Fig. 2B and 2C, PrP^{Sc} concentrations increase nearly as the square of titer. In their Fig. 5 the concentration increases more nearly as the square root. Figures 2B and 2C predict that, at infectivity titers of 1, there will be only 8.9×10^{-10} and 5.1×10^{-9} units of PrP^{Sc} respectively, while Figs. 2A and 5 predict values ten orders of magnitude greater. Depending upon the nature of the arbitrary unit of PrP concentration employed, there may be insufficient PrP to account for infectivity at low titers in Figs. 2B and 2C. The lack of a constant correlation as evidenced in the extreme variation in the value of *a* over different experiments suggests methodological problems that place the entire analysis in doubt.

The lack of proportionality between PrP^{Sc} concentrations and infectivity titers does not disprove that PrP^{Sc} is a component of the infectious agent. However, if PrP^{Sc} is a component, then there must also be other forms of PrP^{Sc} that are not.

Once again a simpler explanation for this divergence in behavior is that PrP^{Sc} and infectivity are separate entities that are degraded with different kinetics and partition differently during fractionations.

2.6.4 Ultrasonication

In scrapie-infected brain homogenates, the scrapie isoform of PrP is found largely in the SAF or prion rods; thus, if infectivity is composed largely of PrP^{Sc}, infectivity must also be sequestered in the same structures (DIRINGER et al. 1983a; PRUSINER et al. 1983, 1984; MCKINLEY et al. 1983a). If unit infectivity is small, then each SAF or prion rod must be comprised of a large number of potentially infectious units. Estimates range from 300 prions/rod to 30 000 prions in the typical aggregate seen in crude homogenates (Table 1 in PRUSINER et al. 1983). One might therefore expect that mechanical disruption of SAF would result in large increases in virus titer. Conversely, if intact SAF were required for infectivity, disruption should diminish titer. Neither effect is observed. Ultrasonication of tissue homogenates of SAF preparations have so far failed to demonstrate any appreciable affect on scrapie titers under conditions that readily disrupt filamentous phages (MCKINLEY et al. 1986; BELLINGER-KAWAHARA et al. 1988; ROHWER 1991). The ultrasonication experiment in Fig. 7 (ROHWER 1991) compares the behavior of the scrapie infectivity to that of bacteriophage controls. Filamentous phage fd and tailed phage λ were rapidly inactivated under conditions that slightly increased the titers of ΦX174 and perhaps scrapie infectivity. Phage PM2 was slowly inactivated, suggesting that ultrasonication was reducing the homogenate to particles smaller than PM2 (60 nm diameter) but larger than ΦX174 (24 nm diameter). Since SAF or prion rods are typically 500–1000 nm or greater in length, a reduction to particle sizes of 60 nm or less should increase titer by a factor of ten or more. Since the fibrils are typically seen as aggregates this number could be much larger. Within the limitations of the quantal titration assay, titer may have increased threefold.

It has been suggested that ultrasonication fails to show significant increases in scrapie titer because the high Boltzman temperatures achieved in the

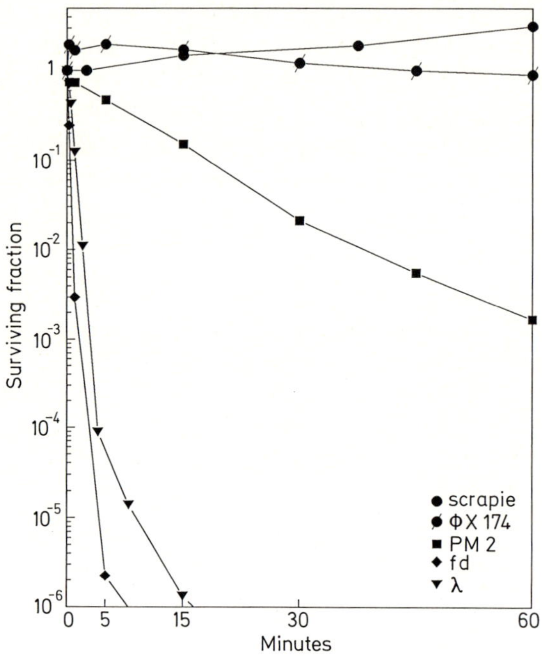

Fig. 7. Inactivation of scrapie infectivity and four bacteriophages by exposure to infense ultrasonication. All sonications were in 10% brain homogenate. Time is cumulative exposure to pulsed sonication. Sample termperature was maintained at <15°C. (From ROHWER 1991)

cavitation event inactivate with the same kinetics as infectivity dispersion (MCKINLEY et al. 1986). The lack of inactivation of the *Φ*X174 control, a virus that, unlike scrapie, is thermolabile, disproves this notion (ROHWER 1984a).

Ultrasonication establishes that the infectious particle must be much smaller than intact SAF, but that there can not be more than one or two infectious particles per SAF as opposed to the hundreds predicted. Either the bulk of the PrP^{Sc} is not directly associated with conferring infectivity, or infectivity is only adventitiously associated with PrP^{Sc}.

Disaggregation of a small spherical virus of the dimensions of *Φ*X174 or smaller from a specific or nonspecific association with SAF would give this same result.

3 The Role of the PrP Protein

From the preceding discussions, it can be concluded that none of the objections to a virus etiology for scrapie disease are sufficiently robust to merit abandoning established viral paradigms for transmissible diseases. There is no good evidence for subviral size. The vast majority of the infectivity inactivates with the kinetics expected for a virus. Agent interactions with the host immune system are circumstantially implicated in the apparent lack of a host immune defense. The agent's resistance to nucleolytic attack and proteolysis are just as might be expected for a virus adapted to survival on pasture while traveling between hosts or that must transit the gut during the infective cycle. The link between PrP^{Sc} and infectivity is weaker than claimed and there is no direct proportionality between the two. Finally, strain variation and mutation are uniquely viral properties.

However, if scrapie is caused by a virus, what then is the role of PrP^{Sc} in the infection? While SAF are well-established pathognomonic indicators of infection by the spongiform encephalopathy agents (P.A. MERZ et al. 1984), and the presence of the protease-resistant scrapie isoform of the PrP protein, PrP^{Sc}, in western blot analyses is diagnostic for these diseases, nevertheless there has not been, until recently, any reason to expect that SAF or prion rods were anything other than pathological endproducts of the degenerating CNS, and as such, only remotely connected to the primary infection. Recent evidence from the genetics of the PrP locus and from transgenic animals bearing foreign PrP genes suggests otherwise.

The PrP locus is closely linked to and may be identical to a host determinant of incubation time. Specific polymorphisms in the PrP gene segregate with the incubation time phenotypes. For incubation time to be controlled by the PrP gene, the gene or its product must participate in the replication of the agent (DICKINSON et al. 1968; DICKINSON and FRASER 1979; CARLSON et al. 1986, 1988; CARP et al. 1987; HUNTER et al. 1987, 1989; WESTAWAY et al. 1987; RACE et al. 1990).

When a large cosmid containing the hamster PrP gene is transfected into mice in multiple copies, the mice become infectible by a hamster-adapted scrapie strain that does not otherwise readily infect mice (SCOTT et al. 1989). This suggests that PrP controls host susceptibility to scrapie. The resultant infection exhibits the short incubation times characteristic of hamster scrapie, 70–75 days, rather than of mouse scrapie, 125–145 days. The change in incubation time provides further evidence for the participation of PrP in replication.

Several different polymorphisms in the human PrP gene are now associated with various pedigrees of familial Creutzfeldt–Jakob disease (CJD) or Gerstmann–Sträussler syndrome (GSS) thus accounting for the genetic component of these diseases in these families (HSIAO et al. 1989; GOLDGABER et al. 1989; OWEN et al. 1990; DOH-URA et al. 1990). Nevertheless, these diseases are also transmissible. This is again consistent with a PrP-determined susceptibility factor in the spongiform encephalopathies. The implication would be that the transmissible agent must either be passed vertically or must be ubiquitous and abundant within the family environments to ensure infection of susceptible individuals.

In the emerging focus of sporadic CJD in Czechoslovakia, all affected individuals share a single polymorphism (codon 200 glutamate to lysine) in common with the familial CJD cases described above (GOLDFARB et al. 1990d). However, these individuals are not closely related. Nor have previous generations been affected (MITROVA 1990). Therefore, the Czechoslovakian focus does not seem to represent familial disease. Furthermore, not all individuals bearing the polymorphism get the disease (GOLDFARB et al. 1990d). Thus the PrP polymorphism may predispose an individual to CJD but is not sufficient to cause disease.

The lack of CJD in previous generations of families of susceptible individuals suggests that the CJD agent has only recently been introduced into the affected Czech community. In other localities these same families might have had long histories of CJD. The current affected individuals may be the founders of new pedigrees of familial CJD. A similar phenomenon may be at work in a focus of CJD among Sephardic Jewish immigrants to Israel (KAHANA et al. 1974; NEUGUT et al. 1979). Affected individuals also carry a PrP polymorphism (GOLDFARB et al. 1990c) which may have put them at risk in a setting of increased exposure to the CJD agent (RADHAKRISHNAN and MOUSA 1988). Thus PrP polymorphisms may predispose families to familial CJD but are not sufficient for its occurrence. Since the disease is distinct from the polymorphism(s), it may some day be possible to break the cycle of familial transmission by identifying the sources of agent contamination within the family environment and protecting family members bearing the susceptibility trait from exposure.

So far, with the exception of the Czech focus, sporadic CJD cases do not usually harbor the polymorphisms associated with the familial disease (GOLDFARB et al. 1990a, b). This again supports the concept that polymorphisms are predisposing rather than necessary conditions for infection.

When a mutation analogous to the human polymorphism at codon 102 that is associated with some GSS families was introduced into the mouse PrP gene and then transfected back to mice in high copy number, some of the animals developed a spontaneous spongiform encephalopathy (HSIAO et al. 1990). This disease has not proved to be transmissible. Nevertheless, this establishes that, far from being a pathogenetic end product of scrapie disease, PrP initiates pathogenesis (HOGAN et al. 1986; BRUCE et al. 1989). It implies that closely linked interactions between the scrapie agent and the PrP gene or its protein are responsible for pathology and disease in the scrapie infection.

Systematic screening of institutionalized patients in a British psychiatric hospital by polymerase chain reaction (PCR) has revealed PrP polymorphisms in patients that would not otherwise have been considered to have CJD or GSS (COLLINGE et al. 1989, 1990). It is presumed by the authors of this study that such cases will prove to be transmissible and that, as a consequence, the incidence of transmissible PrP-related neurological disease may be much greater than indicated by the incidence of CJD. They propose that all such diseases be grouped with CJD and be designated "prion disease". They further speculate that an epidemiology that included this broader spectrum of disease might reveal links to sheep scrapie or to bovine spongiform encephalopathy that have been missed in the past.

It should not be surprising that mutations in a neuronal cell surface protein produce disease in their own right with or without the interactions of an infecting virus. The transgenic experiment described above demonstrates that PrP mutations are sufficient to cause disease. Thus it can not be assumed that all diseases associated with PrP mutations will transmit. Rather, the British cases

(COLLINGE et al. 1989, 1990) may define a new class of entirely genetic disease associated with the PrP locus. As such their designation as "prion disease" is appropriate. However, the inclusion of the transmissible dementias under this heading would be misleading.

This distinction may be clearer in another context. Obstruction of the acetylcholine receptor by autoantibodies causes myasthenia gravis (LINTON and PHILCOX 1990). Presumably there might be mutations that behave similarly. The same cell surface protein is also a rabies virus receptor (LENTZ et al. 1982, 1984). Nevertheless, there would be little point in classifying such diseases together.

4 A Unifying Hypothesis

Many of the properties of the scrapie infection are consistent with PrP being a cell surface receptor for the scrapie virus. Thus, the tenacious association of infectivity with PrP may reflect virus-receptor binding, and the transformation of PrP from the normal protein to the scrapie isoform may be mediated by this interaction. As a receptor, PrP would control susceptibility to the disease and could also regulate replication rates were cell-to-cell spread the rate-limiting step. A vast increase in the expression of the receptor, as in the short incubation transgenics, might increase the efficiency of cell-to-cell spread and shorten the incubation time. Mutations in the receptor might alter susceptibility to the infection or alternatively might disrupt the normal functions of PrP and cause disease outright. However, whatever the role of PrP and however important it may be to the disease process, there is no substantive evidence to suggest that PrP is the infectious agent or is even an obligatory structural component of the infectious agent.

It is instructive to consider the relative dimensions of the scrapie-associated fibril or prion rod and the small isometric viruses (Fig. 8). The most optimistic estimates of the specific infectivity of SAF/prion rod preparations vary from 10 to 100 rods per infectious unit (PRUSINER et al. 1982, 1983). Every electron micrograph ever published of SAF or prion rods, including those of highly purified preparations, contains within the same field, often attached to the fibrils themselves, but sometimes separately, small round particles of dimensions appropriate for small spherical viruses (Fig. 8). However, similar structures are commonly seen in electron micrographs of diverse origins. Thus at ratios of 0.1 or less compared to SAF or prion rods, there is ample opportunity for a small, nondescript virus to have gone unnoticed. In contrast, small, spherical, virus-like structures have been frequently reported in thin sections from the spongiform encephalopathies (Fig. 9) (FIELD and RAINE 1966, 1967; DAVID-FERREIRA et al. 1968; BIGNAMI and PARRY 1970; VERNON et al. 1970; ALLEN et al. 1971; BOTS et al. 1971; LAMPERT et al. 1971a, b; ROY 1972; NARANG 1973, 1974a–c, 1975, 1988; LAMAR et al. 1974; PAYNE and SIBLEY 1975; DE REUCK et al. 1976; ZU RHEIN and VARAKIS

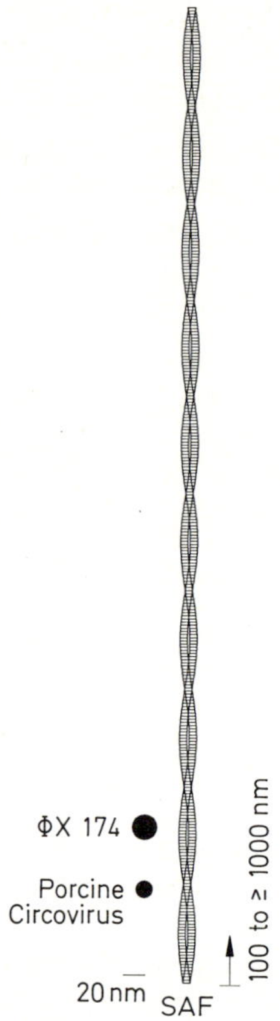

Fig. 8. Relative dimensions of SAF/prion rods and the small spherical viruses. SAF/prion rod lengths are variable but can exceed 1 μm

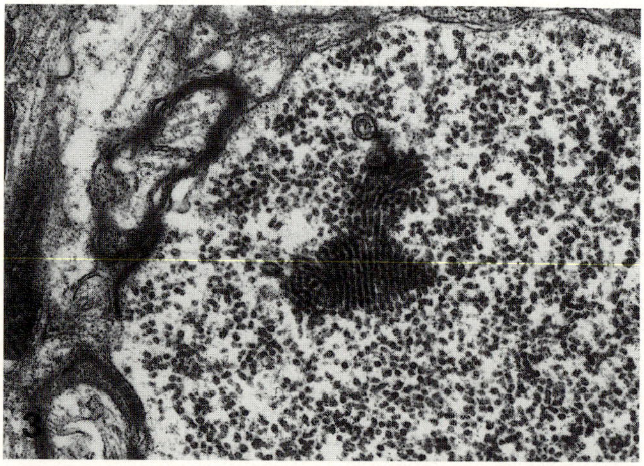

Fig. 9. Electron dense particles from the cortex of a scrapie infected mouse (with permission from BARINGER et al. 1981). The average particle diameter is 23 nm. The tubular array resolves upon tilting into an array of overlapping spherical particles

1976; BARINGER and PRUSINER 1978; BARINGER et al. 1979, 1981; NARANG et al. 1980, 1987a, b, 1988; SHIBAYAMA et al. 1982). However, there is disagreement over whether similar particles are also present in the high-titer hamster scrapie model (BARINGER et al. 1979; NARANG et al. 1987b). Their presence there has seemed a prerequisite for their relevance to the disease.

As originally proposed, "Prions are small proteinaceous infectious particles which are resistant to inactivation by most procedures that modify nucleic acids" (PRUSINER 1982). Depending upon how small was small, this definition subsumed all of virology, and to the extent that it was intended to mean that "the prion is whatever scrapie turns out to be", there is little point in trying to prove or disprove it. More recently this definition has been constrained somewhat by an additional requirement that PrP^{Sc} be a "major and necessary component" of the agent (PRUSINER 1989), a tenuous condition with respect to the scrapie agent. Rather, it is likely that a family of small viruses of low virulence and relatively wide distribution has yet to be discovered.

References

Adams DH (1970) The nature of the scrapie agent. A review of recent progress. Pathol Biol (Paris) 18: 559–577

Aiken JM, Williamson JL, Marsh RF (1989) Evidence of mitochondrial involvement in scrapie infection. J Virol 63: 1686–1694

Aiken JM, Williamson JL, Borchardt LM, Marsh RF (1990) Presence of mitochondrial D-loop DNA in scrapie-infected brain preparations enriched for the prion protein. J Virol 64: 3265–3268

Allen IV, Dermott E, Connolly JH, Hurwitz LJ (1971) A study of a patient with the amyotrophic form of Creutzfeldt–Jakob disease. Brain 94: 715–724

Alper T (1985) Scrapie agent unlike viruses in size and susceptibility to inactivation by ionizing or ultraviolet radiation. Nature 317: 750 [letter]

Alper T (1987) Radio- and photobiological techniques in the investigation of prions. In: Prusiner SB, McKinley MP (eds) Prions. Academic, Berkeley, pp113–148

Alper T, Haig DA (1968) Protection by anoxia of the scrapie agent and some DNA and RNA viruses irradiated as dry preparations. J Gen Virol 3: 157–166

Alper T, Haig DA, Clarke MC (1966) The exceptionally small size of the scrapie agent. Biochem Biophys Res Commun 22: 278–284

Alper T, Cramp WA, Haig DA, Clarke MC (1967) Does the agent of scrapie replicate without nucleic acid? Nature 214: 764–766

Alper T, Haig DA, Clarke MC (1978) The scrapie agent: evidence against its dependence for replication on intrinsic nucleic acid. J Gen Virol 41: 503–516

Bailey JN, McAllister WT (1980) Mapping of promoter sites utilized by T3 polymerase on T3 DNA. Nucleic Acid Res 8: 5071–5088

Baringer JR, Prusiner SB (1978) Experimental scrapie in mice: ultrastructural observations. Ann Neurol 4: 205–211

Baringer JR, Prusiner SB, Wong JS (1981) Scrapie-associated particles in postsynaptic processes: Further ultrastructural studies. J Neuropathol Exp Neurol 40: 281–288

Baringer JR, Wong J, Klassen T, Prusiner SB (1979) Further observations on the neuropathology of experimental scrapie in mouse and hamster. In: Prusiner SB, Hadlow WJ (eds) Slow transmissible diseases of the nervous system, vol 2. Academic, New York, pp 111–121

Barlow, RM, Middleton DJ (1990) Dietary transmission of bovine spongiform encephalopathy to mice. Vet Rec 126: 111–112

Barry RA, McKinley MP, Bendheim PE, Lewis GK, DeArmond SJ, Prusiner SB (1985) Antibodies to the scrapie protein decorate prion rods. J Immunol 135: 603–613

Basler K, Oesch B, Scott M, Westaway D, Wälchli M, Groth DF, McKinley MP, Prusiner SB, Weissmann C (1986) Scrapie and cellular PrP isoforms are encoded by the same chromosomal gene. Cell 46: 417–428
Bellinger-Kawahara C, Cleaver JE, Diener TO, Prusiner SB (1987a) Purified scrapie prions resist inactivation by UV irradiation. J Virol 61: 159–166
Bellinger-Kawahara C, Diener TO, McKinley MP, Groth DF, Smith DR, Prusiner SB (1987b) Purified scrapie prions resist inactivation by procedures that hydrolyze, modify, or shear nucleic acids. Virology 160: 271–274
Bellinger-Kawahara CG, Kempner E, Groth D, Gabizon R, Prusiner SB (1988) Scrapie prion liposomes and rods exhibit target sizes of 55,000 Da. Virology 164: 537–541
Bendheim PE, Barry RA, DeArmond SJ, Stites DP, Prusiner SB (1984) Antibodies to a scrapie prion protein. Nature 310: 418–421
Berg G (ed) (1967) Transmission of viruses by the water route. Wiley Interscience, New York
Bergold GH, Weibel J (1962) Demonstration of the yellow fever virus with the electron microscope. Virology 17: 554–562
Bignami A, Parry HB (1970) Electron microscopic observations in slow virus infections of the central nervous system. VIth International Congress of Neuropathology. Masson. Paris, pp 933–934
Block SS (1983) Disinfection, sterilization and preservation, 3rd edn. Lea and Febiger, Philadelphia
Bolton DC, Bendheim PE (1988) A modified host protein model of scrapie. Ciba Found Symp 135: 164–181
Bolton DC, McKinley MP, Prusiner SB (1982) Identification of a protein that purifies with the scrapie prion. Science 218: 1309–1311
Bolton DC, McKinley MP, Prusiner SB (1984) Molecular characteristics of the major scrapie prion protein. Biochemistry 23: 5898–5906
Bolton DC, Bendheim PE, Marmorstein AD, Potempska A (1987) Isolation and structural studies of the intact scrapie agent protein. Arch Biochem Biophys 258: 579–590
Bots G Th AM, De Man JCH, Verjaal A (1971) Virus-like particles in brain tissue from two patients with Creutzfeldt–Jakob disease. Acta Neuropathol (Berl) 18: 267–270
Braig HR, Diringer H (1985) Scrapie: concept of a virus-induced amyloidosis of the brain. EMBO J 4: 2309–2312
Brown P (1990) The phantasmagoric immunology of transmissible spongiform encephalopathy. Res Publ Assoc Res Nerv Ment Dis 68: 305–313
Brown P, Rohwer RG, Green EM, Gajdusek DC (1982) Effect of chemicals, heat, and histopathologic processing on high-infectivity hamster-adapted scrapie virus. J Infect Dis 145: 683–687
Brown P, Rohwer RG, Gajdusek DC (1986) Newer data on the inactvation of scrapie virus or Creutzfeldt–Jakob disease virus in brain tissue. J Infect Dis 153: 1145–1148
Brown P, Liberski PP, Wolff A, Gajdusek DC (1990a) Resistance of scrapie infectivity to steam autoclaving after formaldehyde fixation and limited survival after ashing at 360° C: practical and theoretical implications. J Infect Dis 161: 467–472
Brown P, Liberski PP, Wolff A, Gajdusek DC (1990b) Conservation of infectivity in purified fibrillary extracts of scrapie-infected hamster brain after sequential enzymatic digestion or polyacrylamide gel electrophoresis. Proc Natl Acad Sci USA 87: 7240–7244
Browne H, Smith G, Beck S, Minson T (1990) A complex between the MHC class I homologue encoded by human cytomegalovirus and beta microglobulin. Nature 347: 770–772
Bruce ME, Dickinson AG, (1979) Biological stability of different classes of scrapie agent. In: Prusiner SB, Hadlow WJ (eds) Slow transmissible diseases of the nervous system, vol 2. Academic, New York, pp 71–86
Bruce ME, Dickinson AG (1980) Asymmetrical foci of vacuolation and agent strain instability in scrapie. 1. Biological aspects. Neuropathol Appl Neurobiol 6: 234
Bruce ME, Dickinson AG (1987) Biological evidence that scrapie agent has an independent genome. J Gen Virol 68: 79–89
Bruce ME, McBride PA, Farquhar CF (1989) Precise targeting of the pathology of the sialoglycoprotein, PrP, and vacuolar degeneration in mouse scrapie. Neurosci Lett 102: 1–6
Buyukmihci NC, Goehring-Harmon F, Marsh RF (1987) Photoreceptor degeneration during infection with various strains of the scrapie agent in hamsters. Exp Neurol 97: 201–206
Carlson GA, Kingsbury DT, Goodman PA, Coleman S, Marshall ST, DeArmond S, Westaway D, Prusiner SB (1986) Linkage of prion protein and scrapie incubation time genes. Cell 46: 503–511
Carlson GA, Goodman PA, Lovett M, Taylor BA, Marshall ST, Peterson–Torchia M, Westaway D, Prusiner SB (1988) Genetics and polymorphism of the mouse prion gene complex: control of scrapie incubation time. Mol Cell Biol 8: 5528–5540

Carp RI, Callahan SM (1981) In vitro interaction of scrapie agent and mouse peritoneal macrophages. Intervirology 16: 8–13

Carp RI, Moretz RC, Natelli M, Dickinson AG (1987) Genetic control of scrapie: incubation period and plaque formation in I mice. J Gen Virol 68: 401–407

Cashman NR, Loertscher R, Nalbantoglu J, Shaw I, Kascsak RJ, Bolton DC, Bendheim PE (1990) Cellular isoform of the scrapie agent protein participates in lymphocyte activation. Cell 61: 185–192

Cates W Jr, (1990) Acquired immunodeficiency syndrome, sexually transmitted diseases, and epidemiology. Past lessons, present knowledge, and future opportunities. Am J Epidemiol 131: 749–758

Caughey B, Neary K, Buller R, Ernst D, Perry LL, Chesebro B, Race RE (1990) Normal and scrapie-associated forms of prion protein differ in their sensitivities to phospholipase and proteases in intact neuroblastoma cells. J Virol 64: 1093–1101

Chesebro B, Race R, Wehrly K, Nishio J, Bloom M, Lechner D, Bergstrom S, Robbins K, Mayer L, Keith JM, Garon C, Haase A (1985) Identification of scrapie prion protein-specific mRNA in scrapie-infected and uninfected brain. Nature 315: 331–333

Cho HJ (1980) Requirement of a protein component for scrapie infectivity. Intervirology 14: 213–216

Cho HJ (1983) Inactivation of the scrapie agent by pronase. Can J Comp Med 47: 494–496

Cho HJ (1986) Antibody to scrapie-associated fibril protein identifies a cellular antigen. J Gen Virol 67: 243–253

Coeroli-Vigouroux C, Bert J, Gordes J-M, Nicoli J (1983) Obtention de fractions subcellulaires hautement infectieuses dans la tremblante du mouton. Etude par chromatographie hydrophobe et ultrafiltration. In: Court LA, Cathala F (eds) Virus non conventionnels et affections du systeme nerveux central. Masson, Paris, pp 114–124

Collinge J, Harding AE, Owen F, Poulter M, Lofthouse R, Boughey AM, Shah T, Crow TJ (1989) Diagnosis of Gerstmann–Sträussler syndrome in familial dementia with prion protein gene analysis. Lancet ii: 15–17

Collinge J, Owen F, Poulter M, Leach M, Crow TJ, Rossor MN, Hardy J, Mullan MJ, Janota I, Lantos PL (1990) Prion dementia without characteristic pathology. Lancet 336: 7–9

Cuille J, Chelle PL (1936) La maladie dite tremblante du mouton, est-elle inoculable? CR Seances Acad Sci (Paris) 203: 1552–1554

Cuille J, Chelle PL (1938) La tremblante du mouton. Son etiologie. (Recherches cliniques et experimentales.). Rev Pathol Comp Hyg Gen 38: 1358–1372

David-Ferriera JF, David-Ferreira KL, Gibbs CJ Jr, Morris JA (1968) Scrapie in mice: ultrastructural observations in the cerebral cortex. Proc Soc Exp Biol Med 127: 313–320

Decker C, Guir J, Kirn A (1969) Infectivity and capacity for DNA replication of vaccinia virus irradiated by gamma-rays. J Gen Virol 4: 221–227

De Reuck J, De Coster W, Otte G, Vander Eecken H, (1976) Papova virus-like particles in a nigral type of Creutzfeldt–Jakob disease. J Neurol 213: 179–188

Dickinson AG, Fraser H (1975) Scrapie: pathogenesis in inbred mice: an assessment of host control and response involving many strains of agent. In: ter Meulen V, Katz M (eds) Slow virus infections of the central nervous system. Springer, Berlin Heidelberg New York, pp 3–14

Dicknson AG, Fraser H (1979) An assessment of the genetics of scrapie in sheep and mice. In: Prusiner SB, Hadlow WJ (eds) Slow transmissible diseases of the nervous system, vol 1. Academic, New York, pp 367–385

Dickinson AG, Meikle VM (1969) A comparison of some biological characteristics of the mouse-passaged scrapie agents, 22A and ME7. Genet Res 13: 213–225

Dickinson AG, Outram GW (1979) The scrapie replication-site hypothesis and its implications for pathogenesis. In: Prusiner SB, Hadlow WJ (eds) Slow transimissible diseases of the nervous system, vol 2. Academic, New York, pp 13–31

Dickinson AG, Outram GW (1988) Genetic aspects of unconventional virus infections: the basis of the virino hypothesis. Ciba Found Symp 135: 63–83

Dickinson AG, Meikle VM, Fraser H (1968) Identification of a gene which controls the incubation period of some strains of scrapie agent in mice. J Comp Pathol 78: 293–299

Dickinson AG, Bruce ME, Outram GW, Kimberlin RH (1984) Scrapie strain differences: the implications of stability and mutation. In: Tateishi J (ed) Proceedings of workshop on slow transmissible diseases. Research Committee on Slow Virus Infection, 31 Aug 1984. Japanese Minsitry of Health and Welfare Tokyo, pp 105–118

Diener TO, McKinley MP, Prusiner SB (1982) Viroids and prions Proc Natl Acad Sci USA 79: 5220–5224

Diringer H, Kimberlin RH (1983) Infectious scrapie agent is apparently not as small as recent claims suggest. Biosci Rep 3: 563–568

Diringer H, Gelderblom H, Hilmert H, Ozel M, Edelbluth C, Kimberlin RH (1983a) Scrapie infectivity, fibrils and low molecular weight protein. Nature 306: 476–478

Diringer H, Hilmert H, Simon D, Werner E, Ehlers B (1983b) Towards purifications of the scrapie agent. Eur J Biochem 134: 555–560

Diringer H, Rahn HC, Bode L (1984) Antibodies to protein of scrapie-associated fibrils. Lancet ii: 345 (letter)

Doh-ura IK, Tateishi J, Kitamoto T, Sasaki H, Sakaki Y (1990) Creutzfeldt–Jakob disease patients with congophilic kuru plaques have the missense variant prion protein common to Gerstmann–Sträussler syndrome. Ann Neurol 27: 121–126

Duguid JR, Rohwer RG, Seed B (1988) Isolation of cDNAs of scrapie-modulated RNAs by subtractive hybridization of a cDNA library. Proc Natl Acad Sci USA 85: 5738–5742

Duguid JR, Bohmont CW, Liu NG, Tourtellotte WW (1989) Changes in brain gene expression shared by scrapie and Alzheimer disease. Proc Natl Acad Sci USA 86: 7260–7264

Ekert B, Muel B, Latarjet R (1970) Irradiation de ribosomes par des rayons ultraviolets: spectre d'action de l'inactivation de la fonction de synthese. Biochem Biophys Acta 204: 275–277

Eklund CM, Hadlow WJ, Kennedy RC (1963) Some properties of the scrapie agent and its behavior in mice. Proc Soc Exp Biol Med 112: 974–979

Field EJ, Raine CS (1966) Observations on "dense-body" structure in nerve cells with special reference to scrapie. Res Vet Sci 7: 292–295

Field EJ, Raine CS (1967) Lattice fibrillar or herringbone structures in kuru. Vet Rec 81: 639–640

Field EJ, Farmer F, Caspary EA, Joyce G (1969) Susceptibility of scrapie agent to ionizing radiation. Nature 222: 90–91

Fraenkel-Conrat H (1956) The role of the nucleic acid in the reconstitution of active tobacco mosaic virus. J Am Chem Soc 78: 882–883

Fraenkel-Conrat H, Wagner RR (1974) Comprehensive virology vol 1, Descriptive catalogue of viruses. Plenum, New York, pp 191

Fraser H (1979) Neuropathology of scrapie: the precision of the lesions and their diversity. In: Prusiner SB, Hadlow WJ (eds) Slow transmissible diseases of the nervous system, volume 1. Academic, New York, pp 387–406

Fraser H, McConnell I, Wells GA, Dawson M (1988) Transmission of bovine spongiform encephalopathy to mice. Vet Rec 123: 472

Fraser H, Bruce ME, McBride PA, Scott JR (1989) The molecular pathology of scrapie and the biological basis of lesion targeting. Prog Clin Biol Res 317: 637–644

Gabizon R, McKinley MP, Groth D, Prusiner SB (1988) Immunoaffinity purification and neutralization of scrapie prion infectivity. Proc Natl Acad Sci USA 85: 6617–6621

Gabizon R, McKinley MP, Groth D, Westaway D, DeArmond SJ, Carlson GA, Prusiner SB (1989) Immunoaffinity purification and neutralization of scrapie prions. Prog Clin Biol Res 317: 583–600

Gajdusek DC (1990) Subacute spongiform encephalopathies: transmissible cerebral amyloidoses caused by unconventional viruses. In: Fields BN, Knipe DM et al. (eds) Virology, 2nd ed Raven, New York, pp 2289–2324

Gajdusek DC, Gibbs CJ Jr (1990) Brain amyloidoses, precursor proteins and the amyloids of transmissible and nontransmissible dementias: scrapie-kuru-CJD viruses as infectious polypeptides or amyloid-enhancing factors. In: Goldstein AL (ed) Biomedical advances in aging. Plenum, New York, pp 3–24

GenBank (1990) Genetic Sequence Data Bank, Computer Systems Division, Bolt, Beranek and Newman Inc., 10 Moulton Street, Cambridge, MA 02238, USA

Gibbs, CJ Jr (1967) Search for infectious etiology in chronic and subacute degenerative diseases of the central nervous system. Curr Top Microbiol Immunol 40: 44–58

Gibbs CJ Jr, Gajdusek DC, Latarjet R (1978) Unusual resistance to ionizing radiation of the viruses of kuru, Creutzfeldt–Jakob disease, and scrapie. Proc Natl Acad Sci USA 75: 6268–6270

Gibbs CJ Jr, Gajdusek DC, Amyx HL (1979) Strain variation in the viruses of Creutzfeldt–Jakob disease and kuru. In: Prusiner SB, Hadlow WJ (eds) Slow transmissible diseases of the nervous system, vol 2. Academic, New York, pp 87–110

Ginoza W (1963) Radiosensitive molecular weight of single-stranded virus nucleic acids. Nature 199: 453–456.

Ginoza W (1968) Inactivation of viruses by ionizing radiation and by heat In: Maramorosch K, Kaprowski H (eds) Methods in virology, vol 4. Academic, London, pp 139–209

Ginoza W, Miller RC (1965) Kinetics of X-ray and heat inactivation of ϕX174 RF-DNA. Proc Natl Acad Sci USA 54: 551–558

Gionza W, Norman A (1957) Radiosensitive molecular weight of tobacco mosaic virus nucleic acid. Nature 179: 520–521

Goldfarb LG, Brown P, Goldgaber D, Asher DM, Rubenstein R, Brown WT, Piccardo P, Kascsak RJ, Boellaard JW, Gajdusek DC (1990a) Creutzfeldt–Jakob disease and kuru patients lack a mutation consistently found in the Gerstmann–Sträussler Scheinker syndrome. Exp Neurol 108: 247–250

Goldfarb LG, Brown P, Goldgaber D, Garruto RM Yanagihara R, Asher DM, Gajdusek DC (1990b) Identical mutation in unrelated patients with Creutzfeldt–Jakob disease. Lancet 336: 174–175 (letter)

Goldfarb LG, Korczyn AD, Brown P, Chapman J, Gajdusek DC (1990c) Mutation in codon 200 of scrapie amyloid precursor gene linked to Creutzfeldt–Jakob disease in Sephardic Jews of Libyan and non-Libyan origin. Lancet 336: 637–638 (letter)

Goldfarb LG, Mitrova E, Brown P, Toh BK, Gajdusek DC (1990d) Mutation in codon 200 of scrapie amyloid protein gene in two clusters of Creutzfeldt–Jakob disease in Slovakia Lancet 336: 514–515 (letter)

Goldgaber D, Goldfarb LG, Brown P, Asher DM, Brown WT, Lin S, Teener JW, Feinstone SM, Rubenstein R, Kascsak RJ, Boellaard JW, Gajdusek DC (1989) Mutations in familial Creutzfeldt–Jakob disease and Gerstmann–Sträussler–Scheinker's syndrome. Exp Neurol 106: 204–206

Gordon WS (1960) Scrapie and scrapie research at the British Agricultural Research Council Field Station, Compton, Berkshire, England. Presented at the Special Meeting on Scrapie, US Department of Agriculture, Nov–Dec 1959. Washington DC, ARS 91–22 (June): pp 1–12

Hadlow WJ, Kennedy RC, Race RE (1982) Natural infection of Suffolk Sheep with scrapie virus. J Infect. Dis 146: 657–664

Haig DA, Clarke MC (1965) Observations on the agent of scrapie. In: Gajdusek DC, Gibbs CJ Jr, Alpers M (eds) Slow, latent, and temperate virus infections. US Government Printing Office, Washington DC pp 215–219 NINDB Monograph no 2, National Institutes of Health, PHS publication no 1378)

Haig DA, Clarke MC Blum E, Alper T (1969) Further studies on the inactivation of the scrapie agent by ultraviolet light. J Gen Virol 5: 455–457

Haig DA, Clarke MC, Millson GC (1977) Failure of fluorocarbon treatment to affect sensitivity of scrapie agent to ultraviolet light. J Comp Pathol 87: 103–107

Hershey AD, Chase M (1952) ndependent functions of viral protein and nucleic acid in growth of bacteriophage. J Gen Physiol 36: 39–56

Hogan RN, Bowman KA, Baringer JR, Prusiner SB (1986) Replication of scrapie prions in hamster eyes precedes reteinal degeneration. Ophthalmic Res 18: 230–235

Hollinger FB (1990a) Hepatitis B virus. In: Fields BN, Knipe DN et al. (eds) Virology, 2nd ed. Raven, New York, pp 2171–2236

Hollinger FB (1990b) Non-A, non-B hepatitis viruses. In: Fields BN, Knipe DN et al. (eds) Virology, 2nd ed. Raven, New York, pp 2239–2273

Hope J, Morton LJ, Farquhar CF, Multhaup G, Beyreuther K, Kimberlin RH (1986) The major polypeptide of scrapie-associated fibrils (SAF) has the same size, charge distribution and N-terminal protein sequence as predicted for the normal brain protein. EMBO J 5: 2591–2597

Hotz G, Muller A (1961) Strahlenbiologische Untersuchungen am Bakteriophagen ϕX174. Z Naturforsch 16b: 282–283

Hradecna Z (1966) X-ray inactivation of lambda phages. I. Damage on adsorption and injection. Int J Radiat Biol 10: 443–449

Hruskova J (1969) Effects of gamma radiation on Western equine encephalomyelitis and tick-borne encephalitis viruses. Acta Virol (Praha) 13: 187–192

Hsiao KK, Baker HF, Crow TJ, Poulter M, Owen F, Terwilliger JD, Westaway D, Ott J, Prusiner SB (1989) Linkage of a prion protein missense variant to Gerstmann–Sträussler syndrome. Nature 338: 342–345

Hsiao KK, Scott M, Foster D, Groth DF, DeArmond SJ, Prusiner SB (1990) Spontaneous neurodegeneration in transgenic mice with mutant prion protein. Science 250: 1587–1590

Hsu D-H, de Waal Malefyt R, Fiorentino DF, Dang M-N, Vieira P, de Vries J, Spits H, Mosmann TR, Moore KW (1990) Expression of interleukin-10 activity by Epstein-barr virus protein BCRFI. Science 250: 830–832

Hunter GD (1965) Progress toward the isolation and characterization of the scrapie agent. In: Gajdusek DC, Gibbs CJ Jr, Alpers M (eds) Slow, latent, and temperate virus infections. US Government Printing Office, Washinton DC pp 259–262 NINDB Monograph no 2, National Institutes of Health, PHS publication no 1378

Hunter GD, Millson GC (1964) Studies on the heat stability and chromatographic behaviour of the scrapie agent. J Gen Microbiol 37: 251–258

Hunter GD, Millson GC (1967) Attempts to release the scrapie agent from tissue debris. J Comp Pathol 77: 301–307

Hunter GD, Gibbons RA, Kimberlin RH, Millson GC (1969) Further studies of the infectivity and stability of extracts and homogenates derived form scrapie affected mouse brains. J Comp Pathol 79: 101–108

Hunter N, Hope J, McConnell I, Dickinson AG (1987) Linkage of the scrapie-associated fibril protein (PrP) gene and *Sinc* using congenic mice and restriction fragment length polymorphism analysis. J Gen Virol 68: 2711–2716

Hunter N. Foster JD, Dickinson AG; Hope J (1989) Linkage of the gene for the scrapie-associated fibril protein (PrP) to the *Sip* gene in Cheviot Sheep. Vet Rec 124: 364–366

Huso DL, Narayan O, Hart GW (1988) Sialic acids on the surface of caprine arthritis-encephalitis virus define the biological properties of the virus. J Virol 62: 1974–1980

Johnson JD (ed) (1975) Disinfection—water and wastewater. Ann Arbor Science, Ann Arbor

Kahana E, Alter M, Braham J, Sofer D (1974) Creutzfeldt–Jakob disease: focus among Libyan Jews in Israel. Science 183: 90–91

Kascsak RJ, Rubenstein R, Merz PA Carp RI, Wisniewski HM, Diringer H (1985) Biochemical differences among scrapie-associated fibrils support the biological diversity of scrapie agents. J Gen Virol 66: 1715–1722

Kascsak RJ, Rubenstein R, Merz PA, Carp RI, Robakis NK, Wisniewski HM, Diringer H (1986) Immunological comparison of scrapie-associated fibrils isolated from animals infected with four different scrapie strains. J Virol 59: 676–683

Kascsak RJ, Rubenstein R, Merz PA, Tonna-DeMasi M, Fersko R, Carp RI, Wisniewski HM, Diringer H (1987) Mouse polyclonal and monoclonal antibody to scrapie-associated fibril proteins. J Virol 61: 3688–3693

Kim KS, Sharp DG (1967a) Multiplicity reactivation of vaccinia virus particles treated with nitrogen mustard J Virol 1: 45–49

Kim KS, Sharp DG, (1967b) Influence of the physical state of formalinized vaccinia virus particles on surviving plaque titer: evidence for multiplicity reactivation. J Immunol 99: 1221–1225

Kim KS, Sharp DG, (1968) Multiplicity reactivation of gamma- and X-irradiated vaccinia virus in L cells. Radiat Res 33: 30–36

Kimberlin RH, Hope J (1987) Genes and genomes in scrapie. Trends Genet 3: 117–118

Kimberlin RH, Walker C (1977) Characteristics of a short incubation model of scrapie in the golden hamster. J Gen Virol 34: 295–304

Kimberlin RH, Walker CA (1978) Evidence that the transmission of one source of scrapie agent to hamsters involves separation of agent strains from a mixture. J Gen Virol 39: 487–496

Kimberlin RH, Millson GC, Hunter GD (1971) An experimental examination of the scrapie agent in cell membrane mixtures 3. Studies of the operational size. J Comp Pathol 81: 383–391

Kimberlin RH, Cole S, Walker CA (1987) Temporary and permanent modifications to a single strain of mouse scrapie on transmission to rats and hamsters. J Gen Virol 68: 1875–1881

Kimberlin RH, Walker CA, Fraser H (1989) The genomic identity of different strains of mouse scrapie is expressed in hamsters and preserved on reisolation in mice. J Gen Virol 70: 2017–2025

Kleczkowski A, Govier DA (1969) Action spectrum for inactivation of the infectivity of potato virus X by UV radiation. Photochem Photobiol 10: 53–59

Kleczkowski A, McLaren AD (1967) Inactivation of infectivity of RNA of tobacco mosaic virus during ultraviolet irradiation of the whole virus at two wavelengths. J Gen Virol 1: 441–448

Kotwal GJ, Isaacs SN, McKenzie R, Frank MM, Moss B (1990) Inhibition of the complement cascade by the major secretory protein of vaccinia virus. Science 250: 827–830

Kroeger AV, Kempf JE (1959) Inactivation of the influenza virus by low-voltage Roentgen rays. J Bacteriol 77: 237–246

Kuroda Y, Gibbs CJ Jr, Amyx HL, Gajdusek DC (1983) Creutzfeldt-Jakob disease in mice: persistent viremia and preferential replication of virus in low-density lymphocytes. Infect Immun 41: 154–161

Lamar CH, Gustafson DP, Krasovich M, Hinsman EJ (1974) Ultrastructural studies of spleens, brains, and brain cell cultures of mice with scrapie. Vet Pathol 11: 13–19

Lampert PW, Gajdusek DC, Gibbs CJ Jr (1971a) Experimental spongiform encephalopathy (Creutzfeldt–Jakob disease) in chimpanzees. Electron microscopic studies. J Neuropathol Exp Neurol 30: 20–32.

Lampert P, Hooks J, Gibbs CJ Jr, Gajdusek DC (1971b) Altered plasma membranes in experimental scrapie. Acta Neuropathol (Berl) 19: 81–93

Latarjet R (1979) Inactivation of the agents of scrapie, Creutzfeldt–Jakob disease, and kuru by radiations. In: Prusiner SB, Hadlow WJ (eds) Slow transmissible diseases of the nervous system, vol 2. Academic, New York, pp 387–407

Latarjet R, Fredericq P (1955) An X-ray study of a colicine and of its relationship to bacteriophage T6. Virology 1: 100–107

Latarjet R, Ephrussi-Taylor H, Rebeyrotte N (1959) On the target size of a transforming factor based on x-ray inactivation. Radiat Res Suppl 1: 417–430

Latarjet R, Muel B, Haig DA, Clarke MC, Alper T (1970) Inactivation of the scrapie agent by near monochromatic ultraviolet light. Nature 227: 1341–1343

Lavelle GC (1973) Multiplicity of scrapie virus in infected mouse spleen cells in vivo. Infect Immun 7: 918–921

Lax AJ, Millson GC, Manning EJ (1983) Involvement of protein in scrapie agent infectivity. Res Vet Sci 34: 155–158

Lea DE (1955) Actions of radiations on living cells, 2nd edn. Cambridge University Press, Cambridge

Lea DE, Salaman MH (1946) Experiments on the inactivation of bacteriophage by radiations, and their bearing on the nature of bacteriophage. Proc R Soc B 133: 434–444

Lea DE, Smith KM (1940) The inactivation of plant viruses by radiation. Parasitology 32: 405–416

Lea DE, Smith KM (1942) The inactivation of plant viruses by radiations. II. The relation between inactivation dose and size of virus. Parasitology 34: 227–237

Lee HW (1982) Korean hemorrhagic fever. Prog Med Virol 28: 96–113

Lee HW, Lee PW, Johnson KJ (1978) Isolation of the etiologic agent of Korean hemorrhagic fever. J Infect Dis 137: 298–308

Lehmann-Grube F, Peralta LM, Bruns M, Lohler J (1983) Persistent infection of mice with the lymphocytic choriomeningitis virus. In: Fraenkal-Conrat H, Wagner RR (eds) Comprehensive virology, vol 18. Plenum, New York, pp 43–103

Lentz TL, Burrage TG, Smith AL, Crick J, Tignor G (1982) Is the acetylcholine receptor a rabies virus receptor? Science 215: 182–184

Lentz TL, Wilson PT, Hawrot E, Speicher DW (1984) Amino acid sequence similarity between rabies virus glycoprotein and snake venom curaremimetic neurotoxins. Science 226: 847–848

Linton DM, Philcox D (1990) Myasthenia gravis. Dis Mon 36: 593–637

Ludwig H, Bode L. Gosztonyi G (1988) Borna disease: a persistent virus infection of the central nervous system. Prog Med Virol 35: 107–151

Lytle CD, Ginoza W (1969) Intracellular development of bacteriophage ϕX174 inactivated by gamma-ray, ultraviolet light, or nitrous acid. Virology 38: 152–165

Majer M (1972) Virus sensitization. Curr Top Microbiol Immunol 58: 69–84

Malone TG, Marsh RF, Hanson RP, Semancik JS (1979) Evidence for the low molecular weight nature of scrapie agent. Nature 278: 575–576

Manuelidis EE, Gorgacs EJ, Manuelidis L (1978) Viremia in experimental Creutzfeldt–Jakob disease. Science 200: 1069–1071

Marsh RF, Hanson RP (1969) Physical and chemical properties of the transmissible mink encephalopathy agent. J Virol 3: 176–180

Massey RJ, Schochetman G (1981) Viral epitopes and monoclonal antibodies: isolation of blocking antibodies that inhibit virus neutralization. Science 213: 447–449

McCrea JF (1960) Ionizing radiation and its effects on animal viruses. Ann NY Acad Sci 83: 692–705

McKinley MP, Bolton DC, Prusiner SB (1983a) A protease-resistant protein is a structural component of the scrapie prion. Cell 35: 57–62

Mckinley MP, Masiarz FR, Isaacs ST, Hearst JE, Prusiner SB (1983b) Resistance of the scrapie agent to inactivation by psoralens. Photochem Photobiol 37: 539–545

McKinley MP, Braunfeld MB, Bellinger CG, Prusiner SB (1986) Molecular characteristics of prion rods purified from scrapie-infected hamster brains. J Infect Dis 154: 110–120

Merz GS, Schwenk V, Schuller-Levis G, Gruca S, Wisniewski HM (1987) Isolation and characterization of macrophages from scrapie-infected mouse brain. Acta Neuropathol (Berl) 72: 240–247

Merz PA (1988) Final general discussion. Ciba Found Symp 135: 261–264

Merz PA, Somerville RA, Wisniewski HM, Iqbal K (1981) Abnormal fibrils from scrapie-infected brain. Acta Neuropathol (Berl) 54: 63–74

Merz PA, Somerville RA, Wisniewski HM (1983) Abnormal fibrils in scrapie and senile dementia of Alzheimer type. In: Court LA, Cathala F (eds) Virus non conventionnels et affections du systeme nerveux central. Masson, Paris, pp 259–281

Merz PA, Rohwer RG, Kascsak R, Wisniewski HM, Somerville RA, Gibbs CJ Jr, Gajdusek DC (1984) Infection-specific particle from the unconventional slow virus diseases. Science 225: 437–440

Meyer RK, McKinley MP, Bowman KA, Braunfeld MB, Barry RA, Prusiner SB (1986) Separation and properties of cellular and scrapie prion proteins. Proc Natl Acad Sci USA 83: 2310–2314
Millson GC, Hunter GD, Kimberlin RH (1976) The physico–chemical nature of the scrapie agent. In: Kimberlin RH (ed) Slow virus diseases of animals and man. North-Holland, Amsterdam, pp 243–266 (North-Holland research monographs, frontiers of biology vol 44)
Mitrova E (1990) Analytical epidemiology and risk factors of CJD. In: Court LA, Dormont D, Brown P, Kingbury DT (eds) Unconventional virus diseases of the Central nervous system, Commissariat a l'Energie Atomique, Service de Documentation, Paris, pp 19–29
Mould DL, Dawson AM (1970) The response in mice to heat treated scrapie agent. J Comp Pathol 80: 595–600
Narang HK (1973) Virus-like particles in natural scrapie of the sheep. Res Vet Sci 14: 108–110
Narang HK (1974a) An electron microscopic study of natural scrapie sheep brain: further observations on virus-like particles and paramyxovirus-like tubules. Acta Neuropathol (Berl) 28: 317–329
Narang HK (1974b) Ruthenium red and lanthanum nitrate: a possible tracer and negative stain for scrapie "particles"? Acta Neuropathol (Berl) 29: 37–43
Narang HK (1974c) An electron microscopic study of the scrapie mouse and rat: further observations on virus-like particles with ruthenium red and lanthanum nitrate as possible trace and negative stain. Neurobiology 4: 349–363
Narang HK (1975) Virus-like particles in Creutzfeldt–Jakob biopsy material. Acta Neuropathol (Berl) 32: 163–168
Narang HK (1988) A chronological study of experimental scrapie in mice. Virus Res 9: 293–305
Narang HK, Chandler RL, Anger HS (1980) Further observations on particulate structures in scrapie-affected brain. Neuropathol Appl Neurobiol 6: 23–28
Narang HK, Asher DM, Gajdusek DC (1987a) Tubulofilaments in negatively stained scrapie-infected brains: relationship to scrapie-associated fibrils. Proc Natl Acad Sci USA 84: 7730–7734
Narang HK, Asher DM, Pomeroy KL, Gajdusek DC (1987b) Abnormal tubulovesicular particles in brains of hamsters with scrapie. Proc Soc Exp Biol Med 184: 504–509
Narang HK, Asher DM, Gajdusek DC (1988) Evidence that DNA is present in abnormal tubulofilamentous structures found in scrapie. Proc Natl Acad Sci USA 85: 3575–3579
Nash AA (1985) Tolerance and suppression in virus diseases. Br Med Bull 41: 41–45
Neugut RH, Neugut AI, Kahana E, Stein Z, Alter M (1979) Creutzfeldt–Jakob disease: familial clustering among Libyan-born Israelis. Neurology 29: 225–231
Oesch B, Groth DF, Prusiner SB, Weissmann C (1988) Search for a scrapie-specific nucleic acid: a progress report. Ciba Found Symp 135: 209–223
Oldstone MBA, Ahmed R, Byrne J, Buchmeier MJ, Riviere Y, Southern P (1985) Virus and immune responses: lymphocytic choriomeningitis virus as prototype model of viral pathogenesis. Br Med Bull 41: 70–74
Owen F, Poulter M, Shah T, Collinge J, Lofthouse R, Baker H, Ridley R, McVey J, Crow TJ (1990) An in-frame insertion in the prion protein gene in familial Creutzfeldt–Jakob disease. Mol Brain Res 7: 273–276
Pattison IH, Millson GC (1961) Further experimental observations on scrapie. J Comp Pathol 71: 350–359
Payne CM, Sibley WA (1975) Intranuclear inclusions in a case of Creutzfeldt–Jakob disease: an ultrastructrual study. Acta Neuropathol (Berl) 31: 353–361
Pocchiari M, Macchi G, Peano S, Conz A (1988) Can potential hazard of Creutzfeldt–Jakob disease infectivity be reduced in the production of human growth hormone? Inactivation experiments with the 263K strain of scrapie. Rapid communication. Arch Virol 98: 131–135
Prusiner SB (1982) Novel Proteinaceous infectious particles cause scrapie. Science 216: 136–144
Prusiner SB (1988) Molecular structure, biology, and genetics of prions. Adv Virus Res 35: 83–136
Prusiner SB (1989) Scrapie prions. Annu Rev Microbiol 43: 345–374
Prusiner SB, Hadlow WJ, Eklund CM, Race RE, Cochran SP (1978a) Sedimentation characterisitics of the scrapie agent from murine spleen and brain. Biochemistry 17: 4987–4992
Prusiner SB, Hadlow WJ, Garfin DE, Cochran SP, Baringer JR, Race RE, Eklund CM (1978b) Partial purification and evidence for multiple molecular forms of the scrapie agent. Biochemistry 17: 4993–4999
Prusiner SB, Garfin DE, Cochran SP, McKinley MP, Groth DF, Hadlow WJ, Race RE, Eklund CM (1980a) Experimental scrapie in the mouse: electrophoretic and sedimentation properties of the partially purified agent. J Neurochem 35: 574–582

Prusiner SB, Groth DF, Bildstein C, Masiarz FR, McKinley MP, Cochran SP (1980b) Electrophoretic propeties of the scrapie agent in agarose gels. Proc. Natl Acad Sci USA 77: 2984–2988

Prusiner SB, Groth DF, Cochran SP, Masiarz FR, McKinley MP, Martinez HM (1980c) Molecular properties, partial purification, and assay by incubation period measurements of the hamster scrapie agent. Biochemsitry 19: 4883–4891

Prusiner SB, Groth DF, Cochran SP, McKinley MP, Masiarz FR (1980d) Gel electrophoresis and glass permeation chromatography of the hamster scrapie agent after enzymatic digestion and detergent extraction. Biochemistry 19: 4892–4898

Prusiner SB, Groth DF, McKinley MP, Cochran SP, Bowman KA, Kasper KC (1981a) Thiocyanate and hydroxyl ions inactivate the scrapie agent. Proc Natl Acad Sci USA 78: 4606–4610

Prusiner SB, McKinley MP, Groth DF, Bowman KA, Mock NI, Cochran SP, Masiarz FR (1981b) Scrapie agent contains a hydrophobic protein. Proc Natl Acad Sci USA 78: 6675–6679

Prusiner SB, Bolton DC, Groth DF, Bowman KA, Cochran SP, McKinley MP (1982) Further purification and characterization of scrapie prions. Biochemistry 21: 6942–6950

Prusiner SB, McKinley MP, Bowman KA, Bolton DC, Bendheim PE, Groth DF, Glenner GG (1983) Scrapie prions aggregate to form amyloid-like birefringent rods. Cell 35: 349–358

Prusiner SB, Groth DF, Bolton DC, Kent SB, Hood LE (1984) Purification and structural studies of a major scrapie prion protein. Cell 38: 127–134

Prusiner SB, Scott M, Foster D, Pan K-M, Groth D, Mirenda C, Torchia M, Yang S-L, Serban D, Carlson GA, Hoppe PC, Westaway D, DeArmond SJ (1990) Transgenic studies implicate interactions between homologous PrP isoforms in scrapie prion replication. Cell 63: 673–686

Race RE, Graham K, Ernst D, Caughey B, Chesebro B (1990) Analysis of linkage between scrapie incubation period and the prion protein gene in mice. J Gen Virol 71: 493–497

Radhakrishnan K, Mousa ME (1988) Creutzfeldt–Jakob disease in Benghazi, Libya. Neuroepidemiology 7: 42–43 (letter)

Ramsay N, Ritchie DA (1980) A physical map of the permuted genome of bacteriophage T1. Mol Gen Genetics 179: 669–675

Rice CM, Lenches EM, Eddy SR, Shin SJ, Sheets RL, Strauss JH, (1985) Nucleotide sequence of yellow fever virus: implications for flavivirus gene expression and evolution. Science 229: 726–735

Robakis NK, Swah PR, Wolfe GC, Rubenstein R, Carp RI, Innis MA (1986) Isolation of a cDNA clone encoding the leader peptide of prion protein and expression of the homologous gene in various tissues. Proc Natl Acad Sci USA 83: 6377–6381

Rohwer RG (1983) Scrapie inactivation kinetics: an explanation for scrapie's apparent resistance to inactivation-a reevaluation of estimates of its small size. In: Court LA, Cathala F (eds) Virus non conventionnels et affections du système Nerveux central. Masson, Paris, pp 84–113

Rohwer RG(1984a) Virus-like sensitivity of the scrapie agent to heat inactivation. Science 223: 600–602

Rohwer RG (1984b) Scrapie infectious agent is virus-like in size and susceptibility to inactivation. Nature 308: 658–662

Rohwer RG (1984c) Scrapie-associated fibrils. Lancet ii: 36 (letter)

Rohwer RG (1986) Estimation of scrapie nucleic acid MW from standard curves for virus sensitivity to ionizing radiation. Nature 320: 381 (letter)

Rohwer RG (1991) Resistance of scrapie infectivity to inactivation by intense ultrasonication is similar to that of small spherical viruses (in preparation)

Rohwer RG, Gajdusek DC (1980) Scrapie virus or viroid-the case for a virus. In: Boese A (ed) Search for the cause of multiple sclerosis and other chronic diseases of the central nervous system. Proceedings of the 1st International Symposion of Hertie Foundation, Frankfurt, Sept 1979. Verlag Chemie, Weinheim, pp 333–355

Roy S, Gupta PC, Sethi U (1972) Creutzfeldt–Jakob disease: an electronmicroscopic study with demonstration of virus-like particles. Neurol India 20: 226–230

Salk JE, Gori JB (1960) A review of theoretical, experimental, and practical considerations in the use of formaldehyde for the inactivation of poliovirus. Ann NYAcad Sci 83: 609–637

Schambra FE, Hutchinson F (1964) The action of fast heavy ions on biological material II. Effects on T1 and ΦX-174 bacteriophage and double-strand and single-strand DNA. Radiat Res 23: 514–526

Scott M, Foster D, Miranda C, Serban D, Coufal F, Wälchli M, Torchia M, Groth D, Carlson G, DeArmond SJ, Westaway D, Prusiner SB (1989) Transgenic mice expressing hamster prion protein produce species-specific scrapie infectivity and amyloid plaques Cell 59: 847–857

Sharp DG, Floyd R, Johnson JD (1975) Nature of the surviving plaque-forming unit of reovirus in water containing bromine. Appl Microbiol 29: 94–101

Shibayama Y, Sakaguchi Y, Nakata K, Goto T, Nakai M, Takai T, Shirakata S (1982) Creutzfeldt–Jakob disease with demonstration of virus-like particles. Acta Pathol Jpn 32: 695–702

Sklaviadis T, Manuelidis L, Manuelidis EE (1986) Characterization of major peptides in Creutzfeldt–Jakob disease and scrapie. Proc Natl Acad Sci USA 83: 6146–6150

Sklaviadis TK, Manuelidis L, Manuelidis EE (1989) Physical properties of the Creutzfeldt–Jakob disease agent. J Virol 63: 1212–1222

Sklaviadis T, Akowitz A, Manuelidis EE, Manuelidis L (1990) Nuclease treatment results in high specific purification of Creutzfeldt–Jakob disease infectivity with a denisty characterisitic of nucleic acid-protein complexes. Arch Virol 112: 215–228

Somerville RA , Ritchie LA (1990) Differential glycosylation of the protein (PrP) forming scrapie-associated fibrils. J Gen Virol 71: 833–839

Stahl N, Borchelt DR, Hsiao K, Prusiner SB (1987) Scrapie prion protein contains a phosphatidylinositol glycolipid. Cell 51: 229–240

Stamp JT, Brotherston JG, Zlotnik I, MacKay JMK, Smith W (1959) Further studies on scrapie. Comp Pathol 69: 268–280

Tateishi J, Kitamoto T, Doh-ura K, Sakaki Y, Steinmetz G, Tranchant C, Warter JM, Heldt N (1990) Immunochemical, molecular genetic, and transmission studies on a case of Gerstmann–Sträussler–Scheinker syndrome. Neurology 40: 1578–1581

Terzi M (1961) Comparative anlaysis of inactivating efficiency of radiation on different organisms. Nature 191: 461–463

Tessman I, Tessman ES, Stent GS (1957) The relative radiosensitivity of bacteriophages S13 and T2. Virology 4: 209–215

Tischer I, Gelderblom H, Vettermann W (1982) A very small porcine virus with circular single-stranded DNA. Nature 295: 64–66

Turk E, Teplow DB, Hood LE, Prusiner SB (1988) Purification and properties of the cellular and scrapie hamster prion proteins. Eur J Biochem 176: 21–30

Vernon ML, Horta-Barbosa L, Fuccillo DA, Sever JL, Baringer JR, Birnbaum G (1970) Virus-like particles and nucleoprotein-type filaments in brain tissue from two patients with Creutzfeldt–Jakob disease. Lancet i: 964–966

Watson JD (1950) The properties of X-ray-inactivated bacteriophage. I. Inactivation by direct effect. J Bacteriol 60: 697–718

Weissmann C (1989) Prions. Sheep disease in human clothing. Nature 338: 298–299

Westaway D, Goodman PA, Mirenda CA, McKinley MP, Carlson GA, Prusiner SB (1987) Distinct prion proteins in short and long incubation period mice. Cell 51: 651–662

Wietgrefe S, Zupancic M, Haase A, Chesebro B, Race R, Frey W 2d, Rustan T, Friedman RL (1985) Cloning of a gene whose expression is increased in scrapie and in senile plaques in human brain. Science 230: 1177–1179

Wills PR (1989) Induced frameshifting mechanism of replication for an information-carrying scrapie prion. Microb Pathog 6: 235–249

Wilson DR, Anderson RD, Smith W (1950) Studies in scrapie. J Comp Pathol 60: 267–282

Woese C, Pollard E (1954) The effect of ionizing radiation on various properties of Newcastle disease virus. Arch Biochem Biophys 50: 354–367

Zlotnik I, Rennie JC (1967) The effect of heat on the scrapie agent in mouse brain. Br J Exp Pathol 48: 171–179

Zu Rhein GM, Varakis J (1976) Subacute spongiform encephalopathies. In: Kimberlin RH (ed) Slow virus diseases of animals and man. North Holland, Amsterdam, pp 359–380

Novel Properties and Biology of Scrapie Prions*

S. B. PRUSINER

1 Introduction

Over the past 40 years, an impressive array of data has been accumulated indicating that many features of scrapie, kuru, Creutzfeldt–Jakob disease (CJD), and Gerstmann–Sträussler syndrome (GSS) are not typical of infectious diseases. Furthermore, the past two decades have witnessed the collection of numerous experimental results which argue strongly that the scrapie agent is a novel pathogen. To distinguish the scrapie and CJD agents from viroids and viruses, the term "prion" was introduced (PRUSINER 1982). Although scrapie can be transmitted to animals by inoculation with extracts of diseased tissue, whether infection features in the spread of natural scrapie is unclear (PARRY 1962, 1983; M'GOWAN 1914; WESTAWAY et al. 1989; HOURRIGAN et al. 1979; GORDON 1966; DICKINSON et al. 1965; DICKINSON 1976). Some investigators have proposed that natural scrapie is a genetic disease and that its transmissibility is incidental

* Portions of this review were adapted from GABIZON and PRUSINER (1990). This work was supported by research grants from the National Institutes of Health (AG02132 and NS14069) and a Senator Jacob Javits Center of Excellence in Neuroscience (NS22786), as well as by gifts from the Sherman Fairchild Foundation, National Medical Enterprises, and the Fairleigh S. Dickinson, Jr. Foundation, Inc.
Departments of Neurology and of Biochemistry and Biophysics, University of California, San Francisco, CA 94143, USA

Current Topics in Microbiology and Immunology, Vol. 172

(PARRY 1962, 1983) while others have argued vehemently that natural scrapie is an infectious disease (DICKINSON et al. 1965; DICKINSON 1976).

The tripartite manifestations of prion diseases—infectious, sporadic and genetic are illustrated by kuru, CJD, and GSS, respectively (PRUSINER 1987). Consistent with some of the observations on natural scrapie noted above, GSS and familial CJD are the only known human diseases which appear to be both inherited and transmissible (HSIAO et al. 1989; OWEN et al. 1989; COLLINGE et al. 1989; GOLDGABER et al. 1989; DOH-URA et al. 1989). Recent genetic linkage studies provide the best evidence, to date, that GSS is a genetic disorder in which infectious prions are produced (HSIAO et al. 1989). An alternative hypothesis, for which there are few supporting data, is that prions are ubiquitous infectious pathogens causing CJD sporadically across the earth at a rate of one case per million population and GSS in all individuals carrying a mutation in one allele of their prion protein (PrP) gene (HSIAO et al. 1989; PRUSINER 1987; RIDLEY et al. 1986; OWEN et al. 1989; WEISSMANN 1989; HARDY 1989). Our genetic linkage results with GSS raise the possibility that sporadic CJD arises from a somatic mutation (WESTAWAY et al. 1989; HSIAO et al. 1989; WEISSMANN 1989).

Over the past 5 years, a large amount of experimental information about the particles causing scrapie has accumulated. Most of the information has been confirmed, and much of it is widely accepted (DIENER 1987). At times, this confirmed body of information has been overshadowed by what appears to be controversy due to the diverse terminology used by different laboratories.

2 On the Definition of Prions

Although the unusual properties of the particle causing scrapie were first appreciated in the 1940s (GORDON 1946), it was not until a mouse bioassay was developed by Chandler (CHANDLER 1961) that sufficient data were obtained to suggest that the scrapie agent might not be a virus, bacterium, fungus, or parasite (ALPER et al. 1966, 1967; PRUSINER et al. 1979). Over the decade from 1965 to 1974, radiobiologic studies offered the most provocative data suggesting that the scrapie agent is fundamentally different from small pathogens such as viruses (ALPER et al. 1966, 1967). The unusual features of the scrapie agent raised the possibility that it might be a replicating membrane fragment (GIBBONS and HUNTER 1967; HUNTER et al. 1968), a viroid (DIENER 1972), a carbohydrate moiety (FIELD 1966, 1967), a protein (GRIFFITH 1967; PATTISON and JONES 1967; LEWIN 1972, 1981), or even a small virus (ROHWER 1984a, b; ADAMS and FIELD 1968; ADAMS 1970). Terms such as "unconventional virus" and "virino" were introduced to distinguish the scrapie agent from viruses (GAJDUSEK 1977; DICKINSON and OUTRAM 1979). Although the term unconventional virus has been used widely, some investigators believe it is counterproductive to adopt such terminology. In surveying the research during the decade 1969–1978 Ian

PATTISON (1988) recently wrote "...with the causal agent of scarpie still obscure, and virologists as adamant as ever that theirs was the only worthwhile point of view. To explain findings that did not fit with a virus hypothesis, they had rechristened the causal agent an 'unconventional virus'. Use of this ingenious cover-up for uncertainty made 'virus' meaningless—for is not a cottage an unconventional castle?" Recently, uncoventional viruses have been suggested to be mineral-protein complexes which multiply by crystallization and have been said to be similar to computer viruses (GAJDUSEK 1986, 1988).

"Virino" was defined as follows: by DICKINSON and OUTRAM (1979): "If the recent experimental results of Marsh and Malone are correct in implicating DNA as a necessary component of the infective unit of scrapie, then an appropriate name for this class of agent would be 'virinos', which (by analogy with neutrinos) are small, immunologically neutral particles with high penetration properties but needing special criteria to detect their presence". Subsequent studies showed that the work of MARSH and colleagues (1978) and MALONE and associates (1979) could not be confirmed (PRUSINER et al. 1980a). The most recent definition of "virino" is "host proteins sequestering the agent genome which may code for no product other than copies of itself" (DICKINSON and OUTRAM 1988).

While several studies raised the possibility that protein might be required for scrapie infectivity (CHO 1980; HUNTER and MILLSON 1967; MILLSON et al. 1976), convincing data showing that a protein molecule is necessary for infectivity were obtained only after an effective protocol for partial purification of the scrapie agent was developed (PRUSINER et al. 1981a). Attempts to use the same protocol to demonstrate that the scrapie agent contains a nucleic acid were unsuccessful. Five procedures that modify or hydrolyze nucleic acids failed to inactivate the scrapie agent, yet these procedures were capable of inactivating numerous viruses as well as viroids (PRUSINER 1982; DIENER et al. 1982). On the basis of these studies, the term "prion" was introduced in order to distinguish the class of particles causing scrapie from those responsible for viral illnesses (PRUSINER 1982).

To avoid prejudging the structure of these infectious particles, prions were defined as "small proteinaceous infectious particles that resist inactivation by procedures which modify nucleic acids" (PRUSINER 1982) and three hypothetical structures for the prion were proposed: (1) proteins surrounding a nucleic acid that encodes them (a virus) (2) proteins surrounding a small noncoding polynucleotide, and (3) a proteinaceous particle devoid of nucleic acid. Data from many laboratories have established that scrapie is not caused by a virus. Although some investigators have chosen to redefine prions as infectious proteins (CARP et al. 1985; HOPE and KIMBERLIN 1987), we have avoided this oversimplification in order not to bias our experimental approaches (CHAMBERLIN 1890).

As described below, a considerable body of data argues persuasively that a protein designated PrP^{Sc} is a major and necessary component of the infectious scrapie particle. Yet unlike viral proteins, PrP^{Sc} is encoded by a chromosomal gene. Based on these observations, the definition of prions (PRUSINER 1989) can be extended to "small proteinaceious infectious particles that resist inactivation by procedures modifying nucleic acids and contain an abnormal isoform of a cellular protein which is major and necessary component."

3 Evidence That Prions Contain PrP^{Sc}

Enriching brain fractions for scrapie infectivity led to the discovery of PrP 27–30 (BOLTON et al. 1982; PRUSINER et al. 1982a). Development of a more rapid and economical bioassay (PRUSINER et al. 1980b, 1982b) greatly facilitated purification of the hamster scrapie agent (BOLTON et al. 1982; PRUSINER et al. 1982a). PrP 27–30 migrates during sodium dodecyl sulfate-polyacrylamide gel electrophoresis (SDS-PAGE) as a broad band with an apparent molecular weight of 27000–30000.

There has been a remarkable convergence of experimental data arguing that PrP^{Sc} (or PrP 27–30) is a component of the infectious prion particle (Table 1). Biochemical purification and characterization studies, molecular cloning and genetic investigations, and neuropathologic studies in animals and humans, as well as immunoaffinity purification and neutralization, all provide data supporting the contention that prions are composed largely, if not entirely, of PrP^{Sc} molecules. PrP 27–30 and the scrapie agent copurify (PRUSINER et al. 1982a, 1983; MCKINLEY et al. 1983; BOLTON et al. 1984) when detergent extractions and limited proteolysis are used to promote aggregation of prions into amyloid rods which are collected by centrifugation. PrP 27–30 is the most abundant macromolecule in purified preparations (PRUSINER et al. 1983). Immunoaffinity purification of scrapie prions was accomplished with PrP monoclonal antibody coupled to protein A-Sepharose. Dispersion of prions into detergent-lipid-protein complexes (DLPCs) was required before meaningful chromatography could be performed (GABIZON et al. 1987, 1988a). Copurification of scrapie prion infectivity and PrP^{Sc} was found in fractions eluted from a PrP monoclonal antibody column (GABIZON et al. 1988b). Indeed, copurification by two different procedures aruges that the molecular structure and properties of PrP^{Sc} (as well as PrP 27–30) and the infectious prion particles must be extremely similar. The PrP 27–30 concentration is proportional to the prion titer (MCKINLEY et al. 1983). PrP^{Sc} is absent in normal, uninfected animals (BARRY and PRUSINER 1986; OESCH et al. 1985). Procedures that denature, hydrolyze, or selectively modify PrP 27–30 also diminish the prion titer (MCKINLEY et al. 1983; BOLTON et al. 1984). The unusual kinetics of PrP 27–30 hydrolysis catalyzed by proteases were found to correlate with the diminution of prion titer. PrP 27–30 and scrapie infectivity copartition into many different forms—membranes, rods, spheres, DLPCs, and liposomes. These dramatically different physical forms all contain PrP 27–30 and high prion titers (BOLTON et al. 1984; PRUSINER et al. 1983; GABIZON et al. 1987; BORCHELT et al. 1990; BARRY et al. 1985; MCKINLEY et al. 1986; MEYER et al. 1986). Rabbit antiserum raised against PrP 27–30 purified by SDS-PAGE was found to

Table 1. Convergence of experimental results arguing that PrP^{Sc} (or PrP 27–30) is a major and necessary component of the prion

A.	*Biochemistry*
1.	Enriching Brain fractions for scrapie infectivity using limited proteolysis, detergent extraction, differential centrifugation, and sedimentation through discontinuous sucrose gradients led to the discovery of PrP 27–30 (BOLTON et al. 1982; PRUSINER et al. 1982a).
2.	Prion titers were found to be proportional to the PrP 27–30 concentration (MCKINELY et al. 1983).
3.	Denaturation, hydrolysis, or selective modification of PrP 27–30 also diminished prion titer (MCKINLEY et al 1983; BOLTON et al. 1984).
4.	Scrapie infectivity and PrP 27–30 copartition into multiple forms—membranes, rods, spheres, DLPCs, and liposomes (PRUSINER et al. 1982a, 1983; GABIZON et al. 1987, 1988a).
B.	*Immunology*
1.	Immunoaffinity purification of scrapie infectivity was accomplished using PrP monoclonal antibodies (GABIZON et al. 1988b).
2.	Rabbit antiserum raised against SDS-PAGE-purified PrP 27–30 neutralized scrapie infectivity in DLPCs (GABIZON et al. 1988b).
C.	*Genetics*
1.	Genetic linkage of the PrP gene to loci controlling the scrapie incubation time (*Prn-i* and *Sinc* (CARLSON et al. 1986; HUNTER et al. 1987). Mice (*Prn-p^b*) with long incubation times exhibit amino acid substitutions of codons 108 and 189 of their PrP gene (WESTAWAY et al. 1987).
2.	Genetic linkage of an amino acid substituion at codon 102 of the human PrP gene with development of GSS (HSIAO et al. 1989).
3.	Tg mice expressing hamster PrP exhibit incubation times, amyloid plaques, hamster PrP^{Sc}, and scrapie infectivity characteristic of hamsters after inoculation with hamster prions (SCOTT et al. 1989).
D.	*Neuropathology*
1.	PrP amyloid plaques are specific to prion diseases in animals and humans (KITAMOTO et al. 1986; ROBERTS et al. 1986; DEARMOND et al. 1985).
2.	Hamster prions inoculated into Tg mice expressing hamster PrP induce cerebral amyloid plaques that are composed of hamster PrP and are distributed in patterns similar to those seen in hamsters with scrapie (SCOTT et al. 1989).
3.	Using a sensitive dot-blot assay for PrP^{Sc} (or PrP^{CJD}), there is excellent correlation between prion diseases confirmed neuropathologically and the presence of protease-resistant PrP (SERBAN et al. 1990).
E.	*Cell Biology*
1.	Cultured cells infected with scrapie prions produce PrP^{Sc} (BUTLER et al. 1988).
2.	PrP^{Sc} is produced more slowly than PrP^{C}; furthermore, PrP^{Sc} accumulates primarily inside ot cells while most of the PrP^{Sc} is transported to the external cell surface (STAHL et al. 1987; TARABOULOS et al. 1990; BORCHELT et al. 1990).

neutralize scrapie infectivity in DLPCs (GABIZON et al 1988b). However, no neutralization of prion infectivity was observed with prions aggregated in amyloid rods (BARRY and PRUSINER 1987; ITO et al. 1988).

The PrP gene (*Prn-p*) in mice is linked to a gene controlling scrapie incubation times (*Prn-i*) (CARLSON et al. 1986). Prolonged incubation periods are a cardinal feature of both scrapie and CJD. The preeminent role of PrP in the pathogenesis of scrapie has been made more compelling by the discovery of a correlation between PrP amino acid sequence and scrapie incubation times (WESTAWAY et al. 1987). An amino acid substitution in the human PrP gene is linked genetically to the development of GSS (HSIAO et al. 1989). GSS and familial CJD are the only known human diseases which are both genetic and infectious.

Recently, transgenic (Tg) mice expressing Syrian golden hamster cellular prion protein isoform (PrP^{C}) have been produced (SCOTT et al. 1989). The Tg mice inoculated with hamster prions synthesize both hamster PrP^{Sc} and infectious hamster prions; these results assert that PrP^{Sc} is a necessary component of the infectious prion. Furthermore, the Tg mice exhibit scrapie incubation times and neuropathology characteristic of hamsters. Numerous amyloid plaques containing hamster PrP were found in the brains of the Tg mice in a distribution similar to that found in hamsters.

Scrapie and CJD prion proteins have been identified only in tissues of animals and humans with transmissible neurodegenerative diseases and not in those with other disorders, such as murine systemic amyloidosis and human Alzheimer's disease, anoxic encephalopathy, or non-neurologic disorders (PRUSINER et al. 1982a, b, 1983; BOLTON et al. 1984; KITAMOTO et al. 1986; ROBERTS et al. 1986; BOCKMAN et al. 1985, 1987; BROWN et al. 1986).

Amyloid plaques in prion disease were first identified in the brains of New Guinea natives dying of kuru (KLATZO et al. 1959). Subsequently, they were reported in scrapie and chronic wasting disease of animals (BECK et al. 1964; BRUCE and FRASER 1975; WISNIEWSKI et al. 1975; BAHMANYAR et al 1985; FRASER and BRUCE 1983). With antisera raised against PrP 27–30 isolated from scrapie-infected hamster brain, CNS amyloid plaques in experimental scrapie were shown to be composed of filaments containing PrP (DEARMOND et al. 1985; BENDHEIM et al. 1984). Similar studies with brain sections from humans dying of kuru, CJD, and GSS were also found to contain amyloid plaques which react with PrP antibodies (KITAMOTO et al. 1986; ROBERTS et al. 1986). To date, amyloid plaques reacting with PrP antibodies have been found only in prion diseases.

Cultured murine neuroblastoma cells have been infected with both scrapie and CJD prions (BUTLER et al. 1988; KINGSBURY et al. 1984; RACE et al. 1987). Clones of the scrapie-infected cells were found to produce PrP^{Sc}, whereas clones showing no infectivity lacked PrP^{Sc} (BUTLER et al. 1988). In scrapie-infected cultured cells, PrP^{C} is synthesized rapidly ($t_{1/2} < < 2$ h) (CAUGHEY et al. 1989) and degraded with a $t_{1/2}$ of ~ 5 h, while PrP^{Sc} is synthesized slowly ($t_{1/2} \sim 15$ h) and does not appear to be degraded (BORCHELT et al. 1990). Most PrP^{C} is transported to the external cell surface, where it is bound by a glycosyl phosphatidylinositol (GPI) anchor but can be released by digestion with phosphatidylinositol phospholipase C (PIPLC) (STAHL et al. 1987). In contrast, PrP^{Sc} accumulates largely inside of cells, probably within the Golgi apparatus, and is not released from membranes by PIPLC (TARABOULOS et al. 1990). The contrasting properties of PrP^{C} and PrP^{Sc} are listed in Table 2.

Many investigators have confirmed the presence of protease-resistant PrP in brains infected with the scrapie or CJD agent (BROWN et al. 1986; GIBBS et al. 1985; DIRINGER et al. 1983; HOPE et al. 1986; MANUELIDIS et al. 1987) Although the amino acid sequence of PrP has been confirmed and there is agreement that PrP is glycosylated, some investigators have suggested that PrP 27–30 may not be a component of the scrapie agent (BRAIG and DIRINGER 1985; CZUB et al. 1988). One argument revolves around the inability of some investigators to detect PrP

Table 2. Properties of cellular and scrapie PrP isoforms

	PrP^{C}	PrP^{Sc}
Normal cells	~1–5 μg/g	—
Scrapie-infected cells	~1–5 μg/g	~10–20 μg/g
Purified prions	—	+[a]
Protease resistance	—	+[b]
Amyloid rods	—	+[c]
Subcellular localization	Cell surface	Primarily intracellular
PIPLC releases from membranes	+	—
Synthesis ($t_{1/2}$)	<<2 h	~15 h
Turnover ($t_{1/2}$)	~5 h	>>24 h

[a] Copurification of PrP^{Sc} and prion infectivity demonstrated by two protocols: *(1)* detergent extraction, sedimentation protease digestion, and *(2)* PrP 27–30 monoclonal antibody affinity chromatography

[b] Limited proteinase K digestion of hamster PrP^{Sc} produces PrP 27–30

[c] After limited proteolysis of PrP^{Sc} to produce PrP27–30 and detergent extraction, amyloid rods form; except for length, the rods are indistinguishable from amyloid filaments forming plaques

mRNA or PrP^{Sc} in spleens of scrapie-infected rodents (Chesebro et al 1985; Czub et al. 1986); however, investigators in several laboratories have clearly shown that both PrP mRNA and PrP^{Sc} are present in spleen tissue (Oesch et al. 1985; McKinley et al. 1987; Rubenstein et al. 1986; Shinagawa et al. 1986). Indeed, recent studies have shown an excellent correlation between the concentration of PrP mRNA in scrapie-infected rodent tissues and prion titer (McKinley et al. 1987). Another argument centers on experiments that demonstrate the loss of CJD infectivity in fractions following lectin chromatography. Neither denaturation of the PrP^{CJD} nor neutralization of CJD infectivity by immobilization on the lectin column was considered as an explanation (Manuelidis et al. 1987). Denaturation of PrP^{Sc} has been demonstrated to be accompanied by a loss of scrapie infectivity (Bolton et al. 1984; Prusiner et al. 1983). Still another argument involves the inability of some investigators to detect PrP 27–30 in hamster brains just prior to clinical signs of scrapie even though high prion titers are found at this time (Czub et al. 1988). Unfortunately, these investigators determined prion titers from homogenates but measured PrP 27–30 in partially purified fractions. Other investigators have measured both prion titers and PrP 27–30 in crude extracts and have found an excellent correlation throughout the course of infection (R. E. Race, personal communication). Indeed, no experimental studies have been reported where fractions with high levels of scrapie infectivity were found to contain less than one PrP^{Sc} (or PrP 27–30) molecule per infectious unit.

4 Structure and Organization of PrP Genes

Oligonucleotides corresponding to a portion of the *N*-terminal amino acid sequence of PrP 27–30 (PRUSINER et al. 1984) were synthesized and used to identify a PrP cDNA (OESCH et al. 1985; CHESEBRO et al. 1985). Southern blotting with PrP cDNA revealed a single-copy gene with the same restriction patterns as those in normal and scrapie-infected DNA from hamster brains. Unexpectedly, PrP mRNA was found at similar levels in both normal and scrapie-infected hamster brains (OESCH et al. 1985; KRETZSCHMAR et al. 1986a).

Molecular cloning of the hamster PrP cDNA led to elucidation of the organization and structure of the hamster PrP gene; the entire open reading frame (ORF) or protein coding region is contained within a single exon (BASLER et al. 1986). The 5′ end of the PrP gene contains multiple G:C-rich initiation sites. An intron of ~10 kb separates exons I and II. Six lines of evidence argue that the two PrP isoforms have the same amino acid sequence: (1) No evidence for rearrangement of the PrP gene in scrapie has been found (OESCH et al. 1985; BASLER et al. 1986). (2) The organization of the PrP gene provides no possibility for alternative splicing within the ORF (BASLER et al. 1986). (3) Only one PrP mRNA of 2.1 kb has been detected, and its concentration does not change throughout the course of scrapie infection in hamster brain (OESCH et al. 1985). (4) Some 70% of PrP^{Sc} and 87% of PrP 27–30 have been sequenced by gas-phase protein sequencing (D. Teplow, D. Groth, L. Hood, S.B. Prusiner, unpublished data). These sequences correspond precisely with the translated genomic DNA sequence. (5) Both scrapie hamster brain and mouse PrP cDNAs have translated ORF sequences which are identical with the corresponding translated ORF sequences (WESTAWAY et al. 1987; OESCH et al. 1985; BASLER et al. 1986; LOCHT et al. 1986). (6) Both PrP isoforms have the same *N*-terminal amino acid sequence as determined by *N*-terminal gas-phase sequencing (TURK et al. 1988). Presumably, the difference in the properties of the two prion proteins is due to a post-translational event (BORCHELT et al. 1990, BASLER et al. 1986); though seemingly remote, the possibility that differences in the amino acid sequences of the two PrP isoforms arising through RNA editing must continue to be considered (CHEN et al. 1987; FEAGIN et al. 1988; POWELL et al. 1987; SHAW et al. 1988).

That PrP genes existed prior to the speciation of mammals is supported by the finding that human and mouse PrP genes are located on homologous chromosomes 20 and 2 (SPARKES et al. 1986). The ORF of PrP genes from humans (KRETZSCHMAR et al. 1986b), Syrian hamsters (BASLER et al. 1986), Chinese hamsters (LOWENSTEIN et al. 1990), Armenian hamsters (LOWENSTEIN et al. 1990), mice (WESTAWAY et al. 1987; LOCHT et al. 1986), rats (LIAO et al. 1987), and sheep (GOLDMAN et al. 1988) have been sequenced and all encode prion proteins of approximately 250 amino acids with *N*-terminal signal peptides. All of these PrP ORFs encode *C*-terminal hydrophobic peptides which are presumably removed upon GPI anchor addition and all contain two conscensus sites for

Asn-linked glycosylation as well as two cysteines within the *C*-terminal half of the molecule. All of the ORFs also possess a series of Gly rich repeats in the *N*-terminal portion of the PrP molecule. Whether eukaryotes other than mammals have authentic PrP genes remains to be established (WESTAWAY and PRUSINER 1986; BOLTON et al. 1985).

5 Post-Translational Modifications of Prion Proteins

There is evidence for at least eight post-translational modifications of PrP (Fig 1). The first modification to be identified was glycosylation; PrP 27–30 is a sialoglycoprotein (PRUSINER et al. 1984). All translated PrP gene ORFs sequenced to date contain two consensus sites for Asn-linked glycosylation. In Syrian hamster PrP, these consensus sites are codons 181 and 197 (OESCH et al. 1985; BASLER et al. 1986). Both the cellular and scrapie PrP isoforms are resistant to digestion by endoglycosidase H but possess *N*-linked oligosaccharides that can be removed by digestion with peptide: *N*-glycosidase F (HARAGUCHI et al. 1989). Hydrazinolysis liberated ~2 mol oligosaccharides/PrP 27–30 (ENDO et al. 1989). The released oligosaccharides were found to be a mixture of bi-, tri-, and tetra-antennary complex-type sugar chains with Manαl→6 (GlcNAcβl→4) (Manαl→3) Manβl→4GlcNAcβl→4(Fucαl→6) GlcNAc as their cores. Variation is produced by combinations of the following oligosaccharides: Galβl→4GlcNAcβl→, Galβl→4 (Fucαl→3) GlcNAcβl→, Neu5Acα2→3Galβl→4GlcNAcβ, and Neu5Acα2→6Galβl→4GlcNAcβl→in their outer chains. Whether the Asn-linked oligosaccharides attached to PrP^{Sc} differ significantly in structure from those attached to PrP^{Sc} remains to be determined.

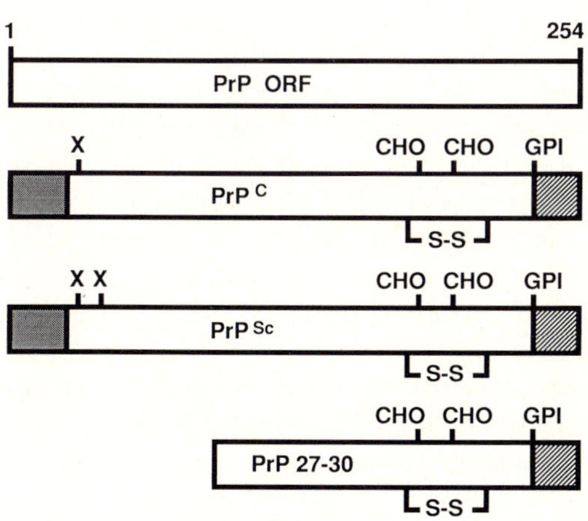

Fig. 1. Prion protein structure. Hamster PrP gene encodes a protein of 254 amino acids. The *N*-terminal signal peptide of 22 amino acids (stippled pattern) is cleaved during maturation of PrP^{C} and PrP^{Sc}. Modified Arg residues (×) at codons 25 and 37 have been identified, as well as Asn-linked oligosaccharides (CHO) attached at codons 181 and 197 within a loop formed by a disulfide bond which joins Cys at codons 179 and 214. Upon removal of a *C*-terminal hydrophobic peptide of ~20 amino acids (diagonal line pattern), a glycosyl phosphatidylinositol (GPI) anchor is added.
(Reproduced, with permission, from Annual Review of Microbiology 43. Copyright 1989 by Annual Reviews Inc.)

Both mature PrP isoforms have two cysteine residues at codons 179 and 214 that are linked by an intramolecular disulfide bond (TURK et al. 1988). This disulfide bond forms a loop which contains both Asn-linked oligosaccharides as well as a variant amino acid at codon 189 found in long incubation period mice (*Prn-p^b*).

A 22-amino-acid signal peptide is found at the *N*-terminus of PrP. Like all other signal peptides, it contains a hydrophobić core and a consensus cleavage site (BASLER et al. 1986; ROBAKIS et al. 1986; BAZAN et al. 1987). Cell-free translation studies have demonstrated the cleavage of the PrP signal peptide (HAY et al. 1987) and *N*-terminal amino acid sequencing of PrP^C and PrP^{Sc} confirm the cleavage (HOPE et al. 1986; TURK et al. 1988; BOLTON et al. 1987).

The *C*-terminal hydrophobic segment of both PrP isoforms appears to be removed during maturation (OESCH et al. 1985) and a GPI anchor added (STAHL et al. 1987) Both PrP^C and PrP^{Sc} have GPI anchors. Endopeptidase digestions combined with amino acid analysis and mass spectrometric studies have shown that the GPI anchor of PrP^{Sc} is attached to Ser_{231} (STAHL et al. 1990) as previously predicted (FERGUSON and WILLIAMS 1988).

Arg_{25} and Arg_{37} in PrP^{Sc} contain an unknown modification which prevents their detection by gas-phase sequencing using Edman degradation (HOPE et al. 1988a; TURK et al. 1988). Arg_{25} and PrP^C contains a similar modification (TURK et al. 1988). Whether Arg_{37} also contains this modification is uncertain; however, Arg_{48} of PrP^{Sc} does not contain the modification.

The results of these genetic linkage and ORF sequencing studies in humans and mice suggest that amino acid substitutions in PrP may modulate the development of prion diseases. Recent studies have demonstrated that GSS is an autosomal dominant disorder and that a Pro→Leu substitution at codon 102 of the PrP gene is genetically linked to the development of GSS (HSIAO et al. 1989). Earlier investigations showed genetic linkage between an incubation time gene (*Prn-i*) and the PrP gene in inbred mice (CARLSON et al. 1986). Mice with long incubation times have PrP genes (*Prn-p^b*) with Leu→Phe and Thr→Val substitutions at codons 108 and 189, respectively (WESTAWAY et al. 1987). Several discordant mice in genetic crosses raised the possibility that the PrP gene and *Prn-i* are separate but tightly linked (CARLSON et al. 1988), but recombinant capture studies (G. CARLSON, M. TORCHIA, S.B. PRUSINER, unpublished results) and Tg mouse studies described below argue for the congruency of the PrP and *Prn-i* genes. Recent studies suggest that these differences in murine PrP sequences produce distinct isolates of prions which exhibit significantly different scrapie incubation times in a single host (CARLSON et al. 1989).

Studies with Tg mice that express foreign PrP genes have brought a wealth of new knowledge about prions (SCOTT et al. 1989). Inoculation of Tg (HaPrP) 81 mice expressing the Syrian hamster (Ha) prion protein with Ha prions caused scrapie in ~75 days; hamsters inoculated with Ha prions had the same incubation period, while non-Tg control mice failed to develop scrapie after > 500 days. Inoculation of a second line, Tg (HaPrP) 71 mice, with Ha prions caused scrapie in ~170 days, similar to the time required for developing scrapie after inoculation with mouse prions. Tg 71 and Tg 81 mice inoculated with Ha prions exhibited spongiform degeneration and reactive astrocytic gliosis, while Tg 81 mice also showed HaPrP amyloid plaques which are characteristic of hamster scrapie. Both Tg 71 and Tg 81 mice produced the hamster PrP^{Sc} in their brains after inoculation with Ha prions. The brains of Tg 81 mice with scrapie were found to contain $\sim 10^9$ ID_{50} units/g of Ha prions based on Ha bioassays. Tg mice carrying a HaPrP minigene failed to express $HaPrP^C$ and had scrapie incubation times after inoculation with Ha prions similar to those observed for non-Tg controls. These findings argue that the PrP gene modulates scrapie susceptibility, the course of the disease, and the neuropathology. Our studies yield the first demonstration of the synthesis of infectious scrapie prions programmed by a recombinant DNA molecule. Indeed, the use of Tg mice expressing foreign PrP genes should bring about a renaissance in prion research.

It now seems feasible to determine first the domains of PrP and later the precise amino acids which are required for production of infectious scrapie prions, modulate the susceptibility of animals to prion diseases, and influence scrapie incubation times. Deciphering how specific amino acid changes in PrP affect both the initiation of the disease and the course of CNS degeneration may have wider implication for our understanding of a variety of neurodegenerative diseases, as noted above.

6 Infectious, Sporadic, and Genetic Prion Diseases

The human prion diseases illustrate that three different manifestations of CNS degeneration may arise: slow infection, sporadic disease, and genetic disorder (RIDLEY et al. 1986; ALPERS 1987). That the three human prion diseases can be transmitted to laboratory animals by inoculation is well documented (GIBBS et al. 1968; GAJDUSEK et al. 1966; MASTERS et al. 1981a). Kuru is thought to have been spread exclusively through a slow infectious mechanism by means of ritualistic cannibalism (ALPERS 1987).

Although a few cases of CJD have been traced to inoculation with prions [e.g., injections of human growth hormone (BROWN 1988; GIBBS et al. 1985; GOUJARD et al. 1988), transplantation of corneas, and implantation of cerebral electrodes], most appear to be sporadic despite considerable effort to implicate scrapie-infected sheep as an exogenous source (GAJDUSEK 1977; KOVANEN and HALTIA 1988). It is possible, although unlikely, that sporadic CJD results from prions that are ubiquitous in humans but have a very low efficiency of infection. In hamsters, scrapie infection by the oral route has been found to be 9 times less efficient intracerebral inoculation (PRUSINER et al. 1985).

Pedigree studies suggest that GSS and familial CJD may be inherited as autosomal dominant disorders (PRUSINER et al. 1985; BAKER et al. 1985; MASTERS et al. 1981b). The significance of this observation was uncertain until recently, when

GSS was demonstrated to be a genetic disease and its development was linked to a missense variant of the prion protein (HSIAO et al. 1989). Molecular cloning studies demonstrated a cytosine to thymine substitution in the second position of codon 102, which probably results from deamination of a methylated cytosine situated 5′ to guanine (BARKER et al. 1984; BIRD 1986). This mutation creates a DdeI restriction site, which was used to demonstrate genetic linkage between the PrP codon 102 amino acid substitution (Leu → Pro) and the development of GSS. Whether this substitution is the cause of GSS remains to be established. About 10%–15% of cases of CJD are familial, while most cases of GSS are inherited (MASTER et al. 1981a; RIDLEY et al. 1986; BAKER et al. 1985). GSS and familial CJD are the only known human diseases which are both genetic and infectious.

The genetic linkage of a missense variant of PrP with GSS constrains further the possible structural models for the prion. If prions contain a small, as yet undetected nucleic acid, then this molecule must be widespread throughout the world to explain the incidence of sporadic CJD, yet it must segregate in rare GSS families with the PrP mutation. It is noteworthy that Huntington's disease, another autosomal dominant genetic disorder, does not manifest as clinical illness until the fifth decade in most cases. The delayed onset of both Huntington's disease and GSS is not understood.

The genetic linkage results also suggest an interesting mechanism for sporadic CJD if prions contain only PrP^{Sc} (or PrP^{CJD}). A somatic mutation (HANSEN and CAVENEE 1987) in the PrP gene or even an RNA editing error (CHEN et al. 1987; FEAGIN et al. 1988; POWELL et al. 1987; HHAW et al. 1988) in a single cell might lead to the generation of PrP^{CJD} in that cell; prions would then spread to neighboring cells upon exit of the PrP^{CJD} molecules. In contrast, the PrP mutation of GSS is a germline mutation, consistent with its genetic transmission.

While scrapie is readily transmitted by inoculation and has been transmitted orally in an experimental setting (PRUSINER et al. 1985; PATTISON et al. 1972; PATTISON and MILLSON 1961), it is unclear how the disease spreads naturally among sheep and goats. Some early investigators favored infection, while others were proponents of a genetic mechanism (PARRY 1962, 1983). In retrospect, both points of view were probably correct. Transmissible mink encephalopathy (TME) probably occurs in domesticated mink after they have consumed scrapie-infected sheep meat (HADLOW and KARSTAD 1968, HARTSOUGH and BURGER 1965). Scrapie prions administered intranasally to mink produce a disease indistinguishable from TME (HADLOW and RACE 1986; HADLOW et al. 1987). Interestingly, neither scrapie in mink nor TME can be transmitted to mice, which suggests a significant species barrier for mink prions (MARSH and KIMBERLIN 1975; W. Hadlow, undpublished data). Chronic wasting disease (CWD) in captive herds of mule deer and elk may occur after oral consumption of prion-contaminated sheep by-products (WILLIAMS and YOUNG 1980). Whether CWD is a naturally occurring disease is uncertain. Over the past 6 years, over 28000 cases of a spongiform degeneration have been reported in dairy cattle in Great Britain (WELLS et al. 1987; WILESMITH 1988). Recent studies have shown that bovine spongiform encephalopathy (BSE) is transmissible to mice after a prolonged incubation period (FRASER et al. 1988) and that a protease-resistant prion protein accumulates in the brains of infected cattle (HOPE et al. 1988b). Epidemiological studies suggest that BSE probably began around the year 1981, when bone meal fed to dairy cattle as a protein supplement was produced by a new process in which the rendering temperature was reduced to increase the yield of sheep by-products (WILESMITH et al. 1988). Whether BSE occurs naturally or is due exclusively to the oral consumption of sheep scrapie prions is unknown. Although the *N*-terminal sequence of PrP recovered from cattle brains is similar to that of hamster PrP (HOPE et al. 1988b), it will be important to learn about the PrP genes from a variety of cattle. Equally interesting will be the PrP sequences from many breeds of sheep as well as mink, mule deer, and elk.

7 Search For a Scrapie-Specific Nucleic Acid

The transfer of scrapie prion infectivity from rods to liposomes provided a new method by which we could search for a hidden or cryptic nucleic acid within the prion. Treatment of DLPCs as well as liposomes with nucleases or Zn^{2+} failed to alter scrapie infectivity (GABIZON et al. 1987). In control experiments, nucleic acids were added to DLPCs at a concentration of one molcule per infectious unit. Under the conditions of our experiments, nucleases or Zn^{2+} degraded the exogenously added DNA or RNA molecules to $< 1\%$ of their initial concentration. In the experiments described above, nucleases were added after the DLPCs and liposomes were formed. In subsequent studies, the nucleases were present during the formation of the DLPCs (GABIZON et al. 1988a). As before, no change in prion titer was observed after prolonged incubations.

Irradiation of the DLPCs at 254 nm produced an inactivation curve virtually identical to that observed for the rods in the experiment reported here and earlier studies (Fig. 2a) (GABIZON et al. 1988a). The kinetics of inactivation of scrapie infectivity by ultraviolet irradiation yielded exponential survival curves characteristic of a single-hit process. From these ultraviolet inactivation curves, we estimated that the D_{37} values for the DLPCs and rods were 17 and 24 J/m^2, respectively. In another study, we found a D_{37} value of 17-22 kJ/m^2 for purified prion rods (BELLINGER–KAWAHARA et al. 1987). The resistance of scrapie infectivity to inactivation by irradiation at 254 nm suggests that if prions contain a nucleic acid, it will be $\sim$5 bases in length for a single-stranded molecule or 30-45 bp for a double-stranded molecule (BELLINGER–KAWAHARA et al. 1987). The D_{37} values for purified prion rods and DLPCs are in good agreement with values reported two decades earlier for murine scrapie agent in brain homogenates (ALPER et al. 1967).

Prolonged exposure of the rods and DLPCs to proteolytic digestion resulted in a significant decrease in scrapie infectivity (GABIZON et al. 1988a). Inactivation of prion rods and DLPCs by proteinase K was a function of both the time of

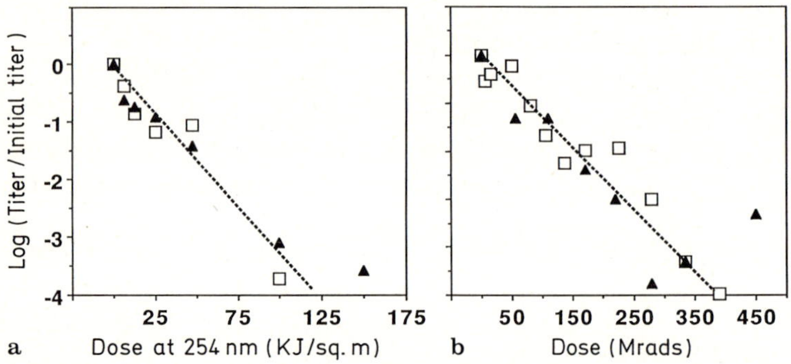

Fig. 2 a, b. Inactivation of scrapie prion rods and liposomes by irradiation. Amyloid rods (—□—) continuing ~30 μg/ml PrP 27–30 or dispersed PrP 27–30 (—▲—) in 20 mg/ml PC as DLPCs were prepared as previously described (PRUSINER et al. 1983; GABIZON et al. 1987; BARRY and PRUSINER 1986; BELLINGER-KAWAHARA et al. 1987). After irradiation studies, samples were inoculated intracerebrally into Syrian hamsters for scrapie bioassay (PRUSINER et al. 1982b). Log Values (titer of irradiated samples/titer of unirradiated samples) were plotted as a function of the irradiation dose. Curves were visually fitted. **a** Ultraviolet irradiation of prions. Fractions were irradiated with ultraviolet light with a 5-GE germicidal lamp for increasing periods of time. **b** Ionizing radiation of prions. All samples were frozen in ethanol dry ice baths prior to storage at −70 °C. The samples were irradiated with 13-MV electrons at −133 °C (BELLINGER-KAWAHARA et al. 1987; HARMON et al. 1985). Controls receiving no irradiation were subjected to the same protocol. (Reproduced, with permission, from *Biochemical Journal*, 266: 1–14, 1990)

digestion and the concentration of protease, as reported earlier for both partially purified (PRUSINER et al. 1981b) and more extensively purified samples (MCKINLEY et al. 1983).

One explanation for the protease resistance of PrP 27-30 and prion infectivity has been the aggregation of PrP 27-30 into amyloid rods (PRUSINER et al. 1983). Dissociation of PrP 27-30 in DLPCs allowed us to test this hypothesis. The kinetics of scrapie infectivity loss as a function of proteinase K digestion were similar for both the DLPCs and rods (GABIZON et al. 1988a). After 24 h digestion at 37°C with 100 μg/ml of proteinase K, the prion titer for both DLPCs and rods decreased by $>10^2$ ID_{50} units/ml; with 500 μg/ml of proteinase K, the titer decreased by $>10^4$ ID_{50} units/ml.

These experiments, as well as studies on immunoaffinity-purified PrP^{Sc}, argue that resistance of PrP 27-30 to enzyme-catalyzed proteolysis is an intrinsic property of the scrapie PrP isoform and not merely a consequence of its polymerization into amyloid rods.

Early radiobiological investigations on the scrapie agent suggested that it might have a molecular weight of <150 000 Da and might not possess a nucleic acid (ALPER et al. 1966, 1967). While the results of these studies were often quoted, their implications have been frequently ignored and more recently their interpretation seriously questioned (ROHWER 1984a, 1986; ROHWER and GAJDUSEK 1980; ALPER 1985). Prion rods and liposomes were irradiated with increasing doses up to 400 Mrad. The inactivation curves (Fig. 2b) were virtually identical to those seen with microsomal fractions isolated from scrapie-infected hamster brain before and after extraction with detergent (BELLINGER-KAWAHARA et al. 1988). The survival curves for all forms were exponential over a range of more than 10^5 infectious units, which is characteristic of a single-hit process.

Results from 14 separate experiments with various types of preparations, each irradiated as a frozen solution, yielded a mean target size for the scrapie agent of 55 000 $\pm$ 9000 Da (BELLINGER-KAWAHARA et al. 1988). The early radiation studies (ALPER et al. 1966; FIELD et al. 1969; LATARJET 1979; GIBBS et al. 1978), corrected for temperature effects (KEMPNER et al. 1986a, b), yielded target sizes between 60 000 and 150 000 Da. An equally important finding of our studies is that the target size for scrapie infectivity appears to be independent of the prion form (BELLINGER–KAWAHARA et al. 1988). Samples containing aggregates in the form of amyloid rods yielded an inactivation curve indistinguishable from those curves observed with microsomes, crude brain homogenates, or liposomes. Clearly the various forms of the scrapie prion do not influence its ionizing radiation target size.

Although there is no chemical or physical evidence for a scrapie-specific nucleic acid within the prion, reports of scrapie "strains" or isolates are interpreted by some investigators as evidence for a scrapie agent genome (BRUCE and DICKINSON 1987; KIMBERLIN et al. 1987, 1989). Studies with *Prn-p*^{a} and *Prn-p*^{b} mice which encode PrP^{Sc} allotypes PrP^{Sc}-A and PrP^{Sc}-B differing by two amino acids (WESTAWAY et al. 1987) are forcing a reevaluation of the significance and molecular basis of scrapie "strains" (CARLSON et al. 1989). Passaging prion in *Prn-p*^{a} (NZW and Swiss), *Prn-p*^{b} (I/Ln) and (NZW x I/Ln)F_1 mice, as well as mice derived in a (NZW x I/Ln)F_2 cross, established that these mice produce prion isolates with different properties which correlate with their *Prn-p* genotype. This is an important point, since C57BL and VM mice were used to derive many distinguishable scrapie isolates (BRUCE and DICKINSON 1987). C57BL mice are *Prn-p*^{a} and VM mice are *Prn-p*^{b}. Compared to prions from NZW or Swiss mice, a single passage of prions through I/Ln mice dramatically shortened incubation times in recipient I/Ln mice with no further change in a second syngeneic passage. Conversely, inoculum obtained from a single passage in I/Ln mice delayed the onset of clinical scrapie in recipient NZW mice compared to inocula previously passaged through NZW and Swiss mice.

The results of experiments with prion isolates derived from *Prn-P*^{a} and *Prn-p*^{b} mice appear to reflect an inbred mouse strain barrier defined by the *Prn* genotype (CARLSON et al. 1989). The *Prn-p*^{a}/*Prn-p*^{b} allotype barrier may be analogous to the "species barrier" described by PATTISON (1965), where a marked prolongation of the scrapie incubation time is observed upon interspecies passage but the minimum incubation time is then achieved upon a single intraspecies passage. Because all species susceptible to scrapie examined, to date, encode distinct prion proteins (WESTAWAY et al. 1987; BASLER et al. 1986; LOCHT et al. 1986; KRETZSCHMAR et al. 1986b; LOWENSTEIN et al. 1990; LIAO et al. 1987; GOLDMAN et al. 1988), we hypothesized that the PrP genotype may be responsible for the "species barrier" (WESTWAY et al. 1989). In support of this hypothesis, studies with Tg mice harboring HaPrP genes showed scrapie incubation times of ~75 days in these Tg mice inoculated with Ha prions similar to those seen with hamsters (SCOTT et al. 1989). In contrast, non-Tg mice inoculated with Ha prions have incubation times >500 days.

While prion isolates with true breeding yet different incubation times in the same inbred host are more compelling as bona fide "strains", there are no biochemical or fluctuation test (LURIA and DELBRUCK 1943) data to equate genesis of these scrapie prion isolates with nucleic acid mutation. Clearly, the molecular basis for genesis of scrapie "strains" is an unresolved issue.

8 Multiple Forms of Prions

The solubilization of prions using phospholipids has helped to clarify the issues concerning the heterogeneity of the infectious particle with respect to size, density, and charge. The apparent heterogeneity of prions is well documented and has plagued scrapie research over the last two decades (PRUSINER et al. 1979). The ability of prions to form rods, DLPCs, and liposomes, each having high prion titers, provided definitive evidence that prions may exist in multiple forms (GABIZON et al. 1987). The results of such studies argue persuasively that hydrophobic interactions are responsible for the extreme apparent heterogeneity of prions as suggested ove a decade ago (PRUSINER et al. 1978, 1979).

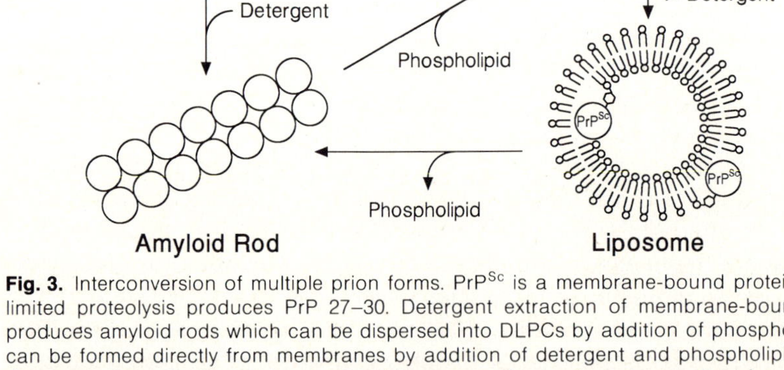

Fig. 3. Interconversion of multiple prion forms. PrP^{Sc} is a membrane-bound protein which upon limited proteolysis produces PrP 27–30. Detergent extraction of membrane-bound PrP 27–30 produces amyloid rods which can be dispersed into DLPCs by addition of phospholipids. DLPCs can be formed directly from membranes by addition of detergent and phospholipid. Removal of detergents from DLPCs produces closed liposomes. Removal of phospholipid from liposomes by organic solvent extraction ($CHCl_3$: methanol) regenerates the rods. All of the prion forms shown here possess high levels of scrapie infectivity (GABIZON et al. 1988a) (Reproduced, with permission, from *Annual Review of Microbiology 43*. Copyright 1989 by Annual Reviews Inc.)

Interconversion of the multiple forms of prions depicted in Fig. 3 depends upon the relative levels of detergent and lipid (GABIZON et al. 1988a). As described above, both detergent extraction and limited proteolysis of PrP^{Sc} are required for amyloid rod formation. Dispersion of the rods into DLPCs was accomplished by the addition of detergent and phospholipid while subsequent removal of detergent produced closed liposomes (GABIZON et al. 1987). Removal of the phospholipid resulted in the reformation of rods with full retention of scrapie infectivity (GABIZON et al. 1988a). As illustrated, dispersion of microsomal membranes into DLPCs bypassed rod formation; this approach was used for the immunoaffinity purification studies.

The foregoing studies demonstrate that prions may exist in multiple forms, and, to date, the only molecule identified in all of these forms is PrP^{Sc}. The rapid interconversion of various prion forms argues, but does *not* prove, that prions are likely to be composed only of a single type of macromolecule. It is difficult to envision how a highly charged, water-soluble macromolecule such as a nucleic acid could partition with PrP^{Sc} throughout these interconversion steps and at the same time remain so highly protected that it has eluded all attempts at detection.

9 Prions and Viruses

Although both prions and viruses multiply, their properties, structures, and modes of replication seem to be fundamentally different. Viruses contain nucleic acid genomes that encode progeny viruses; proteins necessary for producing infectious viruses are encoded by the viral genome. In contrast, prions contain little or no nucleic acid, and PrP is encoded by a cellular gene (OESCH et al. 1985). Unlike viruses, newly synthesized prions possess PrP^{Sc} molecules reflecting the resident (or host) PrP gene and not necessarily the sequence of the PrP^{Sc} molecules found within the infecting prion (BOCKMAN et al. 1987). Although viruses usually evoke an immune response during some phase of their infection, prions do not appear to do so (PRUSINER 1982; OESCH et al. 1985). Variations in PrP amino sequence are linked genetically to altered prion disease phenotypes in both humans and mice.

The convergence of experimental results indicating a pivotal role for PrP^{Sc} (or PrP^{CJD}) in animal and human prion diseases is both impressive and persuasive. The wide variety of independent disciplines, including protein chemistry, molecular genetics, immunochemistry, neuropathology, and experimental neurology, employed to obtain these experimental data greatly strengthens the assertion that PrP^{Sc} has a central role in prion diseases.

That PrP is encoded by a cellular gene and not by a putative nucleic acid carried within the prion is a major feature which distinguishes prions from viruses. This discovery, coupled with recent genetic studies showing linkage between a PrP missense variant and GSS, demands that scrapie and CJD no longer be considered virological disorders. Although prion diseases resemble viral illnesses in some respects, the structure, cell biology, and genetics of prions clearly separate them from viruses.

Whether prions are composed only of an abnormal isoform of the prion protein or they contain some additional molecule is uncertain. Many lines of evidence argue that PrP^{Sc} is the sole component of prions: (1) multiple forms of prions are infectious—membranes, rods, spheres, DLPCs, and liposomes; (2) many attempts to demonstrate the dependence of scrapie infectivity on a nucleic acid have been unsuccessful; (3) the ionizing radiation target size of the prion is 55 000 Da; (4) PrP^{Sc} is encoded by a cellular gene; (5) mice with short and long incubation times have different PrP genes which encode distinct prion proteins and produce prions with distinct properties; and (6) GSS is linked to a mutation in the PrP gene. The partitioning of infectivity in a wide variety of different forms argues for a single component but does not eliminate the possibility of a second macromolecule. Numerous attempts to inactivate scrapie prion infectivity by procedures that hydrolyze or modify nucleic acids have been consistently unsuccessful. As reviewed above, ultraviolet irradiation of membranes, rods, and DLPCs suggests that if prions have an essential nucleic acid, then it will be < 5 bases if single stranded or 30–45 bp if double stranded (GABIZON et al. 1988a; BELLINGER-KAWAHARA et al. 1987). Ionizing radiation studies give a target size too small to protect a large nucleic acid but do not rule out some other macromolecule (BELLINGER-KAWAHARA et al. 1988).

In summary, a wealth of recently acquired experimental data have established that prions are novel and unprecedented pathogens. The uniqueness of prions promises to open many new and unexpected avenues of research. Indeed, the search for other prion diseases should begin in earnest once the events in PrP^{Sc} formation are elucidated. It seems likely that some of the pathogenic mechanisms operative in prion diseases will give new insights into central nervous system disorders of older people such as Alzheimer's disease, Parkinson's disease, and amyotrophic lateral sclerosis. Already familial Alzheimer's disease pedigrees are being reclassified as prion diseases based upon the protein deposits within their amyloid plaques (NOCHLIN et al. 1989; GHETTI et al. 1988, 1989)

References

Adams DH (1970) The nature of the scrapie agent: a recent progress. Pathol Biol 18: 559–577

Adams DH, Field EJ (1968) The infective process in scrapie. Lancet ii: 714–716

Alper T (1985) Scrapie agent unlike viruses in size and susceptibility to inactivation by ionizing or ultraviolet radiation. Nature 317: 750

Alper T, Haig DA, Clarke MC (1966) The exceptionally small size of the scrapie agent. Biochem Biophys Res Commun 22: 278–284

Alper T, Cramp WA, Haig DA, Clarke MC (1967) Does the agent of scrapie replicate without nucleic acid? Nature 214: 764–766

Alpers M (1987) Epidemiology and clinical aspects of kuru. In: Prusiner SB, McKinley MP (eds) Prions-novel infectious pathogens causing scrapie and Creutzfeldt–Jakob disease. Academic, Orlando, PP 451-465

Bahmanyar S, Williams ES, Johnson FB, Young S, Gajdusek DC (1985) Amyloid plaques in spongiform encephalopathy of mule deer. J Comp Pathol 95: 1–5

Baker HF, Ridley RM, Crowe TJ (1985) Experimental transmission of an autosomal dominant spongiform encephalopathy: does the infectious agent originate in the human genome? Br Med J 291: 299–302

Barker D, Schafer M, White R (1984) Restriction sites containing CpG show a higher frequency of polymorphism in human DNA. 36: 131–138

Barry RA, Prusiner SB (1986) Monoclonal antibodies to the cellular and scrapie prion proteins. J Infect Dis 154: 518–521

Barry RA, Prusiner SB (1987) Immunology of prions. In: Prusiner SB, McKinley MP (eds) Prions novel infectious pathogens causing scrapie and Creutzfeldt–Jakob disease. Academic, Orlando, pp 239–275

Barry RA, McKinley MP, Bendheim PE, Lewis GK, DeArmond SJ, Prusiner SB (1985) Antibodies to the scrapie protein decorate prion rods. J Immunol 1 135: 603–613

Basler K, Oesch B, Scott M, Westaway D, Wälchli M, Groth DF, McKinley MP, Prusiner SB, Weissmann C (1986) Scrapie and cellular PrP isoforms are encoded by the same chromosomal gene. Cell 46:417–428

Bazan JF, Fletterick RJ, McKinley MP, Prusiner SB (1987) Predicted secondary structure and membrane topology of the scrapie prion. Protein Engineering 1: 125–135

Beck E, Daniel PM, Parry HB (1964) Degeneration of the cerebellar and hypothalamo-neurohypophysial systems in sheep with scrapie and its relationship to human system degenerations. Brain 87: 153–176

Bellinger-Kawahara C, Cleaver JE, Diener TO, Prusiner SB (1987) Purified scrapie prions resist inactivation by UV irradiation. J Virol 61: 159–166

Bellinger-Kawahara CG, Kempner E, Groth D, Gabizon R, Prusiner SB (1988) Scrapie prion liposomes and rods ehibit target sizes of 55,000 Da. Virology 164: 537–541

Bendheim PE, Barry RA, DeArmond SJ, Stites DP, Prusiner SB (1984) Antibodies to a scrapie prion protein. Nature 310: 418–421

Bird AP (1986) CpG-rich islands and the function of DNA methylation. Nature 321: 209–213

Bockman JM, Kingsbury DT, McKinley MP, Bendheim PE, Prusiner SB (1985) Creutzfeldt–Jakob disease proteins in human brains. N Engl J Med 312: 73–78

Bockman JM, Prusiner SB, Tateishi J, Kingsbury DT (1987) Immunoblotting of Creutzfeldt–Jakob disease prion proteins: host species-specific epitopes. Ann Neurol 21: 589–595

Bolton DC, Bendheim PE, Marmorstein AD, Potempska A (1987) Isolation and structural studies of the intact scrapie agent protein. Arch Biochem Biophys 258: 579–590

Bolton DC, McKinley MP, Prusiner SB (1982) Identification of a protein that purifies with the scrapie prion. Science 218: 1309–1311

Boltion DC, Mckinley MP, Prusiner SB (1984) Molecular characteristics of the major scrapie prion protein. Biochemistry 23: 5898–5906

Bolton DC, Meyer RK, Prusiner SB (1985) Scrapie PrP 27–30 is a sialoglycoprotein J Virol 53: 596–606

Borchelt DR, Scott M, Taraboulos A, Stahl N, Prusiner SB (1990) Scrapie and cellular prion proteins differ in their kinetics of synthesis and topology in culture cells. J Cell Biochem 14C: 72

Braig H, Diringer H (1985) Scrapie: concept of a virus-induced amyloidosis of the brain. EMBO J 4: 2309–2312

Brown P (1988) The decline and fall of Creutzfeldt–Jakob disease associated with human growth hormone therapy. Neurology 38: 1135–1137

Brown P, Coker-Vann M, Pomeroy K, Franko M, Asher DM, Gibbs CJ Jr, Gajdusek DC (1986) Diagnosis of Creutzfeldt–Jakob disease by Western blot identification of maker protein in human brain tissue. N Engl J Med 314: 547–551

Bruce ME, Dickinson AG (1987) Biological evidence that the scrapie agent has an independent genome. J Gen Virol 68: 79–89

Bruce ME, Fraser H (1975) Amyloid plaques in the brains of mice infected with scrapie: morphological variation and staining properties. Neuropathol Appl Neurol 1: 189–202

Butler DA, Scott MRD, Bockman JM, Borchelt DR, Taraboulos A, Hsiao KK, Kingsbury DT, Prusiner SB (1988) Scrapie-infected murine neuroblastoma cells produce protease-resistant prion proteins. J Virol 62: 1558–1564

Carlson GA, Kingsbury DT, Goodman PA, Coleman S, Marshall ST, DeArmond SJ, Westaway D, Prusiner SB (1986) Linkage of prion protein and scrapie incubation time genes. Cell 46: 503–511
Carlson GA, Goodman PA, Lovett M, Taylor BA, Marshall ST, Peterson-Torchia M, Westaway D, Prusiner SB (1988) Genetics and polymorphism of the mouse prion gene complex: the control of scrapie incubation time. Mol Cell Biol 8: 5528–5540
Carlson GA, Westaway D, DeArmond SJ, Peterson-Torchia M, Prusiner SB (1989) Primary structure of prion protein may modify scrapie isolate properties. Proc Natl Acad Sci USA 86: 7475–7479
Carp RI, Merz PA, Kascsak RJ, Merz GS, Wisniewski HM (1985) Nature of the scrapie agent: current status of facts and hypotheses. J Gen Virol 66: 1357–1368
Caughey B, Race RE, Ernst D, Buchmeier MJ, Chesebro B (1989) Prion protein biosynthesis in scrapie-infected and uninfected neuroblastoma cells. J Virol 63: 175–181
Chamberlin TC (1890) The method of multiple working hypotheses. Science [Old Series] 15: 92–97, reprinted in Science (1965) 148: 754–759
Chandler RL (1961) Encephalopathy in mice produced by inoculation with scrapie brain material. Lancet i: 1378–1379
Chen S-H, Habib G, Yang C-Y, Gu Z-W, Lee BR, Weng S-A, Silberman SR, Cai SJ, Deslypere JP Rosseneu M, Gotto AM, Li W-H, Chan L (1987) Apolipoprotein B-48 is the product of a messenger RNA with an organ-specific in-frame stop codon. Science 238: 363–366
Chesebro B, Race R, Wehrly K, Nishio J, Bloom M, Lechner D, Bergstrom S, Robbins K, Mayer L, Keith JM, Garon C, Haase A (1985) Identification of scrapie prion protein-specific mRNA in scrapie-infected and uninfected brain. Nature 315: 331–333
Cho HJ (1980) Requirement of a protein component for scrapie infectivity. Intervirology 14: 213–216
Collinge J, Harding AE, Owen F, Poulter M, Lofthouse R, Boughey AM, Shah T, Crow TJ (1989) Diagnosis of Gerstmann–Sträussler syndrome in familial dementia with prion protein gene analysis. Lancet ii: 15–17
Czub M, Braig HR, Blode H, Diringer H (1986) The major protein of SAF is absent from spleen and thus not an essential part of the scrapie agent. Arch Virol 91: 383–386
Czub M, Braig HR, Diringer H (1988) Replication of the scrapie agent in hamsters infected intracerebrally confirms the pathogenesis of an amyloid-inducing virosis. J Gen Virol 69: 1753–1756
DeArmond SJ, McKinley MP, Barry RA, Braunfeld MB, McColloch JR, Prusiner SB (1985) Identification of prion amyloid filaments in scrapie-infected brain. Cell 41: 221–235
Dickinson AG (1976) Scrapie in sheep and goats. In: Kimberlin RH (ed) Slow virus diseases of animals and man. North-Holland, Amsterdam, pp 209–241
Dickinson AG, Outram GW (1979) The scrapie replication-site hypothesis and its implications for pathogenesis. In: Prusiner SB, Hadlow WJ (eds) Slow transmissible diseases of the nervous system vol 2. Academic, New York, pp 13–31
Dickinson AG, Outram GW (1988) Genetic aspects of unconventional virus infections: the basis of the virino hypothesis. In: Bock G, Marsh J (eds) Novel infectious agents and the central Nervous system. Wiley, Chichester, pp 63–83 (Ciba Found symp 135)
Dickinson AG, Young GB, Stamp JT, Renwich CC (1965) An analysis of natural scrapie in Suffolk sheep. Heredity 20: 485–503
Diener TO (1972) Is the scrapie agent a viroid? Nature 235: 218–219
Diener TO (1987) PrP and the nature of the scrapie agent. Cell 49: 719–721
Diener TO, McKinley MP, Prusiner SB (1982) Viroids and prions. Proc Natl Acad Sci USA 79: 5220–5224
Diringer H, Gelderblom H, Hilmert H, Ozel M, Edelbluth C, Kimberlin RH (1983) Scrapie infectivity, fibrils and low molecular weight protein. Nature 306: 476–478
Doh-ura K, Tateishi J, Sasaki H, Kitamoto T, Sakaki Y (1989) Pro → Leu change at position 102 of prion protein is the most common but not the sole mutation related to Gerstmann–Sträussler syndrome. Biochem Biophys Res Commun 163: 974–979
Endo T, Groth D, Prusiner SB, Kobata A (1989) Diversity of oligosaccharide structures linked to asparagines of the scrapie prion protein. Biochemistry 28: 8380–8388
Feagin JE, Abraham JM, Stuart K (1988) Extensive editing of the cytochrome C oxidase III transcript in *Trypanosoma brucei*. Call 53: 413–422
Ferguson MAJ, Williams AF (1988) Cell-surface anchoring of protein via glycosyl-phosphatidylinositol structures. Annu Rev Biochem 57: 285–320
Field EJ (1966) Transmission experiments with multiple sclerosis: an interim report. Br Med J 2: 564–565

Field EJ (1967) The significance of astroglial hypertrophy in scrapie, kuru, multiple sclerosis and old age together with a note on the possible nature of the scrapie agent. Dtsch Z Nervenheilkd 192: 265–274

Field EJ, Farmer F, Caspary EA, Joyce G (1969) Susceptibility of scrapie agent to ionizing radiation. Nature 222: 90–91

Fraser H, Bruce ME (1983) Experimental control of cerebral amyloid in scrapie in mice. In: Behan PO et al. (eds) Immunology of nervous system infections Elsevier, Amsterdam, pp 281–290 (progress in brain research, vol 59)

Fraser H, McConnell I, Wells GAH, Dawson M (1988) Transmission of bovine spongiform encephalopathy to mice. Vet Rec 123: 472

Gabizon R, Prusiner SB (1990) Prion liposomes. Biochem J 266: 1–14

Gabizon R, McKinley MP, Groth DF, Kenaga L, Prusiner SB (1988a) Properties of scrapie prion liposomes. J Biol Chem 263: 4950–4955

Gabizon R, McKinley MP, Groth DF, Prusiner SB (1988b) Immunoaffinity purification and neutralization of scrapie prion infectivity. Proc Natl Acad Sci USA 85: 6617–6621

Gabizon R, McKinley MP, Prusiner SB (1987) Purified prion proteins and scrapie infectivity copartition into liposomes. Proc Natl Acad USA 84: 4017–4021

Gajdusek DC (1977) Unconventional viruses and the origin and disappearance of kuru. Science 197: 943–960

Gajdusek DC (1988) Transmissible and non-transmissible amyloidoses: autocatalytic post-translational conversion of host precursor proteins to beta-pleated sheet configurations. J Neuroimmunol 20: 95–110

Gajdusek DC (1986) Chronic dementia caused by small unconventional viruses apparently containing no nucleic acid. In: Scheibel AB, Wechsler AF, Brazier MAB (eds) The biological substrates of Alzheimer's disease. Academic, Orlando, pp 33–54 (UCLA forum in medical sciences, no 27)

Gajdusek DC, Gibbs CJ Jr, Alpers M (1986) Experimental transmission of a kuru-like syndrome to chimpanzees. Nature 209: 794–796

Ghetti B, Farlow M, Conneally M, Azzarelli B, Masters C, Giaccone G, Tagliavini F, Bugiani O (1988) Amyloid plaques and neurofibrillary tangles of Gerstmann–Sträussler–Scheinker disease. Neurology 38 (Suppl 1): 266

Ghetti B, Tagliavini F, Masters CL, Beyreuther K, Giaccone G, Verga L, Farlow MR, Conneally PM, Dlouhy SR, Azzarelli B, Bugiani O (1989) Gerstmann–Sträussler-Scheinker disease. II. Neurofibrillary tangles and plaques with PrP-amyloid coexist in an affected family. Neurology 39: 1453–1461

Gibbons RA, Hunter GD (1967) Nature of the scrapie agent. Nature 215: 1041–1043

Gibbs CJ Jr, Gajdusek DC, Asher DM, Alpers MP, Beck E, Daniel PM, Matthews WB (1968) Creutzfeldt–Jakob disease (spongiform encephalopathy): transmission to the chimpanzee. Science 161: 388–389

Gibbs CJ Jr, Gajdusek DC, Latarjet R (1978) Unusual resistance to ionizing radiation of the viruses of kuru, Creutzfeldt–Jakob disease and scrapie. Proc Natl Acad Sci USA 75: 6268–6270

Gibbs CJ JR, Joy A, Heffner R, Franko M, Miyazaki M, Asher DM, Parisi JE, Brown PW, Gajdusek DC (1985) Clinical and pathological features and laboratory confirmation of Creutzfeldt–Jakob disease in a recipient of pituitary-derived human growth hormone. N Engl J Med 313: 734–738

Goldgaber D, Goldfarb LG, Brown P, Asher DM, Brown WT, Lin S, Teener JW, Feinstone SM, Rubenstein R, Kascsak RJ, Boellaard JW, Gajdusek DC (1989) Mutation in familial Creutzfeldt–Jakob disease and Gerstmann–Sträussler–Scheinker syndrome. Exp Neurol 106: 204–206

Goldman W, Hunter N, Multhaup G, Salbalm JM, Foster JD, Beyreuther KT, Hope J (1988) The PrP gene in natural scrapie. Alzheimers Dis Assoc Disorders [Abstr suppl]: 330

Gordon WS (1946) Advances in veterinary research. Vet Res 58: 516–520

Gordon WS (1966) Review of work on scrapie at Compton, England, 1952–1964. In: Report of scrapie seminar, ARS 91–53. US Department of Agriculture, Washington DC, pp 19–36

Goujard J, Entat M, Maillard F, Mugnier E, Rappaport R, Job JC (1988) Human pituitary growth hormone (hgH) and Creutzfeldt–Jakob disease: results of an epidemiological survey in France, 1986. Int J Epidemiol 17: 423–427

Griffith JS (1967) Self-replication and scrapie. Nature 215: 1043–1044

Hadlow WJ, Karstad L (1968) Transmissible encephalopathy of mink in Ontario. Can Vet J 9: 193–196

Hadlow WJ, Race RE (1986) Cerebrocortical degeneration in goats inoculatd with mink-passaged scrapie virus. Vet Pathol 23: 543–549

Hadlow WJ, Race RE, Kennedy RC (1987) Temporal distribution of transmissible mink encephalopathy virus in mink inoculated subcutaneously. J Virol 61: 3235–3240
Hansen MF, Cavenee WK (1987) Genetics of cancer predisposition. Cancer Res 47: 5518–5527
Haraguchi T, Fisher S, Olofsson S, Endo T, Groth D Tarantino A, Borchelt DR, Teplow D, Hood L, Burlingame A, Lycke E, Kobata A, Prusiner SB (1989) Asparagine-linked glycosylation of the scrapie and cellular prion proteins. Arch Biochem Biophys 274: 1–13
Hardy J (1989) "Slow virus" dementias: prion gene holds the key. Trends Neurosci 12: 168–169
Harmon JT, Nielsen TB, Kempner ES (1985) Molecular weight determinations from radiation inactivation. Methods Enzymol 117: 65–94
Hartsough GR, Burger D (1965) Encephalopathy of mink. I. Epizootiologic and clinical observations. J Infect Dis 115: 387–392
Hay B, Barry RA, Lieberburg I, Prusiner SB, Lingappa VR (1987) Biogenesis and transmembrane orientation of the cellular isoform of the scrapie prion protein. Mol Cell Biol 7:914–920
Hope J, Kimberlin RH (1987) The molecular biology of scrapie: The last two years. Trends Neurosci 10: 149–151
Hope J, Morton LJD, Farquhar CF, Multhaup G, Beyreuther K, Kimberlin RH (1986) The major polypeptide of scrapie-associated fibrils (SAF) has the same size, charge distribution and N-terminal protein sequence as predicted for the normal brain protein (PrP). EMBO J 5: 2591–2597
Hope J, Multhaup G, Reekie LJD, Kimberlin RH, Beyreuther K (1988a) Molecular pathology of scrapie-associated fibril protein (PrP) in mouse brain affected by the ME7 strain of scrapie. Eur J Biochem 172: 271–277
Hope J, Reekie LJD, Hunter N, Multhaup G, Beyreuther K, White H, Scott AC, Stack MJ, Dawson M, Wells GAH (1988b) Fibrils from brains of cows with new cattle disease contain scrapie-associated protein. Nature 336: 390–392
Hourrigan J, Klingsporn A, Clark WW, de Camp M (1979) Epidemiology of scrapie in the United States. In: Prusiner SB, Hadlow WJ (eds) Slow transmissible diseases of the nervous system vol 1. Academic, New York, pp 331–356
Hsiao K, Baker HF,Crow TJ, Poulter M, Owen F, Terwilliger JD, Westaway D, Ott J, Prusiner SB (1989) Linkage of a prion protein missense variant to Geratmann–Sträussler Syndrome. Nature 338: 342–345
Hunter GD, Millson GC (1967) Attempts to release the scrapie agent from tissue debris. J Comp Pathol 77: 301–327
Hunter GD, Kimberlin RH, Gibbons RA (1968) Scrapie: a modified membrane hypothesis. J Theor Biol 20: 355–357
Hunter N, Hope J, McConnell I, Dickinson AG (1987) Linkage of the scrapie-associated fibril protein (PrP) gene and Sinc using congenic mice and restriction fragment length polymorphism analysis. J Gen Virol 68: 2711–2716
Ito M, Shinagawa M, Doi S, Sasaki S, Isomura H, Takahashi K, Goto H, Sato G (1988) Effects of the antiserum against a fraction enriched in scrapie-associated fibrils on the scrapie incubation period in mice. Microbiol Immunol 32: 749–753
Kempner ES, Miller JH, McCreery MJ (1986a) Radiation target analysis of glycoproteins. Anal Biochem 156: 140–146
Kempner ES, Wood R, Salovey R (1986b) The temperature dependence of radiation sensitivity of large molecules. J Polym Sci Polym Phys Ed 24: 2337–2343
Kimberlin RH, Cole S, Walker CA (1987) Temporary and permanent modifications to a single strain of mouse scrapie on transmission to rats and hamsters. J Gen Virol 68: 1875–1881
Kimberlin RH, Walker CA, Fraser H (1989) The genomic identity of different strains of mouse scrapie is expressed in hamsters and preserved on reisolation in mice, J Gen Virol 70: 2017–2025
Kingsbury DT, Smeltzer D, Bockman J (1984) Purification and properties of the K. Fu. isolate of the agent of Creutzfeldt–Jakob disease. 6 international congress on virology 1–7 Sept 1984, Sendai, p 70 (abstr no W 47–6)
Kitamoto T, Tateishi J, Tashima I, Takeshita J, Barry RA, DeArmond SJ, Prusiner SB (1986) Amyloid plaques in Creutzfeldt-Jakob disease stain with prion protein antibodies. Ann Neurol 20: 204–208
Klatzo I, Gajdusek DC, Zigas V (1959) Pathology of kuru. Lab Invest 8: 799–847
Kovanen J, Haltia M (1988) Descriptive epidemiology of Creutzfeldt–Jakob disease in Finland. Acta Neurol Scand 77: 474–480
Kretzschmar HA, Prusiner SB, Stowring LE, DeArmond SJ (1986a) Scrapie prion proteins are synthesized in neurons. Am J Pathol 122: 1–5

Kretzschmar HA, Stowring LE, Westaway D, Stubblebine WH, Prusiner SB, DeArmond SJ (1986b) Molecular cloning of a human prion protein cDNA. DNA 5: 315–324
Latarjet R (1979) Inactivation of the agents of scrapie, Creutzfeldt–Jakob disease, and kuru by radiations. In: Prusiner SB, Hadlow WJ (eds) Slow transmissible diseases of the nervous system, vol 2. Academic, New York, pp 387–408
Lewin P (1972) Scrapie: an infective peptide? Lancet i: 748
Lewin P (1981) Infectious peptides in slow virus infections: a hypothesis. Can med Assoc J 124: 1436–1437
Liao Y-C, Tokes Z, Lim E, Lackey A, Woo CH, Button JD, Clawson GA (1987) Cloning of rat "prion-related protein" cDNA. Invest 57: 370-374
Locht C, Chesebro B, Race R, Keith JM (1986) Molecular cloning and complete sequence of prion protein cDNA from mouse brain infected with the scrapie agent. Proc Natl Acad Sci USA 83: 6372–6376
Lowenstein DH, Butler DA, Westaway D, McKinley MP, DeArmond SJ, Prusiner SB (1990) Three hamster species with different scrapie incubation times and neuropathological features encode distinct prion proteins. Mol Cell Biol 10: 1153–1163
Luria SE, Delbruck M (1943) Mutations of bacteria from virus sensitivity to virus resistance. Genetics 28: 491–511
Malone TG, Marsh RF, Hanson RP, Semancik JS (1979) Evidence for the low molecular weight nature of the scrapie agent. Nature 278: 575–576
Manuelidis L, Sklaviadis T, Manuelidis EE (1987) Evidence suggesting that PrP is not the infectious agent in Creutzfeldt–Jakob disease. EMBO J 6: 341–347
Marsh RF, Kimberlin RH (1975) Comparison of scrapie and transmissible mink encephalopathy in hamsters. II. Clinical signs, pathology and pathogenesis. J Infect Dis 131: 104–110
Marsh RF, Malone TG, Semancik JS, Lancaster WD, Hanson RP (1978) Evidence for an essential DNA component in the scrapie agent. Nature 275: 146–147
Masters CL, Gajdusek DC, Gibbs C Jr (1981a) Creutzfeldt–Jakob disease virus isolations from the Gerstmann–Sträussler syndrome. Brain 104 : 559–588
Masters CL, Gajdusek DC, Gibbs C Jr (1981b) The familial occurrence of Creutzfeldt–Jakob disease and Alzheimer's disease. Brain 104: 535–558
McKinley MP, Bolton DC, Prusiner SB (1983) A protease-resistant protein is a structural component of the scrapie prion. Cell 35: 57-62
McKinley MP, Braunfeld MB, Bellinger CG, Prusiner SB (1986) Molecular characteristics of prion rods purified from scrapie-infected hamster brains. J Infect Dis 154: 110–120
McKinley MP, Butler DA, Prusiner SB (1987) Prion protein mRNA in hamsters and mice. J Cell Biol 105: 317a
Meyer RK, McKinley MP, Bowman KA, Braunfeld MB, Barry RA, Prusiner SB (1986) Separation and properties of cellular and scrapie prion proteins. Proc Natl Acad Sci USA 83: 2310–2314
M'Gowan JP (1914) Investigation into the disease of sheep called "scrapie". William Blackwood, Edinburgh
Millson GC, Hunter GD, Kimberlin RH (1976) The physico-chemical nature of the scrapie agent. In: Kimberlin RH (ed) slow virus diseases of animals and man. American Elsevier, New York, pp 243–266
Nochlin D, Sumi SM, Bird TD, Snow AD, Leventhal CM, Beyreuther K, Masters CL (1989) Familial dementia with PrP-positive amyloid plaques: a variant of Gerstmann–Sträussler syndrome. Neurology 39: 910–918
Oesch B, Westaway D, Walchli M, McKinley MP, Kent SBH, Aebersold R, Barry RA, Tempst P, Teplow DB, Hood LE, Prusiner SB, Weissmann C (1985) A cellular gene encodes scrapie PrP 27–30 protein. Cell 40: 735–746
Owen F, Poulter M, Lofthouse R, Collinge J, Crow TJ, Risby D, Baker HF, Ridley RM, Hsiao K, Prusiner SB (1989) Insertion in prion protein gene in familial Creutzfeldt–Jakob disease. Lancet i 51–52
Parry HB (1962) Scrapie: a transmissible and hereditary disease of sheep. Heredity 17: 75–105
Parry HB (1983) Oppenheimer DR (ed) Scrapie disease in sheep. Academic, New York
Pattison IH (1965) Experiments with scrapie with special reference to the nature of the agent and the pathology of the disease. In: Gajdusek DC, Gibbs CJ Jr, Alpers MP (eds) Slow, latent and temperate virus infections. U S Government Printing Office, Washington D C, pp 249–257 (National Institute of Neurological and Communicable Diseases and Blindness monograph no 2)

Pattison IH (1988) Fifty years with scrapie: a personal reminiscence. Vet Rec 123: 661–666
Pattison IH, Jones KM (1967) The possible nature of the transmissible agent of scrapie. Vet Rec 80: 1–8
Pattison IH, Millson GC (1961) Experimental transmission of scrapie to goats and sheep by the oral route. J Comp Pathol 71: 171–176
Pattison IH, Hoare MN, Jebbett JN, Watson WA (1972) Spread of scrapie to sheep and goats by oral dosing with foetal membranes from scrapie–affected sheep. Vet Rec 90: 465–468
Powell LM, Wallis SC, Pease RJ, Edwards YH, Knott TJ, Scott J (1987) A novel form of tissue-specific RNA processing produces apolipoprotein B-48 in intestine. Cell 50: 831–840
Prusiner SB (1982) Novel proteinaceous infectious particles cause scrapie. Science 216: 136–144
Prusiner SB (1987) Prions and neurodegenerative diseases. N Engl J Med 317: 1571–1581
Prusiner SB (1989) Scrapie prions. Annu Rev Microbiol 43: 345–374
Prusiner SB, Hadlow WJ, Garfin DE, Cochran SP, Baringer JR, Race RE, Eklund CM (1978) Partial purification and evidence for multiple molecular forms of the scrapie agent. Biochemistry 17: 4993–4997
Prusiner SB, Garfin DE, Baringer JR, Cochran SP, Hadlow WJ, Race RE, Eklund CM (1979) On the partial purification and apparent hydrophobicity of the scrapie agent. In: Prusiner SB, Hadlow WJ (eds) Slow transmissible diseases of the nervous system, vol 2. Academic, New York, pp 425–464
Prusiner SB, Groth DF, Bildstein C, Masiarz FR, McKinley MP, Cochran SP (1980a) Electrophoretic properties of the scrapie agent in agarose gels. Proc Natl Acad Sci USA 77: 2984–2988
Prusiner SB, Groth DF, Cochran SP, Masiarz FR, McKinley MP, Martinez HM (1980b) Molecular properties, partial purifications, and assay by incubation period measurements of the hamster scrapie agent. Biochemistry 19: 4883–4891
Prusiner SB, Groth DF, McKinley MP, Cochran SP, Bowman KA, Kasper KC (1981a) Thiocyanate and hydroxyl ions inactivate the scrapie agent. Proc Natl Acad Sci USA 78: 4606–4610
Prusiner SB, McKinley MP, Groth DF, Bowman KA, Mock NI, Cochran SP, Masiarz FR (1981b) Scrapie agent contains a hydrophobic protein. Proc Natl Acad Sci USA 78: 6675–6679
Prusiner SB, Bolton DC, Groth DF, Bowman KA, Cochran SP, McKinley MP (1982a) Further purification and characterization of scrapie prions. Biochemistry 21: 6942–6950
Prusiner SB, Cochran SP, Groth DF, Downey DE, Bowman KA, Martinez HM (1982b) Measurement of the scrapie agent using an incubation time interval assay. Ann Neurol 11: 353–358
Prusiner SB, McKinley MP, Bowman KA, Bolton DC, Bendheim PE, Groth DF, Glenner GG (1983) Scrapie prions aggregate to form amyloid-like birefringent rods. Cell 35: 349–358
Prusiner SB, Groth DF, Bolton DC, Kent SB, Hood LE (1984) Purification and structural studies of a major scrapie prion protein. Cell 38: 127–134
Prusiner SB, Cochran SP, Alpers MP (1985) Transmission of scrapie in hamsters. J Infect Dis 152: 971–978
Race RE, Fadness LH, Chesebro B (1987) Characterization of scrapie infection in mouse neuroblastoma cells. J Gen Virol 68: 1391–1399
Ridley RM, Baker HF, Crow TJ (1986) Transmissible and non-transmissible neurodegenerative disease: similarities in age of onset and genetics in relation to aetiology. Psychol Med 16: 199–207
Robakis NK, Sawh PR, Wolfe GC, Rubenstein R, Carp RI, Innis MA (1986) Isolation of a cDNA clone encoding the leader peptide of prion protein and expression of the homologous gene in various tissues. Proc Natl Acad Sci USA 83: 6377–6381
Roberts GW, Lofthouse R, Brown R, Crow TJ, Barry RA, Prusiner SB (1986) Prion-protein immunoreactivity in human transmissible dementias. N Engl J Med 315: 1231–1233
Rohwer RG (1984a) Scrapie infectious agent is virus-like in size and susceptibility to inactivation. Nature 308: 658–662
Rohwer RG (1984b) Virus-like sensitivity of the scrapie agent to heat inactivation. Science 223: 600–602
Rohwer RG (1986) Estimation of scrapie nucleic acid molecular weight from standard curves for virus sensitivity to ionizing radiation. Nature. 320: 381
Rohwer RG, Gajdusek DC (1980) Scrapie-virus or viroid, the case for a virus. In: Boese A (ed) Search for the cause of multiple sclerosis and other chronic diseases of the central nervous system. Verlag Chemie, Weinheim, pp 333–355
Rubenstein R, Kascsak RJ, Merz PA, Papini MC, Carp RI, Robakis NK, Wisniewski HM (1986) Detection of scrapie associated fibril (SAF) proteins using anti-SAF antibody in non-purified tissued preparations. J Gen Virol 67: 671–681

Scott M, Foster D, Mirenda C, Serban D, Coufal F, Walchli M, Torchia M, Groth D, Carlson G, DeArmond SJ, Westaway D, Prusiner SB (1989) Transgenic mice expressing hamster prion protein produce species-specific scrapie infectivity and amyloid plaques. Cell 59: 847–857

Serban D, Taraboulos A, DeArmond SJ, Prusiner SB (1990) Rapid detection of Creutzfeldt-Jakob disease and scrapie prion proteins. Neurology 40: 110–117

Shaw JM, Feagin JE, Stuart K, Simpson L (1988) Editing of kinetoplastid mitochondrial mRNAs by uridine addition and deletion generates conserved amino acid sequences and AUG initiation codons. Cell 53: 401–411

Shinagawa M, Munekata E, Doi S, Takahashi K, Goto H, Sato G (1986) Immunoreactivity of a synthetic pentadecapeptide corresponding to the N-terminal region of the scrapie prion protein. J Gen Virol 67: 1745–1750

Sparkes RS, Simon M, Cohn VH, Fournier REK, Lem J, Klisak I, Heinzmann C, Blatt C, Lucero M, Mohandas T, DeArmond SJ, Westaway D, Prusiner SB (1986) Assignment of the human and mouse prion protein genes to homologous chromosomes. Proc Natl Acad Sci USA 83:7358–7362

Stahl N, Borchelt DR, Hsiao K, Prusiner SB (1987) Scrapie prion protein contains a phosphatidylinositol glycolipid. Cell 51: 229–240

Stahl N, Borchelt DR, Prusiner SB (1990) Differential release of cellular and scrapie prion proteins from cellular membranes by phosphatidylinositol-specific phospholipase C. Biochemistry 29: 5405–5412

Taraboulos A, Serban D, Prusiner SB (1990) Scrapie prion proteins accumulate in the cytoplasm of persistently infected cultured cells. J Cell Biol 110: 2117–2132

Turk E, Teplow DB, Hood LE, Prusiner SB (1988) Purification and properties of the cellular and scrapie hamster prion proteins. Eur J Biochem 176: 21–30

Weissmann C (1989) Prions: sheep disease in human clothing. Nature 338: 298–299

Wells GAH, Scott AC, Johnson CT, Gunning RF, Hancock RD, Jeffrey M, Dawson M, Bradley R (1987) A novel progressive spongiform encephalopathy in cattle. Vet Rec 121: 419–420

Westaway D, Prusiner SB (1986) Conservation of the cellular gene encoding the scrapie prion protein. Nucleic Acids Res 14: 2035–2044

Westaway D, Goodman PA, Mirenda CA, McKinley MP, Carlson GA, Prusiner SB (1987) Distinct prion proteins in short and long scrapie incubation period mice. Cell 51: 651–662

Westaway D, Carlson GA, Prusiner SB (1989) Unraveling prion diseases through molecular genetics. Trends Neurosci 12: 221–227

Wilesmith JW (1988) Bovine spongiform encephalopathy. Vet Rec 122: 614

Wilesmith JW, Wells GAH, Cranwell MP, Ryan JBM (1988) Bovine spongiform encephalopathy: epidemiological studies. Vet Rec 123: 638–644

Williams ES, Young S (1980) Chronic wasting disease of captive mule deer: a spongiform encephalopathy. J Wildl Dis 16: 89–98

Wisniewski HM, Bruce ME, Fraser H (1975) Infectious etiology of neuritic (senile) plaques in mice. Science 190: 1108–1110

Identifying and Mapping Changes in Gene Expression Involved in the Neuropathology of Scrapie and Alzheimer's Disease*

J. DIEDRICH[1], S. WIETGREFE[1], A. HAASE[1], J. DUGUID[2*], and R. I. CARP[3]

1 Introduction

The pathological hallmarks of scrapie are astrogliosis, vacuolation, and, in some combinations of strains of agent and host, neuritic plaques with amyloid cores which resemble those in senile dementia of the Alzheimer type in man (FRASER 1976; WISNIEWSKI et al. 1975). While the nature of the amyloid differs in scrapie and Alzheimer's disease (AD) (PRUSINER et al. 1983; MASTERS et al. 1985), the analogous pathological lesions of neuritic plaques and astrogliosis in scrapie and AD could imply convergent pathogenetic mechanisms. We have set out to determine not only if this general concept can be substantiated, but, more

* This work was supported by the National Institutes of Health, Veterans Administration, The American Parkinson Disease Association, Whitaker Health Sciences Fund, and the Graduate School of the University of Minnesota.

[1] University of Minnesota Medical School, Department of Microbiology, Box 196 Mayo, Minneapolis, MJ 55455, USA

[2] Bedford V.A. Hospital, 200 Springs Road, Bedford, MA 01730, USA

[3] Institute for Basic Research in Developmental Disabilities, 1050 Forest Hill Road, Staten Island, NY 10314, USA

Current Topics in Microbiology and Immunology, Vol. 172

specifically, whether changes in expression of certain genes form the molecular basis for the neuropathological alterations in both conditions (WIETGREFE et al. 1985; DIEDRICH et al. 1987; HAASE et al. 1986; DUGUID et al. 1988, 1989). This chapter begins with an elaboration of the logic underlying this hypothesis, followed by a description of the methods of differential and subtractive screening of cDNA libraries appropriate to testing the tenets of the theory. We conclude with an account of what has already been learned using this approach to understanding the molecular basis of these neuropathological processes.

2 Gene Expression and the Neuropathology of Scrapie and Azheimer's Disease: Rationale and Technical Strategies

The essential premises of our reconstruction of the molecular mechanisms responsible for the neuropathological changes in scrapie and AD are that:

1. Activation or repression of a defined set of genes is the underlying mechanism of the neurodegenerative phenomena. Consequently, identifying these genes offers a unique opportunity to generate novel hypotheses about the fundamental mechanisms of the disease process. The major advantage of this approach is that it is not limited by the current state of knowledge, which necessarily restricts the focus of contemporary analyses of pathological processes; nor does it require a priori assumptions about the underlying basis for these processes.
2. The similarities in tissue lesions reflect changes in expression in similar sets of genes.
3. The changes in gene expression bear a constant and plausible temporal and spatial relationship to the pathological process. This latter statement is, in a sense, the contemporary equivalent of Koch's postulates applied to pathological processes.

Experimental examination of these premises involves construction of a cDNA library that can be screened by differential or subtractive hybridization (Fig. 1) to detect genes whose expression is modulated in disease. We subsequently determine whether or not the altered state of gene expression is common to scrapie and AD, and identify, if possible, the genes involved by sequencing and computer-assisted homology searches of extant data bases. We then use in situ hybridization (HAASE 1986) and immunocytochemistry (with antibodies to the protein encoded by the identified gene) to define the cellular and tissue distribution of transcripts and protein products of the gene whose expression is modulated. In this way we can assess its relationship to the pathological lesions in AD and scrapie. In scrapie, animal models make it possible to see if the changes in gene expression precede the appearance of the tissue lesions (KIM et al. 1987). In the following sections we describe in some detail

Flow Diagram of Procedures

1. Construct cDNA library from RNA from tissues with abundant pathological lesions

↓

2. +/− screening

↓

a) Differential hybridization using radiolabeled probes from RNA from infected or affected tissues (plus state) vs uninfected or unaffected tissues (minus state)

b) Subtractive hybridization with probes from infected or affected tissues (plus state) vs uninfected or unaffected tissues (minus state) to detect altered levels of mRNAs of low abundance

↓

3. Identify cDNAs of genes whose expression increases or decreases in the pathological (infected) state/confirm by comparing the level of RNAs in the plus and minus state

↓

4. If possible, identify the gene by sequencing and computer-assisted homology search of data bases

↓

5. In situ hybridization and immunocytochemistry to map altered gene expression in order to determine its spatial and temporal relationship to the pathological lesions, the constancy of its association with the lesions, and whether it is activated or repressed in both scrapie and AD

Fig. 1. Flow diagram of differential and subtractive hybridization screening. See text for explanation

the construction and screening of the libraries by differential and subtractive hybridization, the advantages of the latter in detecting altered expression of genes of low abundancy, and the identity of some of the genes modulated in scrapie and AD.

3 Differential Hybridization Screening of cDNA Libraries To Identify Genes Involved in Neurodegenerative Processes

3.1 General Comments

The differential screening approach to understanding the molecular mechanisms of neurodegenerative processes is illustrated for scrapie in Figs. 2 and 3. Messenger RNA, isolated from the brains of infected animals, is reverse transcribed into DNA which is subsequently ligated to the right and left "arms" of

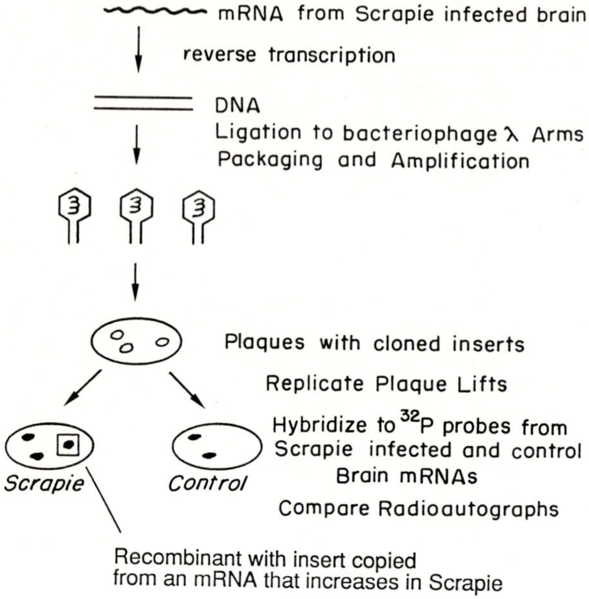

Fig. 2. Schematic diagram of differential screening of a cDNA library to identify an mRNA that increases in scrapie infection. See text for explanation

bacteriophage λ gt10 DNA (HUYNH et al. 1985). After packaging and amplification, the bacteriophages are plated on an *E. coli* host which selects against non-recombinants. Replicate lifts from the agar plates are then hybridized to ^{32}P-labeled cDNA probes reverse transcribed from mRNAs from the brains of infected or uninfected (control) animals. Most of the plaques will contain cloned DNA inserts from mRNAs which do not differ in abundancy in infected and control animals, but there will be some species, in the example illustrated in Figs. 2 and 3, which are increased in scrapie. Since the RNA population contains more copies of this transcript, the ^{32}P-labeled probe will also contain more copies of this sequence and this will increase the number of radioactive molecules bound to the recombinant. By comparing the signals in radioautographs, plaques containing recombinant bacteriophage with DNA copied from mRNAs which are increased in scrapie can be identified as those which appear darker. These plaques are replated for secondary screening to verify the differential hybridization.

Differential screening is a straightforward approach to identifying changes in gene expression for mRNAs that represent about 0.1% or greater of the total mRNA population. Screening by subtractive hybridization, described later in this chapter, facilitates analysis of less abundant mRNAs and is the only way to assess rare mRNAs (0.001%–0.01% of the total mRNA population). The approaches, however, should be viewed as complementary. Differential screening should be done first so that changes in the levels of abundant mRNAs will not be missed by removing these sequences in the subtractive hybridizations.

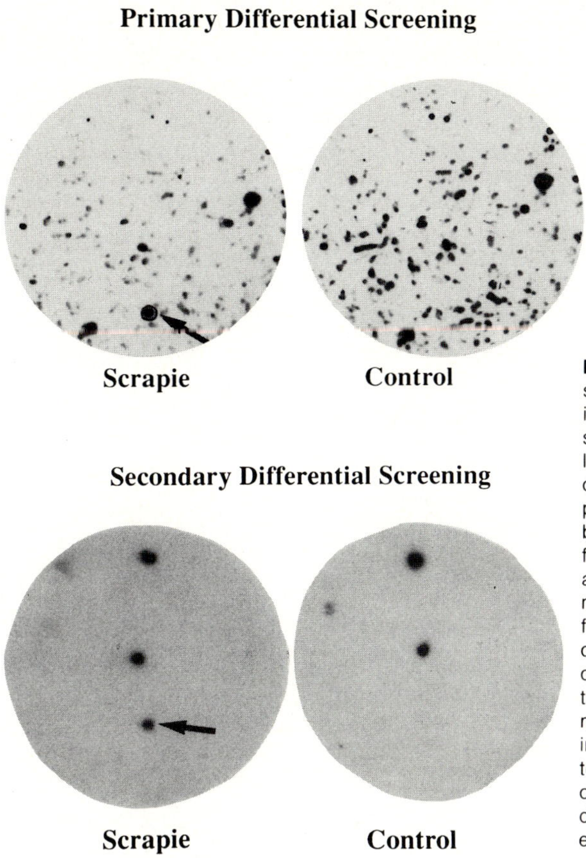

Fig. 3. Illustration of primary and secondary differential screening to identify a mRNA that increases in scrapie infection. Replicate plaque lifts have been hybridized to ^{32}P cDNA reverse transcribed from polyadenylated mRNA from the brains of scrapie-infected or uninfected controls. The *arrow* points to a recombinant with a copy of an mRNA that increases in scrapie infection. Because the control probe contains fewer radiolabeled transcripts of this mRNA (in proportion to the relative abundancy of this mRNA in the population) there is insufficient radioactive probe bound to the corresponding plaque in the control plaque lift to generate a detectable signal for this length of exposure

3.2 Construction of cDNA Libraries

The initial step in construction of the cDNA library is isolation of "relevant" RNA, i.e., RNA from tissues with the tissue lesions which typify the pathological process under investigation. In scrapie and AD this will be brain tissue, which may be from regions selected for a high density of lesions, e.g., hippocampus in AD (KATZMAN 1986). To the extent possible, differences in confounding variables such as the age of the individual, postmortem autolysis times, and fixation conditions should be minimized. Our usual procedure, summarized in MANIATIS et al. (1982), is to isolate total RNA by homogenizing tissues in buffers containing guanidium isothiocyanate as a denaturant to inhibit nucleases. The RNA is separated from DNA and proteins by pelleting it through CsCl, and the polyadenylated population is selected by two cycles of binding and elution from oligo(dT) cellulose. Because about one-half of the brain mRNA population is not polyadenylated, comprehensive analyses of altered gene expression will ultimately require additional studies of the poly A^- mRNA population selected by benzoylated cellulose chromatography (VAN NESS et al. 1979). The isolated RNA is copied into DNA by reverse transcription using oligo(dT) or hexanucleotides to prime first strand synthesis (Fig. 4), RNase H to nick the RNA in the RNA:DNA hybrids, and *E. coli* and T4 DNA polymerases to synthesize the second strand and generate blunt-ended DNA. After internal *Eco*RI sites have been methylated

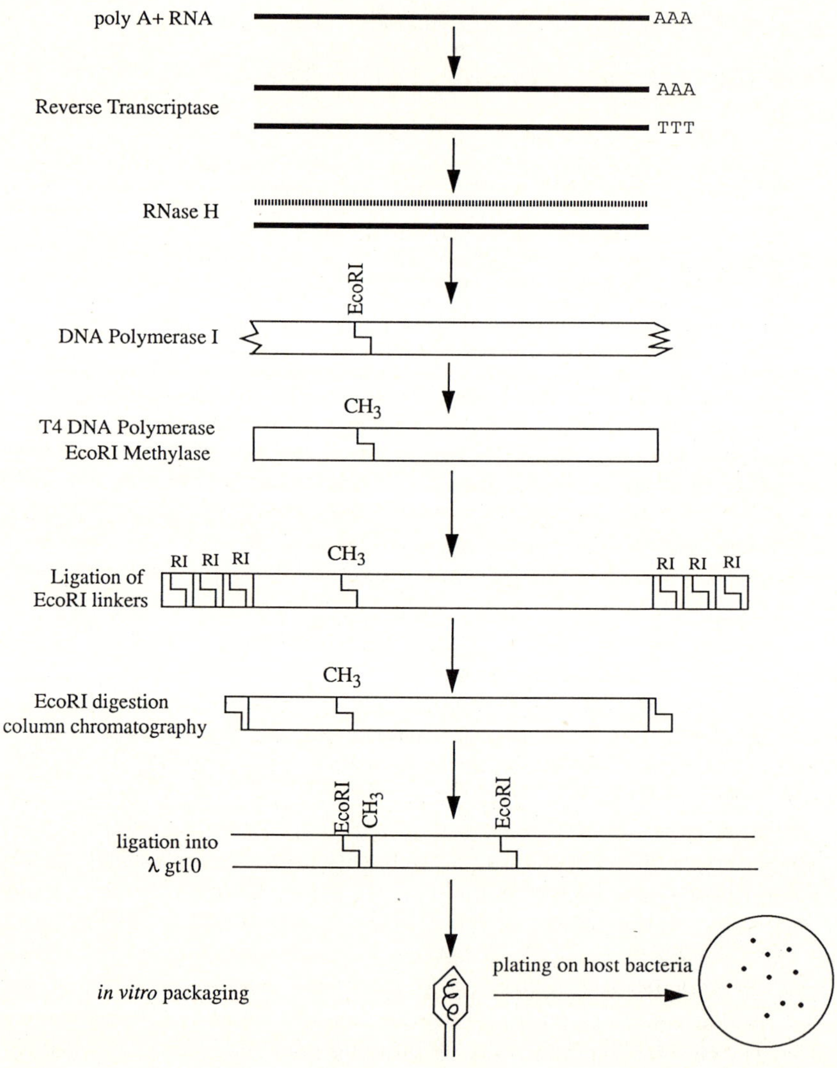

Fig. 4. Schematic diagram of procedures employed to construct a cDNA library in bacteriophage λgt10. See text for explanation

with *Eco*RI methylase, *Eco*RI linkers are ligated to the ends of the DNA. The modified DNA is digested with *Eco*RI, chromatographically separated from excess linkers by passage through a column of sepharose CL–4B, and ligated to dephosphorylated λ gt10 "arms", that are generated by digestion with *Eco*RI which cleaves the vector at a single *Eco*RI site in the repressor gene (cl). After packaging, the bacteriophage preparations are titered, amplified, and screened on *E. coli* carrying a high-frequency lysogeny mutation (hfl) which efficiently suppresses plaque formation by non-recombinant cl^+ bacteriophage. This allows growth of the cl^- λ gt10 with DNA inserts which form clear plaques (HUYNH et al. 1985).

3.3 Screening cDNA Libraries by Differential Hybridization

Libraries can be amplified or screened directly in order to minimize misrepresentation in the library due to variability in growth rates amongst the recombinants. If a large number of individual recombinants are obtained (> 1 million), we screen a small portion of the library directly and amplify and store the remainder. For screening, the library should be plated at a density that allows easy identification of individual plaques (approximately 1000 PFU/150-mm plate). Plates should be prepared several days in advance and allowed to sit unwrapped at ambient temperatures to prevent "smearing" of plaques when the lifts are made.

To screen the library differentially, plaque replicas ("lifts") are prepared by briefly adhering nylon filters to the plates. The lifts are floated on solutions of NaOH and NaCl which lyse the bacteriophage and release the DNA, denature it, and fix it to the filter. Either one plaque lift can be made from the plate which will be hybridized in sequence to the probes transcribed from RNA from normal or diseased tissues, or replicate lifts can be made for hybridization in parallel to the two probes. In our experience the second and third lifts made from each plate often give more consistent and comparable results than the first and second lifts (H. Minnigan, personal communication).

The plaque lifts are hybridized to ^{32}P-labeled probes, prepared by reverse transcribing control and infected scrapie RNA, washed at moderate stringency, dried, and exposed to X-ray film. The autoradiographs are aligned, and the signals are compared to identify plaques that contain DNA from bacteriophage with inserts from genes whose expression increases or decreases in infection. These will hybridize to more or less ^{32}P probe containing, respectively, cDNA copies from more or less abundant mRNAs. The recombinant bacteriophages which apparently contain DNA copies of mRNAs modulated by the disease process are rescreened at very low density to verify the differential level of expression. In secondary and tertiary screening, the plaques should not be plated at high density as only a small fraction of the labeled probe will be complementary to the DNA insert in any recombinant; consequently, dividing these relevant probe transcripts between many such plaques might extinguish the signal.

3.4 Subtractive Hybridization Strategies

In the search for modulated genes in neurodegenerative diseases, a large number of recombinants must be screened because of the sequence complexity of brain mRNA. For example, if there are 30 000 different low-abundance polyadenylated mRNAs in mammalian brain (MILNER and SUTCLIFFE 1983), about 500 000 clones must be screened to identify (with a probability of 0.99) a clone of an mRNA in the 0.001% abundancy class (CLARKE and CARBON 1976). Moreover, with direct differential screening, many false positives will usually be identified, and the search will be made more difficult by intense signals from clones of abundant mRNAs which obscure signals from clones of less abundant modulated RNAs. (The exponential increase in extent of hybridization with increasing concentration of radiolabeled transcripts will disproportionately increase the signal from more abundant species—see discussion of hybridization kinetics below.) These difficulties can be overcome by subtractive hybridization strategies which reduce the sequence complexity of the library and/or probe. This approach has already proved useful in identifying genes expressed differentially in development (ZIMMERMAN et al. 1980; SARGENT and DAWID 1983) and cellular transformation (SUGANO et al. 1987); in identifying genes encoding the T lymphocyte receptor (HEDRICK et al. 1984); and, most recently, in scrapie and AD (DUGUID et al. 1988, 1989).

The methods used to prepare a subtracted scrapie library are shown in Fig. 5. DNA copies of mRNAs isolated from scrapie-infected or control animals are inserted into πH3M, an expression vector developed by ARUFFO and SEED (1987) for propagation in prokaryotic and eukaryotic hosts either as double-stranded plasmid or single-stranded bacteriophage DNA (ss DNA). For subtraction, ss DNA from the control library is biotinylated, sonicated to reduce viscosity, and hybridized in vast excess to ss DNA from the scrapie library. Recombinants with copies of mRNAs equivalently represented in scrapie-infected and control brains will hybridize and can be removed (along with hybridized biotinylated control ss DNAs) by affinity chromatography with avidin bound to Sephacryl. Unhybridized recombinants enriched for scrapie-specific sequences remain in solution. After conversion to double-stranded DNA, a competent *E. coli* host is transformed to generate the subtracted library. This can be screened differentially with scrapie and control cDNA probes to identify modulated genes.

The subtractive hybridization approach of using cloned ss DNA rather than mRNA has a number of advantages over other methods. Systems where only small amounts of tissue are available can be studied and the libraries are also a resource for other studies, e.g., to find genes which are down-regulated in scrapie. In this case the scrapie library is biotinylated and used to subtract the control library. Recombinants of abundant sequence mRNAs such as glial fibrillary acidic protein (GFAP) mRNA can be removed from the subtracted library to make it possible to identify less abundant or very rare recombinants. Generally, the lower limit of abundance of a mRNA that can be detected in an unsubtracted cDNA library hybridization is in the range of about 0.01%–0.05%.

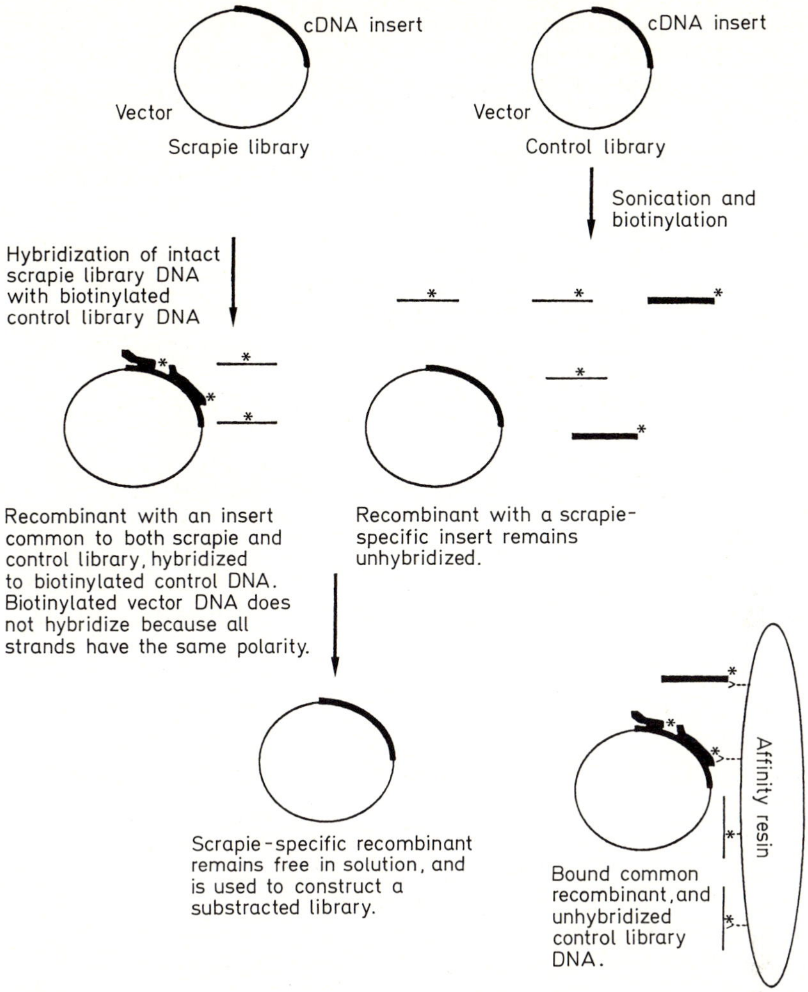

Fig. 5. Schematic diagram of subtractive hybridization cloning. See text for explanation

This limit is imposed by the sequence complexity of the probe; each mRNA is represented in the probe in proportion to its abundancy and accordingly there will be little radiolabeled probe complementary to the cloned DNA of a rare mRNA. This will decrease both the efficiency of hybridization (kinetically unfavorable) and the absolute amount of radioactivity available to generate a detectable signal. Subtraction of sequences common to both libraries with an abundance sufficient for effective hybridization (kinetically favorable) reduces sequence complexity, so that a recombinant that had an abundance of 0.001% in the starting library might have an abundance of 0.01% or greater in the subtracted library. This library could be used to generate a probe that could detect a rare recombinant. To obtain a comparable "minus" probe for differential screening, one would hybridize the subtracting library to itself. This results in a library from which abundant sequences have been removed, and is thereby enriched for low abundance recombinants. Probe generated from this library would serve as a "minus" probe comparable to the "plus" probe generated from the subtracted scrapie library.

Five modulated genes have been identified by subtractive hybridization and characterized by sequencing and homology searches, as discussed in the next section. GFAP, metallothionein, crystallin, sulfated glycoprotein 2, and transferrin are increased in expression in scrapie and AD (the latter is increased in scrapie but not AD). While the latter two are involved in lipid and iron transport, respectively, and might be required to support activation of astrocytes, the sites of transcription in infection and role in the neurodegenerative process of these genes remain matters for speculation and future research.

4 Characterizing Recombinant DNAs of Modulated Genes

Once a recombinant clone of a differentially expressed gene has been identified, the DNA insert can be excised with *Eco*RI purified by preparative agarose gel electrophoresis, and used as a probe to (1) verify and quantitate differential levels of expression and determine the size of the corresponding mRNA(s) in northern blots, and (2) identify recombinants with similar but longer inserts for sequencing by the Sanger dideoxy chain terminator method (SANGER et al. 1977). The sequence is subsequently compared with known sequences by computer-assisted searches of data bases. Finding homology to a known gene is important both because it may suggest novel mechanisms of disease and because it provides valuable information and reagents for the next steps in the analysis. For example, once *Scr-1*, the first modulated gene identified by differential hybridization (WIETGREFE et al. 1985), was found to be homologous to the gene encoding GFAP (DIEDRICH et al. 1987), the major protein of astrocytes, we could obtain complete clones of rodent or human GFAP to assess the changes in mRNA levels in scrapie and AD, thus circumventing the sequence divergence problem (if, for example, the rodent and human genes differed in sequence by more than 30%, it would be difficult to conduct these cross-hybridization comparisons). Moreover, as was the case with GFAP, specific antibodies to the gene product were available for immunocytochemical studies as described in the next section (ENG 1980).

5 Localizing the Transcripts and Protein Products of Modulated Genes in Tissues

The next (critical) step in the analysis is an evaluation of the possible role of the modulated gene in the pathological process. This is assessed by determining whether it is constantly associated with the tissue lesions, and whether the

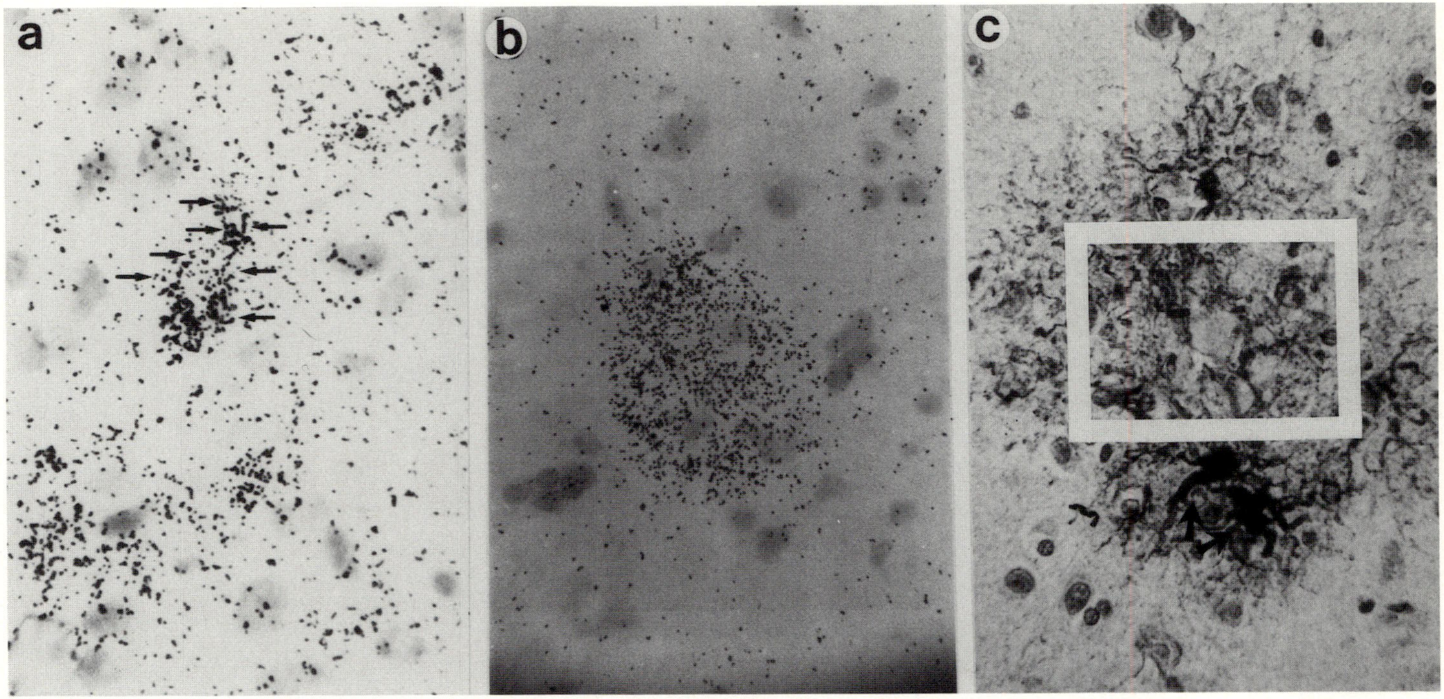

Fig. 6 a–c. GFAP mRNA and protein in astrocytic processes and neutritic plaques. Sections from brains of a scrapie-infected hamster (**a**) and an individual with AD (**b**) were hybridized in situ to an ^{125}I-labeled GFAP-specific probe. In the developed radioautographs, the large number of silver grains over a cell in **a** reflect the increased number of copies of GFAP mRNA in the cell body and processes (marked by *arrows*) in an activated astrocyte in the dentate gyrus. In **b**, the grains are localized to a neuritic plaque identified by staining a subjacent section with thioflavine S. In **c**, the corresponding protein in the processes is revealed by immunohistochemistry using antibodies of GFAP. The darkly stained GFAP-positive fibers infiltrate the plaque from fibrillary astrocytes (*arrows*) at the perimeter of the plaque (*framed*). (Reproduced from DIEDRICH et al. 1987 with permission)

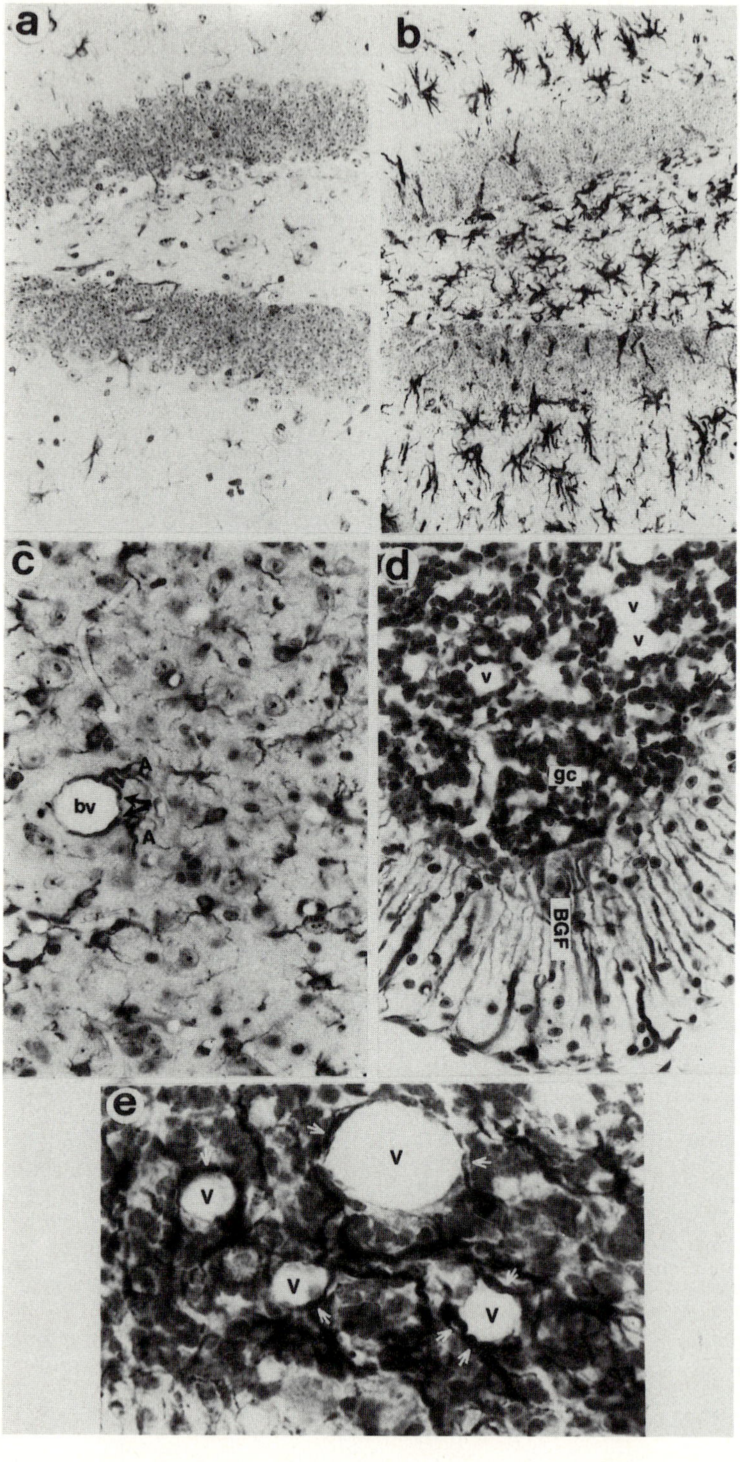
a
b
c
bv
A
A
d
v
v
v
gc
BGF
e
V
V
V
V

change in level of expression bears a plausible temporal and spatial relationship to the formation of tissue lesions. To map changes in gene expression we use in situ hybridization and immunocytochemistry with antibodies to the gene product. To illustrate this approach, Figs. 6 and 7 show examples in scrapie and AD of changes in the amount and distribution of GFAP. Early in the course of infection of rodents with the scrapie agent, astrocytes increase in number and size and their processes also become more numerous and enlarge. This morphological transformation of protoplasmic to fibrillary astrocytes is accompanied by an increase in the amount of GFAP mRNA and protein (DIEDRICH et al. 1987; MACKENZIE 1983; DEARMOND et al. 1987; KRETZSCHMAR et al.1986). In scrapie, the accumulations of GFAP are evident in processes surrounding small blood vessels and the ventricles and in foci in the hippocampus, thalamus, deeper layers of neocortex, septal and brain stem nuclei, and cerebellum. In AD, GFAP mRNA is increased in neuritic plaques, in astrocytic processes infiltrating the plaques from astrocytes at the perimeter. The striking localization of GFAP mRNA in astrocytic processes provides visual evidence of one solution utilized by glial cells to the problem of protein synthesis at sites some distance away from the cell body. Both astrocytes and oligodendrocytes apparently transport mRNAs for GFAP and myelin basic protein respectively to sites of protein synthesis in processes and membranes (DIEDRICH et al. 1987; COLMAN et al. 1982).

The activation of astrocytes in scrapie is a cardinal pathological feature and one of the first manifestations of infection, possibly accounting for the increased rates of DNA synthesis in the brain in the first few weeks of infection (KIMBERLIN 1969). Astrocytosis also parallels increases in infectivity. Stereotaxic inoculation of cerebellum, for example, is followed by regional increases in infectivity and activation of astrocytes at the site of inoculation (KIM et al. 1987, 1990; J. DIEDRICH et al., manuscript in preparation). As the agent spreads to other regions of the brain, there is a corresponding activation of astrocytes accompanied by vacuolation. The terminal stage of infection is characterized by high titers of infectious agent widely distributed throughout the CNS, more extensive astrogliosis and vacuolation, and deposits (plaques) of aggregated PrP^{Sc} which have the staining properties of amyloid (KIMBERLIN and WALKER 1983; COLE and KIMBERLIN 1985; KIM et al. 1987; DEARMOND et al. 1987).

◄

Fig. 7 a–e. Activated astrocytes in scrapie. Formalin-fixed sections of dentate gyrus (**a, b**), cerebral cortex (**c**), and cerebellum (**d, e**) from scrapie-infected or control mice were reacted in sequence with antibody to GFAP, biotinylated secondary antispecies antibody, avidin biotinylated horseradish peroxidase complex, H_2O_2, and diaminobenzidine. Many more GFAP-positive darkly stained fibrillary astrocytes are evident in the photomicrographs from scrapie-infected animals (**b**) than in those from uninfected controls (**a**). Astrocytic processes are also more prominent and numerous, e.g., investing blood vessels (*bv* in **c**) and surrounding vacuoles (**e**) and fibers emanating from Bergmann glia (*BGF*, **d**). *Arrows* indicate processes, A the astrocytes from which they orginate; *v* designates vacuoles in the granule cell layer (*gc*) in cerebellum (**d, e**)

6 "Activated Astrocyte" Vs "Reactive Astrocyte"

One of the central themes that has emerged from these analyses is the importance of astrocyte activation in the formation of lesions. In scrapie, two additional genes encoding apolipoprotein E and cathepsin D have been shown by differential screening to be activated in astrocytes. The increases in gene expression precede overt evidence of neuronal damage and map to sites of histopathological changes (DIEDRICH et al., manuscript in press). Activated astrocytes are also associated with amyloid deposits in scrapie and AD. While it can be argued that amyloid evokes changes in the astrocyte, the recent demonstration that the β-amyloid precursor protein is synthesized in astrocytes in response to injury to hippocampal neurons (SIMAN et al. 1989) makes it just as likely that activation of the astrocyte lies more proximate in the chain of events leading to amyloid deposition. For this reason, we think that the term "activated astrocyte" is preferable to "reactive astrocyte" with the implied restrictions on the role it plays in lesion formation.

7 Conclusion and Prospects

The approaches we have described to identifying and mapping modulated gene expression in the neuropathological processes of scrapie and AD have already revealed early and prominent changes in the astrocyte. It is our hope that, as we continue our investigations of modulated gene expression in neurodegenerative diseases, other common themes and patterns will emerge that will provide deeper insight into the primary causes of these disorders. The addition of methods to enhance the kinetics of subtractive hybridization (TRAVIS and SUTCLIFFE 1988) and to amplify rare sequences using the polymerase chain reaction (AKOWITZ and MANUELIDIS 1989) provide new opportunities to seek clues to the fundamental mechanisms of pathological processes by examining changes in expression of very rare species of mRNAs.

Acknowledgements. We thank Dana Clark and Tim Leonard for preparation of the manuscript and figures and Hal Minnigan and D. Blomster–Hautamaa for helpful discussion and contributions to the projects.

References

Akowitz A, Manuelidis L (1989) A novel cDNA/PCR strategy for efficient cloning of small amounts of undefined RNA. Gene 81: 295–306

Aruffo A, Seed B (1987) Molecular cloning of a CD28 cDNA by a high-efficiency COS cell expression system. Proc Natl Acad Sci USA 84: 8573–8577

Clarke L, Carbon J (1976) A colony bank containing synthetic Col El hybrid plasmids representative of the entire *E. coli* genome. Cell 9: 91–99

Cole S, Kimberlin RH (1985) Pathogenesis of mouse scrapie: Dynamics of vacuolation in brain and spinal cord after intraperitoneal infection. Neuropathol Appl Neurobiol 11: 213–227

Colman DR, Kreibich G, Frey AB, Sabatini DD (1982) Synthesis and incorporation of myelin polypeptides into CNS myelin. J Cell Biol 95: 598–608

DeArmond SJ, Mobley WC, DeMott BA, Barry RA, Beckstead JH, Prusiner SB (1987) Changes in the localization of brain prion proteins during scrapie infection. Neurology 37: 1271–1280

Diedrich J, Wietgrefe S, Zupancic M, Staskus K, Retzel E, Haase AT, Race R (1987) The molecular pathogenesis of astrogliosis in scrapie and Alzheimer's disease. Microbial Pathogenesis 2: 435–442

Diedrich JF, Minnigan H, Carp RI, Whitaker JN, Race R, Frey W, Haase AT (1991) Neuropathological changes in scrapie and Alzheimer's disease are associated with increased expression of apolipoprotein E and cathepsin D in astrocytes. J Virol (in press)

Duguid JR, Rohwer RG, Seed B (1988) Isolation of cDNAs of scrapie-modulated RNAs by subtractive hybridization of a cDNA library. Proc Natl Acad Sci USA 85: 5738–5742

Duguid JR, Bohmont CW, Liu N, Tourtellotte WW (1989) Changes in brain gene expression shared by scrapie and Alzheimer disease. Proc Natl Acad Sci USA 86: 7260–7264

Eng LF (1980) The glial fibrillary acidic (GFA) protein. In: Proteins of the nervous system, 2nd edn. Raven, New York

Fraser H (1976) The pathology of natural and experimental scrapie. In: Slow virus diseases of animals and man. North–Holland, Amsterdam

Haase AT (1986) Analysis of viral infections by in situ hybridization. In: In situ hybridization—applications to neurobiology. Symposium Monograph, Oxford University Press, Fairlawn

Haase AT, Lewis E, Wietgrefe S, Zupancic M, Diedrich J, Minnigan H, Ball MJ (1986) Speculations on the role of transmissible agents in the pathogenesis of Alzheimer's disease. Can J Neurol Sci 13: 449–451

Hedrick SM, Cohen DI, Nielsen EA, Davis MM (1984) Isolation of cDNA clones encoding T cell-specific membrane-associated proteins. Nature 308: 149–153

Huynh TV, Young RA, Davis RW (1985) Constructing and screening cDNA libraries in λgt10 and λgt11. In: Glover DH (ed) DNA cloning, vol 1: a practical approach. IRL, Oxford

Katzman R (1986) Medical progress: Alzheimer's disease. N Engl J Med 314: 964

Kim YS, Carp RI, Callahan SM, Wisniewski HM (1987) Incubation periods and survival times for mice injected stereotaxically with three scrapie strains in different brain regions. J Gen Virol 68: 695–702

Kim YS, Carp RI, Callahan SM, Wisniewski HM (1990) Pathogenesis and pathology of scrapie after stereotaxic injection of strain 22L in intact and bisected cerebella. J Neuropathol Exp Neurol 49: 114–121

Kimberlin RH (1969) Biochemical changes in scrapie affected brain. Proceedings of the 495th Meeting of the Biochemical Society, Universities of Glasgow and Strathclyde, May 29–31, 1969

Kimberlin RH, Walker CA (1983) Invasion of the CNS by scrapie agent and its spread to different parts of the brain. In: Court LA, Cathala F (eds) Virus non-conventionnels et affections du système nerveux central. Masson, Paris, pp 11–13

Kretzschmar HA, Prusiner SB, Stowring LE, DeArmond SJ (1986) Scrapie prion proteins are synthesized in neurons. J Pathol 122: 1–5

Mackenzie A (1983) Immunohistochemical demonstration of glial fibrillary acidic protein in scrapie. J Comp Pathol 93: 251–259

Maniatis T, Fritsch EF, Sambrook J (1982) Molecular cloning (a laboratory manual). Cold Spring Harbor Laboratory. New York

Masters CL, Simms G, Weinman NA, Multhaup G, McDonald BL, Beyreuther K (1985) Amyloid plaque core protein in Alzheimer disease and Down syndrome. Proc Natl Acad Sci USA 82: 4254

Milner RJ, Sutcliffe JG (1983) Gene expression in rat brain. Nucleic Acids Res 11: 5497–5520

Prusiner SB, McKinley MP, Bowman KA, Bolton DC, Bendheim PE, Groth DF, Glenner GG (1983) Scrapie prions aggregate to form amyloid-like birefringent rods. Cell 35: 349

Sanger F, Nicklen S, Coulson AR (1977) DNA sequencing with chain-terminating inhibitors. Proc Natl Acad Sci USA 45: 407–415

Sargent TD, Dawid IB (1983) Differential gene expression in the gastrula of *Xenopus laevis*. Science 222: 135–139

Siman R, Card JP, Nelson RB, Davis LG (1989) Expression of β-amyloid precursor protein in reactive astrocytes following neuronal damage. Neuron 3: 275–285

Sugano S, Stoeckle MY, Hanafusa H (1987) Transformation by Rous sarcoma virus induces a novel gene with homology to a mitogenic platelet protein. Cell 49: 321–328

Travis GH, Sutcliffe JG (1988) Phenol emulsion-enhanced DNA-driven subtractive cDNA cloning: isolation of low-abundance monkey cortex-specific mRNAs. Proc Natl Acad Sci USA 85: 1696–1700

Van Ness J, Maxwell IH, Hahn WE (1979) Complex population of nonpolyadenylated messenger RNA in mouse brain. Cell 18: 1341–1349

Wietgrefe S, Zupancic M, Haase A, Chesebro B, Race R, Frey WII, Rustan T, Friedman RL (1985) Cloning of a gene whose expression is increased in scrapie and in senile plaques in human brain. Science 230: 1177–1179

Wisniewski HM, Bruce ME, Fraser H (1975) Infectious etiology of neuritic (senile) plaques in mice. Science 190: 1108

Zimmermann CR, Orr WC, Leclerc RF, Barnard EC, Timberlake WE (1980) Molecular cloning and selection of genes regulated in aspergillus development. Cell 21: 709–715

Search for a Transmissible Agent in Alzheimer's Disease: Studies of Human Buffy Coat*

E. E. MANUELIDIS and L. MANUELIDIS

The etiology of Alzheimer's disease (AD) is totally unknown, despite the vigorous efforts in several laboratories during recent years to solve this mystery. The lack of success in delineating one or more underlying etiologies for AD may be due to the fact that there are no animal models directly derived from this disease. Further, almost all investigative studies of AD concentrate on end stages of disease where fully developed lesions are apparent. In searching for potential etiologies of AD, such fully developed cases might represent the wrong end of the disease spectrum and be relatively uninformative with respect to original inciting events (E.E. MANUELIDIS 1985). In autopsy brains of individuals with AD, there is clear evidence of a neuronal cemetery, with neuronal devastation and gliosis in selected brain regions. Tangles and more widespread plaques and amyloid deposits are the remaining skeletons and bushes associated with end stage neuronal degeneration. We have postulated that these very important cardinal lesions are the end result rather than the originating causes of AD. Although these sequelae and their associated biochemical processes may play a role in the progression of AD, the initial inciting events are not known. Without underestimating the value of the data generated by our colleagues on the biochemistry and histology of these lesions, we concentrate, rightly or wrongly, on the concept that a transmissible agent may underlie AD during very early stages of disease. At such early stages, humans may have mild or unrecognized neurological symptoms. Should we prove successful in demonstrating a transmissible agent, our findings would not necessarily be in conflict with the results of our colleagues who have different working hypotheses.

In considering of the possibility of a transmissible agent as one etiology of AD, we were greatly influenced by our investigations with Creutzfeldt–Jakob Disease (CJD), a rare dementia of viral etiology (E.E. MANUELIDIS et al. 1976; 1978a; E.E. MANUELIDIS and L. MANUELIDIS 1989; L. MANUELIDIS and E.E. MANUELIDIS 1986, 1989). Although the nature of the infectious agent is controversial, infectivity in CJD copurifies with nucleic acid and displays physical properties that are most consistent with a viral structure (AKOWITZ et al. 1990; MURDOCH et al. 1990; SKLAVIADIS et al. 1989, 1990). We felt that events set in motion during the

* This work was supported by NIH grants NS 12674 and AG 03106
Section of Neuropathology, Yale University School of Medicine, New Haven, CT 06510, USA

Current Topics in Microbiology and Immunology, Vol. 172

pathogenesis of a rare dementia such as CJD might be applicable to, or shed some light on, the pathogenesis of AD, a common and widespread dementia with the well-known devastating effects upon the patients, their families, and the community in general. Although we believe our hypothesis has merit, we do not want to minimize the findings of others who entirely disagree with us.

Numerous attempts to demonstrate transmission of AD to laboratory animals using end-stage human tissues were unsuccessful in other laboratories (GOUDSMIT et al. 1980). A few attempts in our own laboratory using autopsy brain tissue from AD patients also failed to demonstrate any transmissible agent (E.E. MANUELIDIS 1985). We therefore decided to utilize the buffy coat of human blood, and chose for initial pilot studies healthy relatives of AD patients. In this study we searched for a putative transmissible agent. Relatives of AD patients were selected because this group might include individuals with early stages of AD. The buffy coat assay was chosen because this sample is readily obtained and, several years ago, using buffy coat of blood, we were able to demonstrate viremia in CJD. Viremia was first shown in experimental guinea pigs (E.E. MANUELIDIS et al. 1978c) and subsequently was also detected in humans (E.E. MANUELIDIS et al. 1985). Our original findings have been confirmed in other laboratories in the USA and elsewhere (DIRINGER 1984; KURODA et al. 1983; TATEISHI et al. 1980). Viremia is a very important event in the pathogenesis of classical viral infections (JOHNSON and MIMS 1968). Even in neurotropic viral disease such as rabies, where there is viral spread to the CNS along neuronal pathways, viremia has been demonstrable.

In our pilot study (E.E. MANUELIDIS et al. 1988), 11 buffy coat samples were evaluated. The techniques we used were similar to those used in investigating viremia in CJD and are published in detail elsewhere (E.E. MANUELIDIS et al. 1978c). Essentially, 0.05 ml of buffy coat homogenate was injected intracerebrally into each of six young (6 to 8-week-old) LVG outbred hamsters per sample. The inoculated animals were kept in our P3 animal facility and were observed daily, including weekends. Positive transmissions were judged on histological criteria and, most importantly, were verified by successful transmission (in a second passage) from first hamster passage brain material. Findings with these 11 samples are summarized in Table 1. Additional data on this pilot experiment and repeat samplings of buffy coat are presented here.

In the first passage the buffy coat of five samples produced a spongiform encephalopathy in the brain of the inoculated hamsters between 196 and 517 days after inoculation (Table 1). The sacrificed hamsters showed only very subtle neurological signs or no clinical signs at all prior to sacrifice. This situation is similar to that seen in first-passage animals inoculated with human CJD material. For the second passage, brain from first-passage hamsters was sterilely removed at autopsy and these brain homogenates were intracerebrally inoculated into each of six healthy young outbred LVG hamsters per sample. Four out of five positive samples from the first passage produced a spongiform encephalopathy in the second passage (Table 1). The spongiform changes in the second passage were, however, considerably more pronounced than in the

Table 1. Transmission studies with human buffy coat in serial passages and in a repeated sampling

Sample	Patient's age	Sex	Passage 1	Passage 2	Passage 3	Repeat
1	51	F	–	NA	NA	–
2	42	F	+	+	+	+
3	59	F	±	–	–	–
4	46	F	–	–	–	–
5	68	M	±	–	–	Died
6	57	F	±	–	–	±
7	37	F	+	+	Pending	+
8	77	F	+	+	+	NA
9	44	F	+	–	–	+
10	50	F	–	–	–	NA
11	53	F	+	+	+	NA

NA, sample not available; +, positive spongiform changes; –, negative

first passage. All hamsters which were negative during the first passage remained negative during the second passage (Figs. 1, 2). A third passage to hamsters from the second passage material is almost completed, and the findings are very similar to those seen in the brains of second-passage-positive hamsters (Table 1). At least three of the initial positive cases were serially transmitted through several passages.

The spongiform encephalopathy in affected animals was characterized by small, well-demarcated vacuoles in the neuropil, as well as mild astrocytosis and

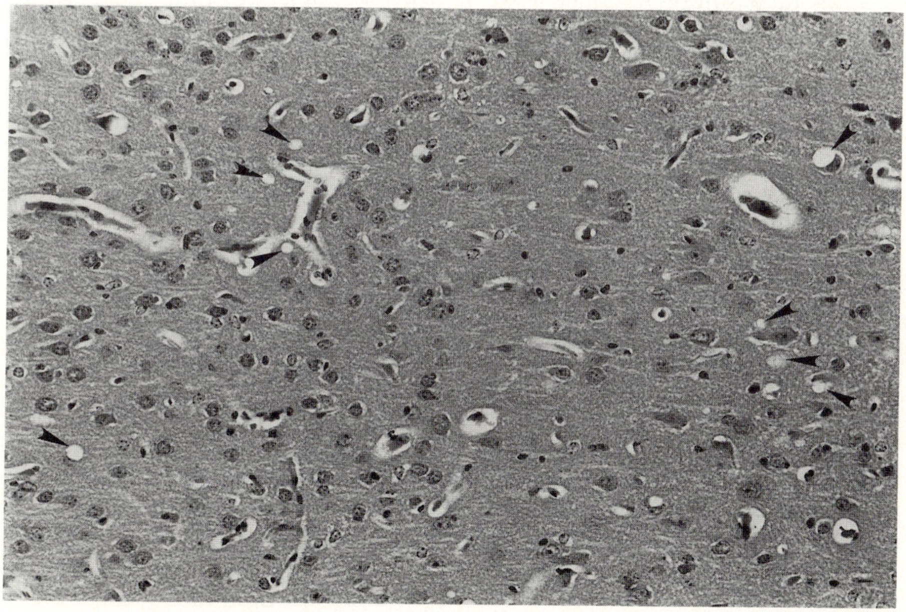

Fig. 1. Passage 1, buffy coat sample 11, hamster no. 2 (Table 1). Cortex shows isolated vacuoles in the neuropil (*arrowheads*). Hematoxylin-eosin, × mal 184

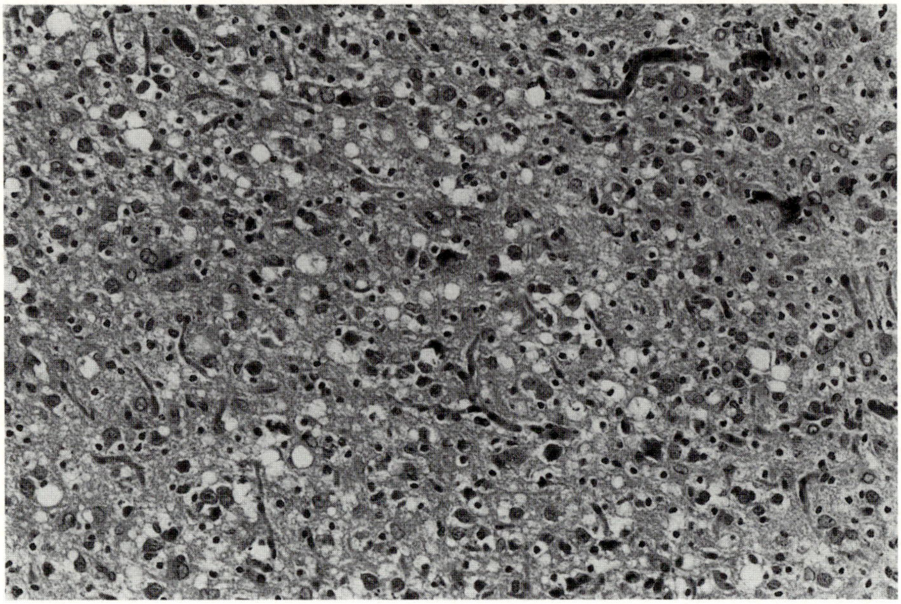

Fig. 2. Passage 2, buffy coat sample 2, hamster no. 6 (Table 1). Cortex shows extensive vacuolization in the neuropil. Hematoxylin-eosin, ×mal 184

neuronal loss. These changes were indistinguishable from the histological findings in experimental CJD. No neurofibrillary tangles, plaques, or AD-like amyloid deposits were detected in any of the inoculated hamsters.

In order to verify the data for these individuals, we subsequently collected fresh blood samples from the original volunteers. In the interim period, 2-3 years, one volunteer (original sample 5) had died of AD and three others (samples 8, 10 and 11) had left town and were not available. Thus the seven remaining "random" patient samples were reassessed. We purchased new homogenizers to rule out contamination by human or rodent CJD. Brand new instruments were also used for each repeat sample. The findings of this repeat experiment are listed in Table 1. Three of the original five positive volunteers remained positive in the ensuing years, and four volunteers who were initially negative remained negative. This almost absolute correspondence of initial primary and repeat primary passage findings is very unlikely to be due to chance.

The unusual and totally unexpected findings of our initial pilot study were received, with very few exceptions, with skepticism, disbelief, and on occasion ridicule and/or hostility from the cognoscenti of the AD club, despite our own acknowledgement of the study's limitations. A number of justified criticisms may be answered by this repeat study and by other ongoing larger studies of buffy coat.

Since we work with CJD, the question of cross-contamination, despite fastidious cleaning of homogenizers, is always an issue, even though we have had plenty of negative results when we have attempted to transmit unrelated human disease to rodents. In the pilot experiment we also had a fair proportion of "internal controls", because not all buffy coats from AD family members resulted in positive transmission, either during the first passage or in subsequent passages. Furthermore, not all six hamsters used per sample were positive. The decontamination and sterilization procedures used in our laboratory are those routinely used in laboratories throughout the world by investigators experimenting with spongiform encephalopathies. Nevertheless, we purposely repeated our first pilot transmission with fresh buffy coats using new instruments and homogenizers never used before in our laboratory for each human sample tested. The results of the new "repeat" experiments do not contradict, but rather confirm, our first positive passage results in virtually all available samples. This work is time-consuming, and it is the rule in our laboratory to wait for 1–2 years before we declare an experiment or an animal negative.

One criticism voiced to us was that if AD is a transmissible disease, why then do spouses not infect each other? This criticism was not taken seriously, because not all transmissible viral infections are "contagious". Numerous examples exist in classical virology. Furthermore, no one doubts that CJD is transmissible, yet spouses and relatives are not infected. Indeed, CJD-like agents are known not to be highly contagious.

A serious question which arose in our minds was that of phanerosis of an endogenous virus in healthy hamsters elicited by experimental manipulations and injection of experimental inocula. That, however, is most unlikely, judging from our own experimental data. Not all hamsters developed an encephalopathy when injected with human buffy coat (internal control). Furthermore, when homologous tissue was used as an inoculum during the second and third passages, hamster brain inoculum did not evoke a phanerosis of an endogenous virus when injected into the brain of healthy hamsters (see passages 2 and 3, Table 1). Numerous electron micrographs from positive passage animals also failed to show any conventional viral particles.

One very serious criticism in our published pilot study concerned the small number of control samples from people who had no relatives with AD. In ongoing recent experiments we have therefore obtained buffy coat from healthy normal volunteers who did not have AD or any other dementing illness by clinical history in their ancestry. Arbitrarily, three decades of human life were chosen for further investigation, viz. 20–30, 40–50 and 60–70 years of age. The buffy coat from 11, 11 and 7 volunteers respectively was inoculated into hamsters. This experiment is still in progress, and a long time will be needed before it is completed with accompanying verifying serial passages. However, several preliminary findings are worth mentioning. As in the original study, none of the first-passage animals have yet shown any detectable clinical signs of disease. As part of our revised protocol, to avoid losing animals between 200 and 400 days from unknown causes, we sacrificed random hamsters from each sample after 200 days of inoculation in primary passage. In hamsters inoculated with buffy coat from each of the three chosen decades of life, our preliminary data again show a spongiform encephalopathy during the first passage. Thus, until proven wrong, our working hypothesis is that many humans carry a putative CJD-like transmissible agent and are infected in utero or early in life. We suppose that one or more co-factors, which are currently unknown, may additionally determine whether and when these people develop a dementia.

References

Akowitz A, Sklaviadis T, Manuelidis EE, Manuelidis L (1990) Nuclease resistant polyadenylated RNAs of significant size are detected by PCR in highly purified Creutzfeldt–Jakob disease preparations. Microb Pathog 9: 33–45

Diringer H (1984) Sustained viremia in experimental hamster scrapie. Arch Virol 82: 105–109

Goudsmit J, Morrow CH, Asher DM, Yanagihera RT, Masters CL, Gibbs CJ Jr, Gajdusek DC (1980) Evidence for and against the transmissibility of Alzheimer's disease. Neurology 30: 945–950

Johnson RT, Mims CA (1968) Pathogenesis of viral infections of the nervous system. N Engl J Med 278: 23–30, 84–92

Kuroda Y, Gibbs CJ Jr, Amyx HL, Gajdusek DC (1983) Creutzfeldt–Jakob disease in mice: persistent viremia and preferential replication of virus in low-density lymphocytes. Infect Immun 41: 154–161

Manuelidis EE (1985) Presidential address. Creutzfeldt–Jakob disease. J Neuropathol Exp Neurol 44: 1–17

Manuelidis EE, Manuelidis L (1989) Suggested links between different types of dementias: Creutzfeldt–Jakob disease, Alzheimer's disease and retroviral CNS infections. Alzheimers Dis Relat Disorders 3: 100–109

Manuelidis EE, Angelo JN, Kim JH, Manuelidis L (1976) Serial propagation of Creutzfeldt–Jakob disease in guinea pigs. Proc Natl Acad Sci USA 73: 223–227

Manuelidis EE, Gorgacz EJ, Manuelidis L (1978a) Transmission of Creutzfeldt–Jakob disease with scrapie-like syndromes to mice. Nature 271: 778–779

Manuelidis EE, Gorgacz EJ, Manuelidis L (1978b) Interspecies transmission of Creutzfeldt–Jakob disease to Syrian hamsters with reference to clinical syndromes and strains of agent. Proc Natl Acad Sci USA 75: 3432–3436

Manuelidis EE, Gorgacz EJ, Manuelidis L (1978c) Viremia in experimental Creutzfeldt–Jakob disease. Science 200: 1069–1071

Manuelidis EE, Kim JH, Mericangas JR, Manuelidis L (1985) Transmission of Creutzfeldt–Jakob disease from human blood. Lancet ii: 896–897

Manuelidis EE, deFigueiredo JM, Kim JH, Fritch WW, Manuelidis, L (1988) Transmission studies from blood of Alzheimer's disease patients and healthy relatives. Proc Natl Acad Sci USA 85: 4898–4901

Manuelidis L, Manuelidis EE (1986) Recent developments in scrapie and Creutzfeldt–Jakob disease. Prog Med Virol 3: 78–98

Manuelidis L, Manuelidis EE (1989) Creutzfeldt–Jakob disease and dementias. Microb Pathog 7: 157–164

Murdoch G, Sklaviadis T, Manuelidis EE, Manuelidis L (1990) Potential retroviral RNA transcripts in infectious fractions of Creutzfeldt–Jakob disease. J Virol 64: 1477–1486

Sklaviadis T, Manuelidis L, Manuelidis EE (1989) Physical properties of the Creutzfeldt–Jakob disease agent. J Virol 63: 1212–1222

Sklaviadis T, Akowitz A, Manuelidis EE, Manuelidis L (1990) Nuclease treatment results in high specific purification of Creutzfeldt–Jakob disease infectivity with a density characteristic of nucleic acid-protein complex. Arch Virol 112: 215–228

Tateishi J, Sato Y, Koga M, Doi H, Ohta M (1980) Experimental transmission of human subacute spongiform encephalopathies to small rodents. 1. Clinical and histological observations. Acta Neuropathol 51: 127–134

Subject Index

Current Topics in Microbiology and Immunology

Volumes published since 1986 (and still available)

Vol. 127: **Potter, Michael; Nadeau, Joseph H.; Cancro, Michael P. (Ed.):** The Wild Mouse in Immunology. 1986. 119 figs. XVI, 395 pp. ISBN 3-540-16657-2

Vol. 128: 1986. 12 figs. VII, 122 pp. ISBN 3-540-16621-1

Vol. 129: 1986. 43 figs., VII, 215 pp. ISBN 3-540-16834-6

Vol. 130: **Koprowski, Hilary; Melchers, Fritz (Ed.):** Peptides as Immunogens. 1986. 21 figs. X, 86 pp. ISBN 3-540-16892-3

Vol. 131: **Doerfler, Walter; Böhm, Petra (Ed.):** The Molecular Biology of Baculoviruses. 1986. 44 figs. VIII, 169 pp. ISBN 3-540-17073-1

Vol. 132: **Melchers, Fritz; Potter, Michael (Ed.):** Mechanisms in B-Cell Neoplasia. Workshop at the National Cancer, Institute, National Institutes of Health, Bethesda, MD, USA, March 24–26, 1986. 1986. 156 figs. XII, 374 pp. ISBN 3-540-17048-0

Vol. 133: **Oldstone, Michael B. (Ed.):** Arenaviruses. Genes, Proteins, and Expression. 1987. 39 figs. VII, 116 pp. ISBN 3-540-17246-7

Vol. 134: **Oldstone, Michael B. (Ed.):** Arenaviruses. Biology and Imminotherapy. 1987. 33 figs. VII, 242 pp. ISBN 3-540-14322-6

Vol. 135: **Paige, Christopher J.; Gisler, Roland H. (Ed.):** Differentiation of B Lymphocytes. 1987. 25 figs. IX, 150 pp. ISBN 3-540-17470-2

Vol. 136: **Hobom, Gerd; Rott, Rudolf (Ed.):** The Molecular Biology of Bacterial Virus Systems. 1988. 20 figs. VII, 90 pp. ISBN 3-540-18513-5

Vol. 137: **Mock, Beverly; Potter, Michael (Ed.):** Genetics of Immunological Diseases. 1988. 88 figs. XI, 335 pp. ISBN 3-540-19253-0

Vol. 138: **Goebel, Werner (Ed.):** Intracellular Bacteria. 1988. 18 figs. IX, 179 pp. ISBN 3-540-50001-4

Vol. 139: **Clarke, Adrienne E.; Wilson, Ian A. (Ed.):** Carbohydrate-Protein Interaction. 1988. 35 figs. IX, 152 pp. ISBN 3-540-19378-2

Vol. 140: **Podack, Eckhard R. (Ed.):** Cytotoxic Effector Mechanisms. 1989. 24 figs. VIII, 126 pp. ISBN 3-540-50057-X

Vol. 141: **Potter, Michael; Melchers, Fritz (Ed.):** Mechanisms in B-Cell Neoplasia 1988. Workshop at the National Cancer Institute, National Institutes of Health, Bethesda, MD, USA, March 23–25, 1988. 1988. 122 figs. XIV, 340 pp. ISBN 3-540-50212-2

Vol. 142: **Schüpach, Jörg:** Human Retrovirology. Facts and Concepts. 1989. 24 figs. 115 pp. ISBN 3-540-50455-9

Vol. 143: **Haase, Ashley T.; Oldstone Michael B. A. (Ed.):** In Situ Hybridization 1989. 22 figs. XII, 90 pp. ISBN 3-540-50761-2

Vol. 144: **Knippers, Rolf; Levine, A. J. (Ed.):** Transforming. Proteins of DNA Tumor Viruses. 1989. 85 figs. XIV, 300 pp. ISBN 3-540-50909-7

Vol. 145: **Oldstone, Michael B. A. (Ed.):** Molecular Mimicry. Cross-Reactivity between Microbes and Host Proteins as a Cause of Autoimmunity. 1989. 28 figs. VII, 141 pp. ISBN 3-540-50929-1

Vol. 146: **Mestecky, Jiri; McGhee, Jerry (Ed.):** New Strategies for Oral Immunization. International Symposium at the University of Alabama at Birmingham and Molecular Engineering Associates, Inc. Birmingham, AL, USA, March 21–22, 1988. 1989. 22 figs. IX, 237 pp. ISBN 3-540-50841-4

Vol. 147: **Vogt, Peter K. (Ed.):** Oncogenes. Selected Reviews. 1989. 8 figs. VII, 172 pp. ISBN 3-540-51050-8

Vol. 148: **Vogt, Peter K. (Ed.):** Oncogenes and Retroviruses. Selected Reviews. 1989. XII, 134 pp. ISBN 3-540-51051-6

Vol. 149: **Shen-Ong, Grace L. C.; Potter, Michael; Copeland, Neal G. (Ed.):** Mechanisms in Myeloid Tumorigenesis. Workshop at the National Cancer Institute, National Institutes of Health, Bethesda, MD, USA, March 22, 1988. 1989. 42 figs. X, 172 pp. ISBN 3-540-50968-2

Vol. 150: **Jann, Klaus; Jann, Barbara (Ed.):** Bacterial Capsules. 1989. 33 figs. XII, 176 pp. ISBN 3-540-51049-4

Vol. 151: **Jann, Klaus; Jann, Barbara (Ed.):** Bacterial Adhesins. 1990. 23 figs. XII, 192 pp. ISBN 3-540-51052-4

Vol. 152: **Bosma, Melvin J.; Phillips, Robert A.; Schuler, Walter (Ed.):** The Scid Mouse. Characterization and Potential Uses. EMBO Workshop held at the Basel Institute for Immunology, Basel, Switzerland, February 20–22, 1989. 1989. 72 figs. XII, 263 pp. ISBN 3-540-51512-7

Vol. 153: **Lambris, John D. (Ed.):** The Third Component of Complement. Chemistry and Biology. 1989. 38 figs. X, 251 pp. ISBN 3-540-51513-5

Vol. 154: **McDougall, James K. (Ed.):** Cytomegaloviruses. 1990. 58 figs. IX, 286 pp. ISBN 3-540-51514-3

Vol. 155: **Kaufmann, Stefan H. E. (Ed.):** T-Cell Paradigms in Parasitic and Bacterial Infections. 1990. 24 figs. IX, 162 pp. ISBN 3-540-51515-1

Vol. 156: **Dyrberg, Thomas (Ed.):** The Role of Viruses and the Immune System in Diabetes Mellitus. 1990. 15 figs. XI, 142 pp. ISBN 3-540-51918-1

Vol. 157: **Swanstrom, Ronald; Vogt, Peter K. (Ed.):** Retroviruses. Strategies of Replication. 1990. 40 figs. XII, 260 pp. ISBN 3-540-51895-9

Vol. 158: **Muzyczka, Nicholas (Ed.):** Viral Expression Vectors. 1992. Approx. 20 figs. Approx. XII, 190 pp. ISBN 3-540-52431-2

Vol. 159: **Gray, David; Sprent, Jonathan (Ed.):** Immunological Memory. 1990. 38 figs. XII, 156 pp. ISBN 3-540-51921-1

Vol. 160: **Oldstone, Michael B. A.; Koprowski, Hilary (Ed.):** Retrovirus Infections of the Nervous System. 1990. 16 figs. XII, 176 pp. ISBN 3-540-51939-4

Vol. 161: **Racaniello, Vincent R. (Ed.):** Picornaviruses. 1990. 12 figs. X, 194 pp. ISBN 3-540-52429-0

Vol. 162: **Roy, Polly; Gorman, Barry M. (Ed.):** Bluetongue Viruses. 1990. 37 figs. X, 200 pp. ISBN 3-540-51922-X

Vol. 163: **Turner, Peter C.; Moyer, Richard W. (Ed.):** Poxviruses. 1990. 23 figs. X, 210 pp. ISBN 3-540-52430-4

Vol. 164: **Bækkeskov, Steinnun; Hansen, Bruno (Ed.):** Human Diabetes. 1990. 9 figs. X, 198 pp. ISBN 3-540-52652-8

Vol. 165: **Bothwell, Mark (Ed.):** Neuronal Growth Factors. 1991. 14 figs. IX, 173 pp. ISBN 3-540-52654-4

Vol. 166: **Potter, Michael; Melchers, Fritz (Ed.):** Mechanisms in B-Cell Neoplasia 1990. 143 figs. XIX, 380 pp. ISBN 3-540-52886-5

Vol. 167: **Kaufmann, Stefan H. E. (Ed.):** Heat Shock Proteins and Immune Response. 1991. 18 figs. IX, 214 pp. ISBN 3-540-52857-1

Vol. 168: **Mason, William S.; Seeger, Christoph (Ed.):** Hepadnaviruses. Molecular Biology and Pathogenesis. 1991. 21 figs. X, 206 pp. ISBN 3-540-53060-6

Vol. 169: **Kolakofsky, Daniel (Ed.):** Bunyaviridae. 1991. 34 figs. X, 256 pp. ISBN 3-540-53061-4

Vol. 170: **Compans, Richard W. (Ed.):** Protein Traffic in Eukaryotic Cells. Selected Reviews. 1991. 14 figs. X, 186 pp. ISBN 3-540-53631-0

Vol. 171: **Kung, Hsing-Jien (Ed.):** Retroviral Insertional Mutagenesis and Oncogene. 1991. 18 figs. X, 179 pp. ISBN 3-540-53857-7